Hormone Receptors

Horizons in Biochemistry and Biophysics Series

Editorial Board

*Horizons in Biochemistry
and Biophysics*

Volume 6

Hormone Receptors

Volume Editor
L. D. Kohn
*Laboratory of Biochemical Pharmacology, National Institute of Arthritis,
Metabolism and Digestive Diseases, National Institutes of Health,
Bethesda, Maryland*

Series Editors
E. Quagliariello, *Editor-in-Chief*
and
F. Palmieri, *Managing Editor*
Department of Biochemistry, University of Bari

T. P. Singer, *Consulting Editor*
*Department of Molecular Biology, University of California and School of Medicine,
Veterans Administration Hospital, San Francisco*

A Wiley-Interscience Publication

JOHN WILEY & SONS

Chichester · New York · Brisbane · Toronto · Singapore

British Library Cataloguing in Publication Data:

Hormone receptors.—(Horizons in biochemistry
  and biophysics, ISSN 0096–2708; v. 6)
  1. Hormone receptors
  I. Kohn, L. D. II. Series
  612′.405   QP188.5.H67

  ISBN 0 471 10049 8

Filmset and printed in Northern Ireland at The Universities Press (Belfast) Ltd.,
and bound at the Pitman Press, Bath, Avon.

Contents

List of contributors

ALOJ, S. M.   *Centro di Endocrinologia ed Oncologia Sperimentale del C.N.R., c/o Laboratorio di Patologia Generale L. Califano, 2 Facoltà di Medicina e Chirurgia, Università di Napoli, Naples, Italy*

BONATTI, S.   *2° Istituto di Chimica Biologica, 2° Facoltà di Medicina e Chirurgia, 80131, Napoli, Italy*

COOPER, D. M. F.   *Laboratory of Nutrition and Endocrinology, National Institute of Arthritis, Metabolism and Digestive Diseases, National Institutes of Health, Bethesda, Maryland 20205, U.S.A.*

FAIN, J. N.   *Section of Physiological Chemistry, Division of Biology and Medicine, Brown University, Providence, Rhode Island 02912, U.S.A.*

GABRIEL, O.   *Department of Biochemistry, Schools of Medicine and Dentistry, Georgetown University, Washington, D.C. 20007, U.S.A.*

GARDNER, J. D.   *Digestive Diseases Branch, National Institute of Arthritis, Metabolism and Digestive Diseases, National Institutes of Health, Bethesda, Maryland 20205, U.S.A.*

GILL, D. L.   *Laboratory of Biochemical Pharmacology, National Institute of Arthritis, Metabolism and Digestive Diseases, National Institutes of Health, Bethesda, Maryland 20205, U.S.A.*

GROLLMAN, E. F.   *Laboratory of Biochemical Pharmacology, National Institute of Arthritis, Metabolism and Digestive Diseases, National Institutes of Health, Bethesda, Maryland 20205, U.S.A.*

JENSEN, R. T.   *Digestive Diseases Branch, National Institute of Arthritis, Metabolism and Digestive Diseases, National Institutes of Health, Bethesda, Maryland 20205, U.S.A.*

KOHN, L. D.   *Laboratory of Biochemical Pharmacology, National Institute of Arthritis, Metabolism and Digestive Diseases, National Institutes of Health, Bethesda, Maryland 20205, U.S.A.*

LEWIS, J. A.   *Department of Anatomy and Cell Biology, State University of New York, Downstate Medical Centre, 450 Clarkson Avenue, Brooklyn, New York 11203, U.S.A.*

LONDOS, C.   *Laboratory of Nutrition and Endocrinology, National Institute of Arthritis, Metabolism and Digestive Diseases, National Institutes of Health, Bethesda, Maryland 20205, U.S.A.*

MINTON, A. P.   *Laboratory of Biochemical Pharmacology, National Institute of Arthritis, Metabolism and Digestive Diseases, National Institutes of Health, Bethesda, Maryland 20205, U.S.A.*

RASMUSSEN, H.   *Departments of Cell Biology and Internal Medicine, Yale University, New Haven, Connecticut 06510, U.S.A.*

SHIFRIN, S.   *Division of Cancer Biology and Diagnosis, National Cancer Institute, National Institutes of Health, Bethesda, Maryland 20205, U.S.A.*

SMITH, M. E.   *Laboratory of Biochemical Pharmacology, National Institute of Arthritis, Metabolism and Digestive Diseases, National Institutes of Health, Bethesda, Maryland 20205, U.S.A.*

TORRENCE, P. F.   *Laboratory of Chemistry, National Institute of Arthritis, Metabolism and Digestive Diseases, National Institutes of Health, Bethesda, Maryland 20205, U.S.A.*

YAVIN, E.   *Department of Neurobiology, The Weizmann Institute of Science, Rehovot, Israel.*

ZOON, K. C.   *Bureau of Biologics, Food and Drug Administration, National Institutes of Health, Bethesda, Maryland, U.S.A.*

Preface

With this volume, Horizons in Biochemistry and Biophysics inaugurates a new format. From its inception, this Series has aimed at the presentation of articles which summarized important advances in specialized fields of Biochemistry and Biophysics, but which were readable by and informative to investigators and students not versed in that particular field. Its object was thus a presentation of work in a manner which could be used to cross-fertilize fields and inform readers of current and future horizons in subject areas which could well be important to their own work in the near future. This policy will continue, but in an altered format.

The first volumes of this series grouped single articles in a diversity of fields; thus, for example, individual articles were grouped in a single volume on mitochondrial bioenergetics, mammalian hormone receptors, bacterial transport, suicide substrates for affinity labeling of enzymes, thermodynamic properties of water solvents, etc. In the revised format articles encompassing a general area, but attacking the problem from different directions, will be grouped. The original aim of the series remains; there is, however, the added hope that each volume can serve as a single handy reference or even text source for studies in the area. The editors look forward to this improvement as an even better means of serving the needs of all Biochemists and Biophysicists in an era of increasing specialization.

The specific area taken as the topic for this volume is hormone receptors. In keeping with our aims, the articles in the book are not aimed at an exhaustive or detailed explanation of each hormone receptor but rather have used articles from individual receptor areas to illustrate problems and approaches in the area. Individual articles will vary in their technical detail. Thus some are very general and others more specialized as the intent is to provide both the overall concept and problems in the field and some indication of detailed approaches to the problem.

In the first chapter, by Dr Kohn, the definition of a receptor is explored and the experimental problems in transposing receptor studies from the physiologic realm to the biochemical or biophysical arena will be raised. A single receptor is used as the model, but the problems can be generalized to receptor for many hormones, toxins, and bioactive ligands recognized by cells, i.e. interferon. The complexity of coupling ligand recognition to cell information, i.e. to the process by which the hormone tells the cell what to do, is explored.

In Dr Minton's chapter the kinetic aspects of ligand binding are explored and a new and innovative approach is presented to the problem. The original simple concept of ligand receptor binding is brought to the era of multicomponent receptor structures and approached thermodynamically. Dr. Yavin presents an important future application, the use of monoclonal antibodies to define physiologic receptors for *in vitro* studies of biochemists and biophysicists. The problem of recognition is explained in more detail by Drs Aloj and Gabriel. The former examines the question of lipids as receptor components and modulators, an important area of the future as the lipid membrane milieu of the receptor becomes recognized as more than an innert sac surrounding the cell. The latter explores the key role of carbohydrates as recognition determinants. Finally Dr Bonatti raises the problem of receptor synthesis, insertion in the lipid bilayer, and movement through different membrane compartments.

The next articles explore modes of receptor signalling to the cell. The well recognized 2nd message system, adenylate cyclase and cAMP, is not dealt with in detail since it is already a text book issue. Rather, other means of signalling and regulation of the cAMP system are considered. Drs Grollman, Rasmussen, Gill, Fain, Garder, Jensen, Cooper, and Londos explore this area from several directions. Particular emphasis is given to the role of Ca^{++} in signalling, and its relation to the cAMP system.

Finally the problem of a 'hormone receptor' field in its infancy is presented and the relevance of the receptor approach is viewed through interferon. The problem of purifying and defining a ligand is described by Drs Zoon, Smith, and Torrence. The problem of receptor mediated action integral to a more complex mechanism involving translational or transcriptional control is elaborated by Dr Lewis. These articles should be viewed as a problem area for a receptor approach to which all the other articles are relevant. They also present an informative view of interferon and its biologic and biochemical problems.

Future volumes of the series will take different areas to explore. For example, the next volume will examine gene structure and the expression of genetic information. Topics will include the following.

1. Nucleosomes and chromatine function in eukaryotes
2. Mosaic genes and mosaic RNAs
3. Splicing: theory and facts
4. The non-university of the genetic code
5. Transposons
6. Differential expression of multiple genes in different tissues
7. Expression of eukaryotic genes in prokaryotic cells
8. Immunoglobulins
9. Ribsome structure and function

10. Regulatory mechanisms in eukaryotic initiation of protein synthesis
11. Transport of proteins from the sites of genetic expression to the sites of functional expression
12. The interplay between different genetic systems in eukaryotic cells

As can be seen topic 11 will further expand the area approached in this volume. The continued interrelatedness of volumes will be stressed as the series progresses. The final aim of presenting a horizon view will, however, remain the guiding principle in future series. The editors hope the revised format will improve the usefulness of the series.

E. Quagliariello
F. Palmieri
T. P. Singer

1 Receptor structure and function: an exploratory approach using the thyrotropin receptor as a vehicle

L. D. Kohn and **S. Shifrin**

1 INTRODUCTION

A receptor is that site on the surface of a cell which recognizes a molecule in the environment surrounding the cell. It must be capable of transmitting information derived from that molecule to the interior of the cell or must be able to recognize a molecule as either unimportant or harmful so the cell can either ignore it, destroy it, or avoid it. In its broadest analogy, the receptor on the surface of a cell replaces at the cellular level those senses of our body which we use for recognition of our environment, i.e. sight, touch, smell, hearing and taste. It is the essential feature of our cell membrane which changes it from an inert wall of lipids to a dynamic, living, responsive structure upon which the entire cell depends for its existence.

The purpose of this chapter and book will be to explore a critical period in our fascination with receptors as we move from recognition of their existence, through their characterization, to an understanding of the molecular and chemical basis for function. No longer can we rest by identifying anatomical or physiological properties of a receptor just as we have long since passed the period of identifying only the gross anatomy and physiology of the eye. This in no way means we understand all of the anatomy and physiology but only that we must start using histological, biochemical and physicochemical techniques to explore receptor structure and function. We shall see that this requires new concepts of identifying and defining a receptor which challenge the past. This new basis of understanding is already having a practical fallout in understanding pathological processes causing human disease.

In this particular chapter several general concepts of receptor structure and function will be explored, but it will be done with a personal reference.

Thus the 'Receptor Concept' will be explored through the eyes of a particular receptor, the thyrotropin (TSH)† receptor.

2 DEFINITION OF A RECEPTOR USING THE TSH RECEPTOR AS A MODEL

2.1 Receptor definition

The definition of a receptor is evolving today. The receptor, in its limited sense, can be considered the initial recognition site on the cell surface with which a ligand interacts or binds, i.e. the eye. Alternatively, and in the broadest sense, it can be defined by the measurable response resulting from the interaction, i.e. sight. This is well illustrated in the case of the thyrotropin receptor.

2.2 Thyrotropin

Thyrotropin is a pituitary hormone which is functionally identified by its action on a target tissue, the thyroid.[1] It causes the thyroid to produce and secrete thyroid hormones, triiodothyronine (T_3) and tetraiodothyronine or thyroxine (T_4), which regulate the metabolic state of every cell in the body.[1] Thyrotropin is a glycoprotein composed of two approximately equal-sized subunits, α and β, whose sequence has been defined[2] (Figure 1). It is, however, one of a family of glycoproteins which includes luteinizing hormone (LH), follicle-stimulating hormone (FSH), and human chorionic gonadotropin (hCG). Each of these hormones has an α subunit with nearly an identical primary sequence and a β subunit with a great deal of sequence homology.[2] The carbohydrate structures on each are different, but it is known that the α subunit of LH can be combined with the β subunit of thyrotropin to create a hybrid molecule which behaves like native TSH.[2] Similarly, the α TSH plus β LH hybrid behaves like LH. In short, the β subunit controls target tissue specificity, but the α subunit can be recognized by any target for this hormone family in terms of the functional information it provides.[2]

---

† The abbreviations used are: TSH, thyrotropin; LH, luteinizing hormone; hCG, human chorionic gonadotropin; DNS, dansyl,5-dimethylamino-1-naphthalene sulphonate; G_{M1}, galactosyl-N-acetylgalactosaminyl-[N-acetylneuraminyl]-galactosylglucosylceramide; G_{M2}, N-acetylgalactosaminyl-[N-acetylneuraminyl]-galactosylglucosylceramide; G_{M3}, N-acetylneur-aminylgalactosylglucosylceramide; G_{D1a}, N-acetylneuraminylgalactosyl-N-acetylgalactos-aminyl-[N-acetylneuraminyl]-galactosylglucosylceramide; G_{D1b}, galactosyl-N-acetylgalactos-aminyl-[N-acetyl-neuraminyl-N-acetylneuraminyl]-galactosylglucosylceramide; G_{T1}, N-acetyl-neuraminylgalactosyl - N - acetylgalactosaminyl - N - acetylneuraminyl - N - acetylneuraminyl - galactosylglucosylceramide; T_3, triiodothyronine; T_4, tetraiodothyronine.

Primary Structure of Bovine Thyrotropin

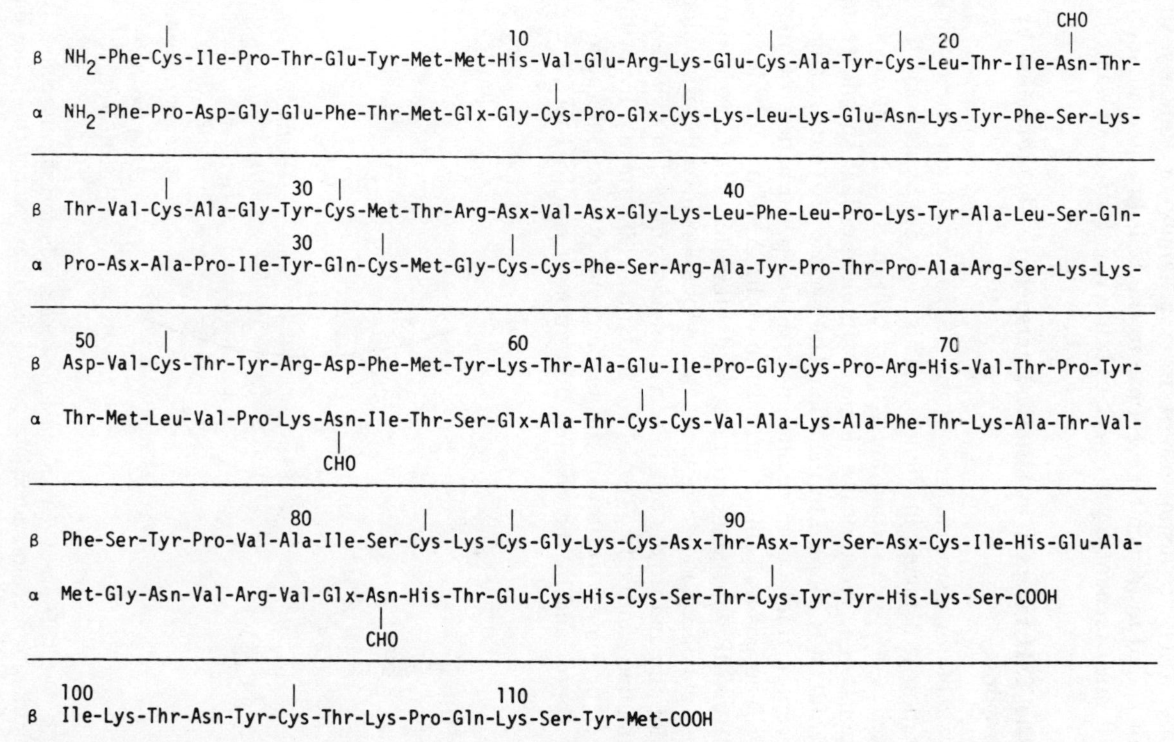

Figure 1. Primary sequence of TSH. CHO stands for carbohydrate units

2.3   The TSH receptor role and problem

The basic role or problem of the TSH receptor on the thyroid cell is to recognize TSH, to distinguish it from a family of similarly structured molecules, and to transmit a response to the cell. In an equal and opposite sense, it must recognize yet reject LH, hCG or FSH as unwanted.

2.4   The TSH receptor in historical perspective—identification by response

The simple or historical view of the TSH receptor and the mechanism of TSH action is depicted in Figure 2. This view has as its basis the concept that cyclic AMP is a second messenger[1] and that the receptor is defined by its response. The receptor is a single component of high specificity unique to the thyroid. It is coupled to the adenylate cyclase enzyme directly, such that the receptor interaction results in a direct perturbation of the cyclase activity. There is a resultant increase in cyclic AMP production; the activation of cAMP-dependent kinases; and ultimately the classic thyroid responses: enhanced iodide uptake, thyroglobulin biosynthesis, iodination of thyroglobulin and thyroglobulin secretion and conversion to T_3 and T_4.[1] An 'autoimmune' antibody (TSAb) in patients with primary hyperthyroidism has been presumed to subvert this mechanism by direct interaction with the TSH receptor and by the ensuing utilization of the TSH message transmission system.[3]

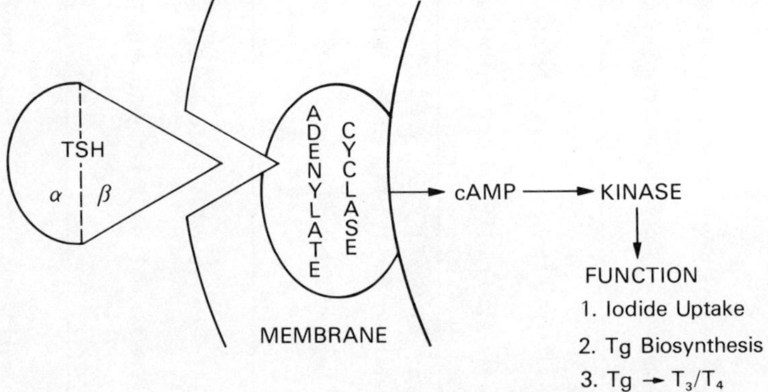

Figure 2. The TSH message transmission process in historical perspective. Biochemical definition of receptor, adenylate cyclase, or other components is not made. The assumption is that the receptor and adenylate cyclase are a single component and that measurements of cAMP levels or adenylate cyclase activity are directly related to the receptor recognition process

2.5 The TSH receptor today—a molecular view

The current view of the TSH receptor is much more complicated and is beginning to invoke biochemically defined components. This view now defines the receptor as a molecular recognition site involving a series of molecular steps analogous to a series of enzymes and enzyme regulators. It distinguishes the receptor from the transmission step or series of transmission steps since these are now known to be biochemically distinct molecules which are not uniquely coupled to the TSH receptor. Thus the TSH receptor is now suggested to be composed of both a glycoprotein and a ganglioside, i.e. a glycolipid containing sialic acid[4-7] (Figure 3).

The glycoprotein receptor component is believed to be a high affinity, primary binding component which is able to distinguish TSH from non-related polypeptide hormones. The ganglioside is believed to carry out the following functions: (i) complete specificity by distinguishing among glycoprotein hormones and related ligands such as tetanus toxin, cholera toxin and interferon; (ii) modulate the apparent affinity and capacity of the glycoprotein receptor component; (iii) induce a conformational change in the hormone believed necessary for subsequent message transmission; (iv) form an anhydrous complex between the hormone and the lipid bilayer of the membrane; (v) allow the ligand to perturb the phospholipid bilayer through alterations in lipid order; and (vi) alter the ion flux across the membrane.

In a sense, therefore, the model proposes that the glycoprotein is high affinity 'fly paper' attaching the hormone to the cell membrane in the same manner as the sperm to the ovum or a virus to the surface of a cell. The ganglioside, in contrast, acts as an emulsifying agent which allows the hormone to penetrate into the lipid bilayer where it can interact with other membrane components in an anhydrous environment. The active, biologically relevant receptor is probably the complex of the glycoprotein with a G_{T1}- or G_{D1b}-like ganglioside. The ganglioside in essence overcomes a major energy barrier by bringing an aqueous-soluble hormone into the lipid environment of the membrane.

2.6 The TSH receptor—a molecular complex which must transmit information

Before turning to the coupling of the receptor to the adenylate cyclase system, it is important to recognize that other informational routes exist. Thus, TSH exerts an effect on the membrane potential of the cell and on ion flux across the membrane independent of the cyclase system.[4,8] The direct relevance of this TSH action to TSH effects on iodide uptake are already

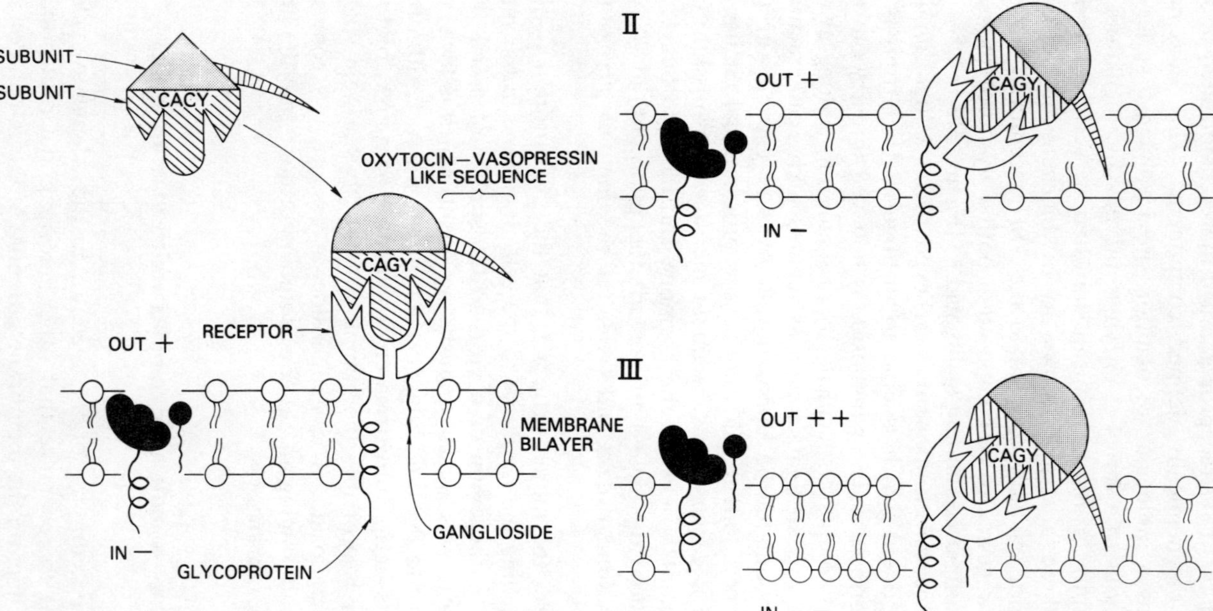

Figure 3. Proposed model of TSH receptor composed of glycoprotein and ganglioside component. After the TSH β subunit interacts with the receptor, the hormone changes its conformation and the α subunit is brought into the bilayer where it interacts with other membrane components. The end result includes a change in organization of the membrane bilayer and a change in the transmembrane electrochemical ion gradient (symbolized by +/−). The membrane change includes changes in lipid array and in the expression of other receptors. The receptor itself, i.e. the recognition site, involves both a glycoprotein (high affinity, low capacity) and ganglioside (low affinity, high capacity) component. An assumption is made that the α subunit contribution to the receptor interaction is indirect, i.e. by virtue of its conformational perturbation of the β subunit[18]

suggestive. Thus proton flux, intracellular pH and sodium cotransport processes have been linked both to iodide uptake and to effects by TSH, or ligands analogous to TSH, in the absence of cAMP.

Further, recent evidence has implicated the importance of Ca^{2+}-dependent kinases in the stimulation of thyroid cell activity. Thus a case has been observed in cultured human cells wherein the cAMP-dependent kinase system was missing but the Ca^{2+}-dependent system was intact.† TSH was able to effect stimulation of the thyroid function in these cells. These data emphasize the potential importance of a non-cAMP-mediated process in certain cases of thyroid stimulation and, in turn, emphasize the importance of calmodulin, the Ca^{2+} binding protein which has been implicated in the stimulation of many Ca^{2+}-dependent kinase systems.[9,10] The 'inter-convertability' of these systems to yield similar end responses is discussed elsewhere in this volume (J. D. Gardner).

In terms of receptor coupling to the adenylate cyclase system, i.e. the classically defined messenger response, it is now clear that the adenylate cyclase system is composed of at least two components, a regulatory unit and a catalytic unit.[11] The mechanism by which these interact is unclear. However, it is clear that there is a GTP binding site on the regulatory subunit. The presence of GTP or a non-hydrolysable analog [Gpp(NH)p] on the regulatory subunit results in its 'activation' and, in turn, the enhancement of adenylate cyclase catalytic activity. A GTPase on the regulatory subunits results in a deactivation process.[11]

Studies which have compared the mechanism of action of TSH with cholera toxin have recently provided additional insight into the mechanism by which TSH may perturb the system. The α subunit of cholera toxin catalyses the ADP ribosylation of the G regulatory protein using NAD as a substrate.[12,13] The ADP ribosylation is believed to alter the GTPase activity, guanine nucleotide exchange, or both, and thereby increases cyclase activity. One study has shown that TSH has no ADP ribosylation activity.[14] However, a second study has shown that TSH can perturb a membrane-associated ADP ribosyltransferase which carries out this same reaction or, in the absence of acceptor, the conversion of NAD to free ADP ribose and nicotinamide.[6,7,15,16] Recent studies have also shown that thyroid membranes, or a solubilized glycoprotein extract of thyroid membranes, have the ability to reverse this action.[15,16] Thus one can ADP ribosylate arginine methyl ester with the α subunit of cholera toxin and then convert the modified acceptor to its non-ADP-ribosylated form by the thyroid membrane preparation.

These data can be explained by the following schema (Figure 4). There

† R. J. Winand, personal communication.

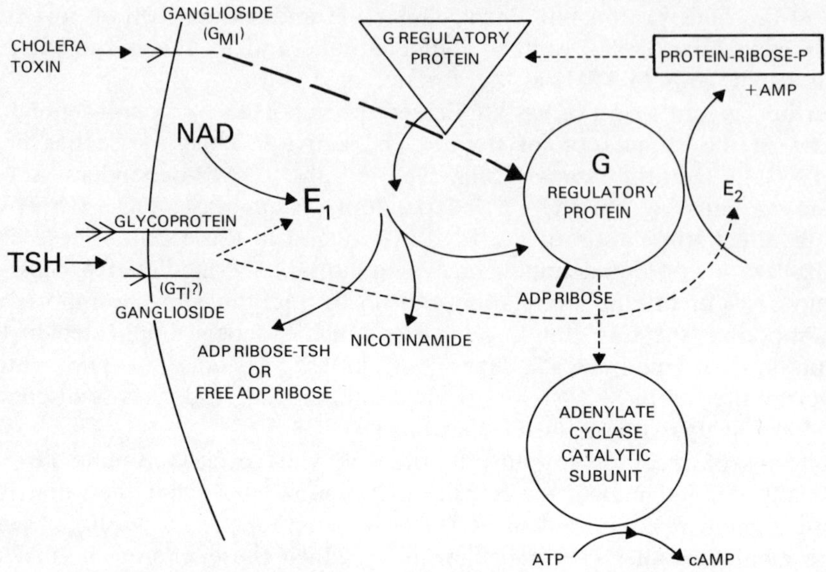

Figure 4. Proposed model wherein enzymes which regulate ADP ribosylation of the regulatory subunit of the adenylate cyclase are the site of hormonal control of the cAMP message transmission process. It is suggested that TSH acts through E_1 or E_2; cholera toxin subverts this through its own intrinsic ADP ribosyltransferase action. It is further proposed that this also affords a regulatory pathway whereby the cell can—with chronic TSH stimulation, maximal G regulatory protein ribosylation, and a stable activated state of adenylate cyclase activity—become desensitized to further TSH stimulation and convert NAD to free ADP ribose and nicotinamide or even ADP ribosylate TSH itself (See below for further discussion)

exists in thyroid membranes two enzymes: E_1 catalyses the ADP ribosylation of the G regulatory protein and E_2 catalyses the reversal of this process. TSH, by directly perturbing E_1 or E_2, causes a net increase in activated ADP-ribosylated G protein. In the absence of an acceptor substrate, i.e. the G protein, the ADP ribosyltransferase activity catalyses the conversion of NAD to ADP ribose and nicotinamide.

In short, TSH, like cholera toxin, would stimulate adenylate cyclase activity by ADP ribosylating the G regulatory subunit of the adenylate cyclase but would use a different means to do it. A cycle would exist whereby the G regulatory protein could be in one of three forms—native, ADP ribosylated, and a form containing ribose-phosphate. Their distribution would determine the stability of the hormone response or even regulate the hormone response by limiting the amount of native acceptor available. This might be the mechanism by which desensitization to a hormone—lack of response after initial stimulation—is achieved or where the stimulatory role of the hormone is blended with the stimulatory effects of the many other adenylate cyclase stimulatory agents.

2.7   The receptor definition today

The question at hand, then, is to define how these different and defined molecular components which can be distinguished as interacting with a ligand in the environment can be joined to create a 'physiologic receptor' and how this receptor can be linked to those molecular events which can be termed informational or 'message-generating'. Questions which require changing concepts are numerous and intuitive in the above models. For example, what will define a receptor component if one separates the binding or recognition process from the response process? What information resides in or is contributed by the ligand or environmental molecule and what resides in the receptor? What is target tissue specificity in molecular or biochemical terms? Can one relate *in vitro* data obtained using subcellular membrane preparations to the *in vivo* state of a cell and its receptors? When we solubilize and purify receptor components, what will be our criteria or assay to relate purified or identified components to the functioning 'physiologically relevant' receptor? Must all receptors be 'high affinity' in properties? What are the regulatory paths? In order to place these and other questions in their perspective, this report will summarize the history and evidence which led to the concepts of the TSH receptor structure and function outlined above. Many questions have been raised, but, as will be seen, few have been clearly resolved. The recognition of the problems may, however, allow new experimental approaches to evolve.

3   THE TSH RECEPTOR—EVIDENCE FOR ITS CURRENT STRUCTURE/FUNCTION MODEL

3.1   Initial studies

The TSH receptor was initially defined by its response, i.e. the ability of TSH to enhance adenylate cyclase activity.[17] The presumption of these studies was, however, that the cyclase response reflected a specific cell surface interaction with a 'receptor' able to discriminate between TSH, LH and hCG, since the latter two hormones did not stimulate the thyroid adenylate cyclase or stimulated it minimally.[17,18]

In order to define this site directly, several laboratories prepared bioactive radiolabelled TSH and directly measured TSH binding to membranes and cells,[19–26] i.e. tried to characterize the properties of the 'receptor' in terms of the initial binding and recognition process. Studies with cells using physiologic salt concentrations[20] showed a very low number of TSH binding sites with a high degree of specificity. Studies with membranes at physiologic salt concentrations or with the assay conditions optimal for measuring adenylate cyclase activity also measured low levels of TSH binding[21,23,25] and a high degree of specificity; however, it was noted that TSH binding could be dramatically improved in low salt buffers and at lower pH values (pH 6.0), i.e. conditions far from the physiologic state.[19,21,22,24,26] These experiments did note, however, that binding curves as a function of hormone concentration were non-linear,[22,24,26] suggesting the existence of 'negative cooperativity' or a high affinity TSH binding component with low capacity and a low affinity component with high capacity. The presence of the low capacity site shifted displacement curves such that specificity could still be demonstrated but not at 'physiologic' levels of hormone.

3.2   The 'high affinity' glycoprotein component of the TSH receptor

Studies assaying TSH binding at low pH (pH 6.0) and at low salt concentrations showed that limited tryptic digestion of functioning thyroid cells resulted in a coincident loss and coincident return of TSH binding and cell function; TSH stimulated adenylate cyclase activity[27] and TSH stimulated changes in membrane potential.[8] Tryptic digestion did not, however, destroy the receptor but rather released the TSH binding component into the media.[22,27,28] The binding component could be radiolabelled by pulsing cultures with [^{14}C]glucosamine;[27] it and the tryptic fragment of this component were purified by applying affinity chromatographic procedures to tryptic digests of bovine thyroid membranes solubilized with lithium diiodosalicylate.[28,29] Antisera prepared against the purified receptor fragment were able to precipitate the TSH binding activity in crude solubilized receptor preparations and reacted with the receptor fragment which was lost from the plasma membrane after trypsinization of the thyroid cells.[28,29] There was no evidence for a change in the ganglioside composition of cells or membranes exposed to trypsin (Figure 5). In sum, then, early evidence pointed to a membrane glycoprotein as the TSH receptor and was uncovered using non-physiologic assay conditions *in vitro*.

More recent evidence supported the view that this component was a part of the physiologically relevant receptor. First, the solubilized glycoprotein component of the TSH receptor could be reconstituted in liposomes and was shown to exhibit many of the characteristics of TSH binding exhibited by plasma membranes.[4,30] Moreover, using liposomes containing a self-quenched fluorescent dye, 6-carboxyfluorescein, TSH could cause a dose-dependent release of 6-carboxyfluorescein which was not duplicated by hCG (Figure 6).[31] This last result was obtained despite the observation that hCG could interact with these liposomes (Table 1) and was a significantly better inhibitor of TSH binding to the solubilized glycoprotein component of the

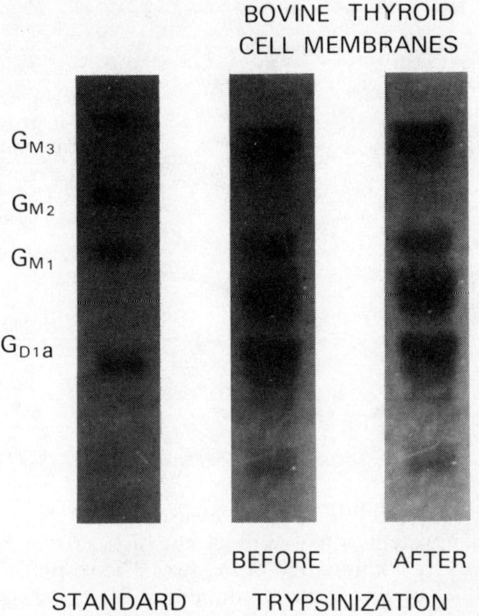

Figure 5. Ganglioside composition of bovine thyroid membranes before and after trypsinization to remove 90 per cent of the ^{125}I-TSH binding activity measured at 1×10^{-12} M to 1×10^{-10} M ^{125}I-TSH concentrations. Gangliosides were extracted and chromatographed as described[41]

TSH receptor than to the intact thyroid cell or membrane.[28,29,31] This observation suggested (as will be discussed below) that a specific TSH interaction with its receptor results (i) in a specific conformational change in both and (ii) in a change in membrane properties which cannot be duplicated even by a structurally similar hormone (hCG) or hormone analogue (cholera toxin) which might be able to bind to the receptor.

Second, a monoclonal antibody has been made against the glycoprotein component of the TSH receptor.[32,33] This antibody inhibits TSH binding to membranes in a competitive manner and is prevented from binding to functional thyroid cells in culture by TSH. It does not bind to non-functional thyroid cells in culture. The antibody can bind to the glycoprotein component of the receptor reconstituted in liposomes as measured either directly or as 6-carboxyfluorescein release but only poorly to gangliosides reconstituted in the liposomes. The monoclonal antibody to the TSH receptor is inactive as a cyclase stimulator; it can, however, inhibit the TSH activation of the cyclase,[6] and TSH stimulated iodide uptake in functioning thyroid cell cultures.[32]

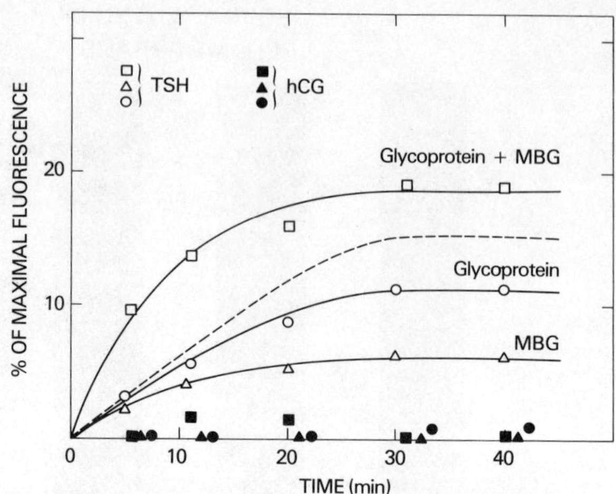

Figure 6. Effect of TSH and hCG on the release of 6-CF from liposomes embedded with either the glycoprotein component of the thyrotropin receptor, mixed brain gangliosides (MBG), or a combination of the two. The response to TSH or hCG were compared using 20 μl of liposomes containing $\sim$20 μg protein ($\bigcirc$, $\bullet$), 200 ng of MBG ($\triangle$, $\blacktriangle$), or a mixture containing one-half of each, i.e. $\sim$10 μg protein and 100 ng MBG ($\square$, $\blacksquare$). The dashed line represents the theoretic results from a mixture containing one-half the amount of glycoprotein, i.e. a simple summation of each activity alone. Liposome preparations and the meaning of the fluorescence values are described in Reference 31. The fluorescence sample contained in a 1-cm^2 quartz cuvette 1.5 ml buffered sucrose and 20 μl of the noted liposome preparation. Fluorescence at 516 nm was measured at 22 °C with the excitation monochromoter set at 480 nm. The measurements were started when TSH was added at a final concentration of 2.8×10^{-6} M

Third, support for the role of a glycoprotein receptor component comes from studies of the relationships between TSH and interferon receptors wherein a fragment of the glycoprotein component was shown to increase interferon bioactivity 50-fold after being incubated with cells in culture.[34] This experiment initially derived from the observation that TSH and interferon could interact with gangliosides *in vitro*[35] and from experiments showing that TSH could inhibit interferon bioactivity.[36] A glycoprotein receptor component was isolated from cells which were sensitive to interferon's antiviral protective effect but not to TSH in terms of either adenylate cyclase stimulation or altered membrane permeability.[37] Based on liposome reconstitution data, the glycoprotein receptor component could account for the TSH binding activity.[37] It was subjected to trypsinization and, using TSH binding as an assay, several glycoprotein receptor fragments were purified by a column chromatographic procedure. Two of these fragments, when incubated with interferon-sensitive cells, caused a 50-fold increase in interferon bioactivity in terms of antiviral protection.[34] The effect could be

Table 1. Ability of hCG to interact with glycoprotein component of TSH receptor as measured by its inhibition of ^{125}I-TSH
binding to 6-CF liposomes[a]

Assay component	^{125}I-TSH bound
	c.p.m. $\times 10^{-3}$
Glycoprotein liposomes	45
+unlabelled TSH 1×10^{-6} M	1.2
1×10^{-7} M	10.8
1×10^{-8} M	30.8
+unlabelled hCG 1×10^{-6} M	1.2
1×10^{-7} M	18.4
1×10^{-8} M	44
Thyroid plasma membranes	84
+unlabelled TSH 1×10^{-6} M	3.8
1×10^{-7} M	20.4
1×10^{-8} M	62
+unlabelled hCG 1×10^{-6} M	58
1×10^{-7} M	76
1×10^{-8} M	84

[a] See Reference 31 for details of assays.

obtained even if cells were preincubated with the receptor fragment and was
associated with binding of the fragment to the cell; no enhanced activity was
seen in cells without gangliosides or after additional proteolytic digestion of
the fragment. In sum, at least one reconstitution type experiment and data
using a monoclonal antibody to the TSH receptor support the biologic
relevance of the glycoprotein receptor component.

A recent study[38] may partially resolve the 'condition problem', whereby
optimal measurement of this receptor component is at non-physiologic
optima. Thus, the tetrameric form of concanavalin A has been shown to
interact with the thyroid membrane at a site distinct from the TSH receptor.
This causes a 2-fold increase in TSH binding *whether* measured in the low
salt, pH 6.0 *in vitro* assay system or the 'more physiologic' pH 7.4 system
with salts. In both cases, i.e. with both *in vitro* assays, the cause of the
increased binding appears to be a 10-fold increase in the affinity of the 'high
affinity' glycoprotein receptor component for TSH (Figure 7) with no change
in the low affinity component or receptor number. That this in both
situations is the 'physiologic' receptor is evident from the data in Figure 8.
Thus, the monoclonal antibody to the TSH receptor which blocks both TSH
binding to cells and the TSH functional response will inhibit the concanavalin A-induced increased TSH binding, will inhibit with no increase in

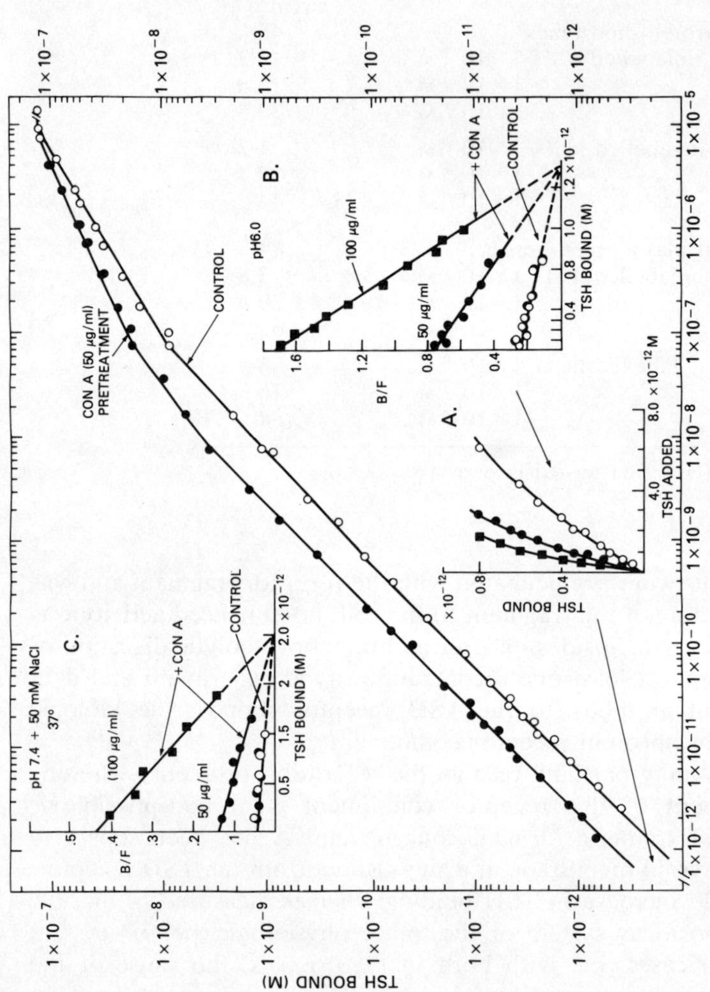

Figure 7. TSH binding to control bovine thyroid membranes (○) or to bovine thyroid membranes pretreated with 50 μg/ml (●) or 100 μg/ml concanavalin A (■) as a function of TSH concentration. The data from the main body of the figure represent experiments with ^{125}I-TSH alone (below 1×10^{-10} M concentrations of added TSH) or with a mixture of ^{125}I-TSH and unlabelled TSH (above 1×10^{-11} M concentrations of added TSH). The binding was performed using 10 μg membrane protein per assay; at pH 6.0 in 0.025 M Tris-acetate; at 0°C to 4°C; and included 0.1 M α-methyl mannoside in the assay. *Insert A* details data below 8×10^{-12} M added TSH; *insert B* transposes the *insert A* data using a Scatchard analysis; and *insert C* represents a Scatchard analysis of binding data at 37°C in 50 mM Trischloride, pH 7.4, containing 50 mM NaCl, i.e. the more physiologic conditions used by the workers of Reference 61. For additional details, see Reference 38

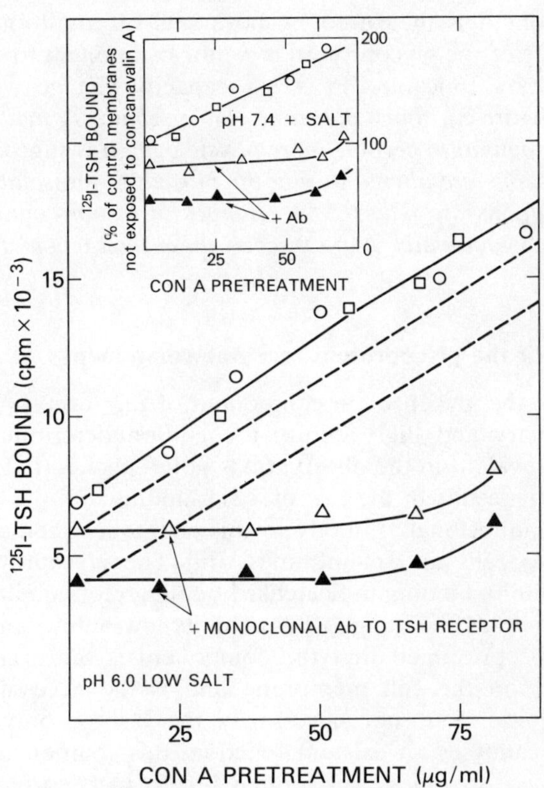

Figure 8. Effects of the monoclonal antibody to the TSH receptor[32] on ^{125}I-TSH binding to bovine thyroid membranes pretreated with the noted concentrations of tetramer concanavalin A. The antibody preparation was an ammonium sulfate precipitate of the media from the T-13 D11 clone;[32] its preparation has been described. The antibody was added at 2 concentrations to achieve approximately 10 per cent (△) and 40 per cent (▲) inhibition, respectively, of ^{125}I-TSH binding to bovine thyroid membranes untreated with concanavalin A. Equal amounts of membranes were present in each assay. The membranes were exposed to the antibody preparation for 1 hour at pH 7.0 and 22 °C; membranes were washed; and ^{125}I-TSH binding was measured at pH 6.0 in the standard way. To insure that this procedure had no effect on the concanavalin A-induced increase in ^{125}I-TSH binding, the experiment was duplicated using an ammonium sulphate preparation of a 'naive' antibody (□) which could interact with the thyroid membrane, but not inhibit ^{125}I-TSH binding (Group I antibody, Reference 32). The insert depicts an identical experiment wherein TSH binding was measured at 37 °C in Tris-chloride, pH 7.4, containing 50 mM NaCl. Further details are in Reference 38

antibody necessary, and will affect inhibition using both physiologic and non-physiologic assay conditions. This last observation shows again that the receptor component is measured by both sets of conditions and that the increased affinity of the glycoprotein receptor component for TSH is the key change elicited by concanavalin A. A capacity change would cause the dashed lines (Figure 8). *Since the same concanavalin A treatment causes the in vitro binding optima to become more physiologic, it is suggested that normal cell–cell interactions are altered during the procedure for isolating membranes, that these alterations are reflected by changes in in vitro optima and expression, and that concanavalin A may revert these interactions toward 'normality' in vitro.*

3.3 The role of the glycoprotein receptor component

In sum, then, the glycoprotein component of the receptor clearly appears to be the primary and high affinity recognition determinant on the cell surface. This is evident in the observations noted above. (i) Trypsinization of cells to release it results in the loss of TSH binding and the TSH-stimulable cyclase. (ii) A monoclonal antibody to this component competitively blocks TSH binding to cells and membranes. (iii) The glycoprotein component exhibits high affinity binding in Scatchard binding plots similar to membrane preparations whereas the ganglioside exhibits low affinity binding.

It can thus be presumed that the component is the original binding or recognition site on the cell membrane and is the necessary first step in receptor recognition but not necessarily the last or only step. Like an enzyme which catalyses an existent reaction, this component can be considered to enhance 'stickiness' by comparison to inert surfaces—plastic and glass—which often can bind ligands at the low concentrations used in these assays. Its existence and properties should, however, involve new questions.

This receptor must recognize the hormone in an aqueous environment where electrostatic interactions dominate; yet the ligand must be moved to the hydrophobic plane of the bilayer where it interacts with the cyclase system. What are the electrostatic forces involved in recognition? Why doesn't the high affinity lock the ligand into a compartment where it cannot function, in analogy to the 'high affinity-antagonist' principle of pharmacologic studies? What intrinsic roles exist for this receptor component in the recognition process of a cell independent of the coupling to cyclase?

It is important to note in this respect that this component exists on non-thyroid cells, can account for TSH binding to these cells, and is biologically relevant. For example, the glycoprotein component of the TSH receptor can account for TSH binding to mouse L cells.[36,37] This binding is functional in the sense that it blocks interferon bioactivity in these cells, albeit in the absence of a TSH-stimulated adenylate cyclase activity.

The last experiment suggests that the glycoprotein component in and of itself is not sufficient for a full biologic response and that other components of the cell membrane must be invoked. In this regard the existence of low affinity sites on thyroid membranes, able to interact with TSH, and apparently related to bioactivity in thyroid tumour experiments, may answer some of the questions as to how the receptor might function in a biologically responsive receptor system on the thyroid cell.

4 THE 'LOW AFFINITY' GANGLIOSIDE COMPONENT OF THE TSH RECEPTOR

4.1 Initial identification

As noted above, studies of TSH binding to thyroid membranes at pH 6.0 and in low salts (non-physiologic conditions) identified a low affinity binding component. Studies of the properties of the glycoprotein component of the TSH receptor indicated that sialic acid might be a receptor determinant,[28,29] i.e. the binding activity of the solubilized fragment was found to be sensitive to neuraminidase. With this result in mind and the evidence that the ganglioside G_{M1} was the receptor for cholera toxin,[39,40] the idea that a ganglioside or ganglioside-like structure could be involved in or be an analog of the glycoprotein receptor component was considered. Accordingly, a variety of these gangliosides were evaluated for their effects on TSH binding to bovine thyroid receptors.

4.2 Gangliosides as a receptor component

The gangliosides G_{D1b} and G_{T1} were found to be inhibitors of TSH binding to its specific membrane receptor by comparison to G_{M1} and G_{D1a}. This inhibition was caused by an interaction of these gangliosides with the hormone rather than the membrane, and the inhibition was hormone-specific.[4-7] The possibility that the ganglioside might actually be a receptor component evolved from the following observations. Thyroid plasma membranes were not only rich in gangliosides with the chromatographic characteristics of G_{D1b}, G_{T1} and G_{M1}, but also in a higher order ganglioside which (i) is an even more potent inhibitor of TSH binding and (ii) does not appear to be present in the brain.[41] A TSH receptor defect in a rat thyroid tumour was correlated with an alteration in the ganglioside content of the membranes of this tumour and with a defect in the biosynthetic enzymes concerned with the synthesis of the higher order gangliosides implicated as components of the TSH receptor structure.[42] Resynthesis of gangliosides in membranes from this rat tumour caused a return in both TSH binding and an effect on adenylate cyclase stimulation (Table 2).[7] Gangliosides could be

Table 2. Improvement in TSH receptor expression of 1-8R thyroid tumour membranes after treatment with a heterologous ganglioside synthesis system

Membrane	^{125}I-TSH binding c.p.m. $\times 10^{-5}$/mg/ membrane protein		TSH-stimulated adenylate cyclase activity picomoles above basal/15 min/mg/ membrane protein	Total cofactor radioactivity in ganglioside from	
	pH 6.0	pH 7.5		^{3}H-CMP AcNeu	^{14}C-UDP GalNAc
				pmoles	pmoles
1-8R tumour	3.8	0.8	0	—	—
1-8R tumour pretreated with chick brain +cofactors[a]	11.4	3.5	110	1.8	48

[a] Incubation system included the following: (i) tumour 1-8R plasma membranes; (ii) chicken brain preparation containing G_{M3}:G_{M2} N-acetylgalactosaminyltransferase activity; and (iii) cofactors: ^{14}C-UDP GalNAc, UDP-Gal and ^{3}H-CMP AcNeu. The G_{M3} acceptor was present in the tunour 1-8R membranes.

reconstituted in liposomes and both bind TSH and respond to TSH with the release of 6-carboxyfluorescein from the liposomes in a manner not significantly different from liposomes containing the glycoprotein component of the TSH receptor.[30,31,43,44]

4.3 The role of the ganglioside may relate to its role in the cholera toxin receptor

Circumstantial support for the role of gangliosides exists if one accepts cholera toxin-G_{M1} as a model and if one accepts that sequence homologies between cholera toxin and TSH or its sister glycoprotein hormones are potentially relevant to similarities in their structure and function.[4–7,37,45] Sequence homologies have been demonstrated to exist between cholera toxin, TSH and the other glycoprotein hormones. One sequence analogy on the β subunit of cholera toxin, TSH and the other glycoprotein hormones is extremely highly preserved and can be argued to be implicated in the cell surface binding reaction since the β subunits of all seem to determine the specificity of the cell surface action. A second sequence homology has been noted between the A protein of cholera toxin and the α subunit of the glycoprotein hormones. This region was also startlingly similar in sequence to the nonapeptide neurohypophyseal hormones, oxytocin and vasopressin.[4–7,37] This sequence has been synthesized and is able to stimulate adenylate cyclase activity in membranes.[45]

4.4　The prime role of the ganglioside in the cholera toxin receptor is to bring the α subunit of the toxin into or through the hydrophobic membrane bilayer

Studies which have compared the mechanism of action of TSH with cholera toxin have recently provided an additional view of the ganglioside role. As noted earlier, there is a regulatory and catalytic subunit in the cyclase complex.[11] The α subunit of cholera toxin catalyses the ADP ribosylation of the G regulatory protein using NAD as a substrate.[12,13] ADP ribosylation is believed to alter GTPase activity, guanyl nucleotide exchange, or both, and thereby increase adenylate cyclase activity.[11–13] TSH can perturb a membrane-associated ADP ribosyltransferase which carries out this same reaction or, in the absence of acceptor, the conversion of NAD to free ADP ribose and nicotinamide.[15,16] If one presumes that the ganglioside receptor is an important component of the cholera toxin action, and that its role is to internalize the α subunit of the toxin to effect the ADP ribosylation reaction, a major presumptive role for the ganglioside in the TSH receptor emerges. *It is important to note, however, that internalization does not mean into the cytoplasm but rather into the hydrophobic plane of the membrane bilayer. Internalization also does not necessarily require subunit separation.*

4.5　*In vitro* studies suggest a multiple role

Understanding of the role of the ganglioside in the receptor structure is complicated by the fact that the *in vitro* interaction between TSH and ganglioside is inhibited by salts.[44] Also disturbing is the observation that they have a low affinity constant in a micellar array ($\sim 1 \times 10^5 \text{ M}^{-1}$)[44] and even in liposomes ($1 \times 10^8 \text{ M}^{-1}$).[43] Thus, although they can account for the existence of a low affinity component on thyroid membranes,[22,43,44] their mechanistic role in a recognition process must be reconciled with these properties.

One approach to understanding function is to reconstitute gangliosides in liposomes and examine their properties with respect to TSH responses. At low levels of gangliosides in liposomes, i.e. at concentrations where the ganglioside alone has no measurable TSH binding, the presence of the ganglioside markedly enhances TSH binding to liposomes containing the glycoprotein receptor component[5,7] (Figure 6). At high ganglioside concentrations, the presence of both components can inhibit TSH binding (Figure 9(a)). This effect is not restricted to the glycoprotein component of the TSH receptor. Thus, thyroid membranes have a glycoprotein component which interacts with immunoglobulins and has characteristics of an F_c receptor (Figure 9(b)). Gangliosides in liposomes have little or no immunoglobulin

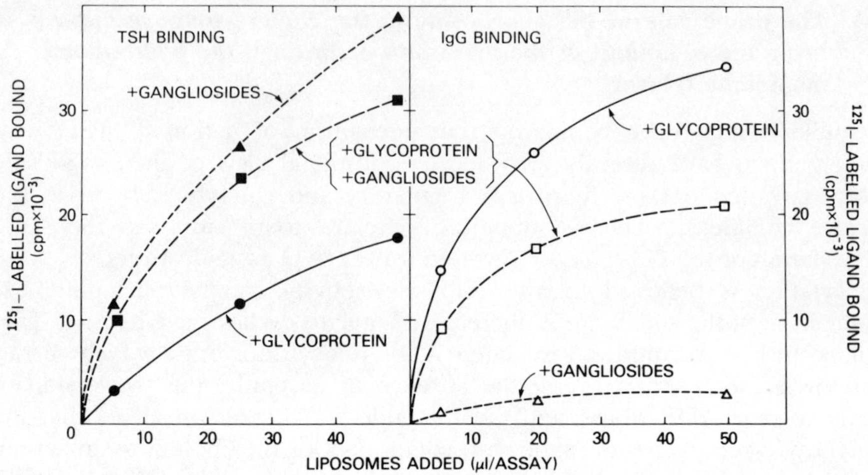

Figure 9. (a) ^{125}I-TSH binding to liposomes containing mixed brain gangliosides (▲), the glycoprotein component of the TSH receptor (●), and an equal mixture of each (■) wherein the ganglioside and glycoprotein receptor concentrations were each equal to the liposomes with only one incorporated. (b) ^{125}I-labelled immunoglobulin G binding to the same liposomes. Details of liposome preparations and assays can be obtained in References 30 and 44

binding activity; however, the presence of the gangliosides at high concentrations can inhibit the binding activity of the glycoprotein serving as the immunoglobulin receptor (Figure 9(b)) as well as that recognizing TSH. *These studies thus suggest that one role of the ganglioside is to modulate binding by the glycoprotein TSH receptor component.* Using 6-carboxy-fluorescein release from liposomes as a more functional assay than simple binding, similar evidence has been obtained[31] (Figure 6).

A most important recent analogy between cholera toxin and TSH ganglioside interactions raised the possibility of the mechanistic basis for the 'internalization' role of the ganglioside. As noted above, salts reduce the interaction of gangliosides with TSH[44] and other glycoprotein hormones.[46] When this was evaluated further,[44] it was noted that a preformed ganglioside-TSH adduct was not significantly salt-sensitive in terms of dissociating the adduct.

The ability of salts to prevent complex formation suggests that the initial association reaction involves electrostatic interactions. The inability of salts to reverse the ganglioside-TSH interaction indicates that short-range interactions, most likely hydrophobic, dominate the kinetic dissociation constant. The hydrophobic nature of the short-range interactions are also indicated by the observation that dansylation of the TSH molecule increased its affinity for gangliosides in all cases and by the observed fluorescence changes of dansyl-TSH in each instance. That *specific* short-range interactions, as opposed to a *generalized 'hydrophobic effect'*, are involved is adduced from the observation that while the addition of Ca^{2+} could not disrupt a pre-formed ganglioside-TSH adduct, it could disrupt a preformed phospholipid-TSH adduct.

Studies of Ca^{2+}-sialic acid adducts offer a potential explanation for the specificity of the short-range interactions insofar as gangliosides are concerned.[47] Ca^{2+} interacts not only with the negatively-charged carboxylate ion in the C1 position but also with the glycerol-like moiety at the C7, C8 and C9 positions of the molecule. In an apolar environment resulting from the approximation of the TSH-ganglioside by short-range hydrophobic forces, the glycerol moiety of the sialic acid might well be involved in hydrogen bonding to amino acid residues of the TSH molecule. The ability of Ca^{2+} to interact with sialic acid would then be significantly reduced; this phenomenon would not be evident with phospholipids.

Gangliosides are amphiphilic molecules which form micelles of 250,000–350,000 molecular weight in aqueous solution at 1×10^{-10} M concentrations.[48] Since the lipid moiety is buried within the micellar matrix, the short-range interaction between ligand and ganglioside must result subsequent to the initial electrostatic interaction involving the negatively-charged polar head groups of the carbohydrate moiety of the ganglioside.

In the context of data indicating that the TSH receptor might involve both a glycoprotein and a ganglioside component, these data would suggest the possibility that the short-range hydrophobic interactions which dominate the ganglioside–TSH dissociation reaction are highly specific; can result in an exclusion of water and salts from approximating surfaces of the TSH and ganglioside; and can bring the ganglioside from the aqueous environment of the glycoprotein receptor component to the hydrophobic environment of the lipid bilayer. The possibility is thus raised that the ganglioside role is to bring a portion of the TSH molecule into the hydrophobic portion of the lipid bilayer where it can interact with other membrane components.

Recent studies of iodide quenching[49] of the tryptophan fluorescence of cholera toxin and of the effect of oligo-G_{M1} on this process support the view that the G_{M1}-cholera toxin interaction functions with exactly the same role. Thus, after an initial interaction involving electrostatic forces, solvent is extruded from between the interacting species; an anhydrous complex is formed; and penetration of the lipid bilayer can occur in the absence of a significant energy barrier.

These results should be viewed with the following perspective. As noted above, current results indicate that the role of G_{M1} is effectively to allow the α subunit of cholera toxin to enter the bilayer, where its α subunit can carry out the ADP ribosyltransferase reaction.[12,13] Extrapolated to TSH, the role of the ganglioside may also be to allow α subunit of TSH to enter the bilayer where it can stimulate the membrane ADP ribosyltransferase and the ADP ribosylation reaction can be initiated. In this regard, it was noted that in analogy to cholera toxin α subunit, the TSH α subunit appears to undergo ADP ribosylation when exposed to membrane preparations.[15,16] It is, however, important to remember that whereas the α subunit of cholera toxin has an intrinsic ADP ribosyltransferase activity, the α subunit of TSH uses a membrane enzyme. *The common denominator is that the role of the ganglioside is to bring the subunit to the location in the bilayer where it can act.*

Before passing further, it is perhaps important to note a potentially important general principle emerging from this model. As noted above, the hormone or ligand is usually in an aqueous environment where its solubility assumes a surface covered with negative and positive charges and sites for hydrogen bonding to water. To bring such a ligand into the oil of the lipid bilayer requires that the water and charge be neutralized or eliminated such that hydrophobic interactions can dominate. The model proposes that the key in this process is a *charged carbohydrate* moiety which can shift from an initial electrostatic to a hydrogen bonding mode, exclude water and neutralize charge, and emulsify the ligand. The model predicts that carbohydrate polymers may be archtypical emulsifying agents.

Vibrational Raman spectroscopy provides a sensitive technique for specifically probing lipid conformations in bilayer assemblies. Raman spectroscopy indicates that G_{M1} is inserted within the bilayer and does not exist as separate micellar aggregates (Figure 10).† Further, G_{M1} increases the intermolecular disorder of the DPPC bilayer while concomitantly increasing its intramolecular order (Figures 10 and 11). Finally, for the G_{M1}-containing liposomes to which cholera toxin has been added, the data demonstrate that whereas the *intermolecular* disorder parameters are not significantly altered, changes arise in the *intramolecular* order characteristics which are best described as an enhanced bilayer order containing localized patches of disordered lipids (Figure 12). Thus, the gangliosides induce a discrete or localized intramolecular and intermolecular chain disordering within the phospholipid matrix around the ganglioside; this relatively small disordering effect is superimposed upon a general increase in intrachain (trans-gauche) order experienced by the lipid chains upon adding either G_{M1} or G_{M1} plus cholera toxin.

If the ganglioside-ligand effects changes in the bilayer structure, ligand-induced membrane permeability changes in cells could be related. *Thus, the possibility exists that another common mechanistic link among all the agents interacting with gangliosides might be related to the ability of this interaction to*

---

† Data in Figures 10 through 12 are derived from a report in preparation by I. R. Hill, I. W. Levin, P. S. Lazo and L. D. Kohn.

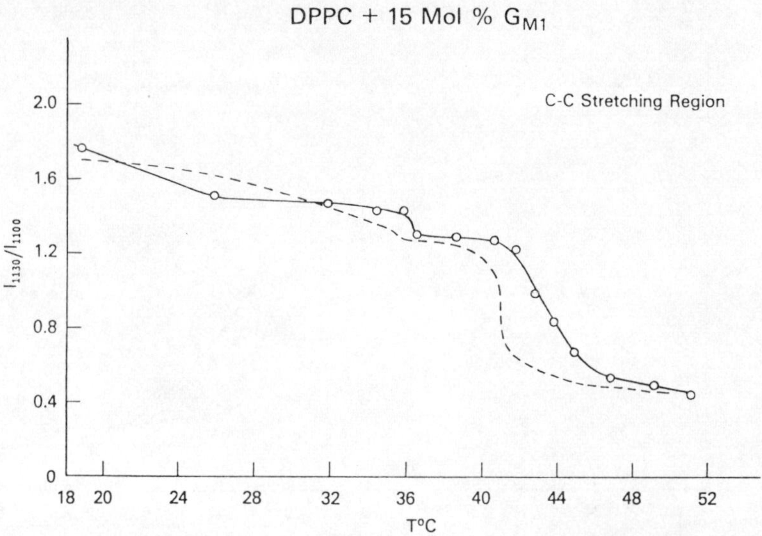

Figure 10. Temperature profile for DPPC-G_{M1} liposomes using the I_{1130}/I_{1100} peak height intensity ratio as a marker (see Figure 12 for details). The solid line represents the DPPC-G_{M1} system, while the dotted line represents the pure DPPC bilayer assembly. The $1000\,cm^{-1}$ to $1170\,cm^{-1}$ region defines the C–C stretching mode region. The decrease in intensity of the $1130\,cm^{-1}$ C–C stretching mode monitors the loss of the all-trans nature of the hydrocarbon chains, while the increase in the $1100\,cm^{-1}$ C–C stretching features implies an increase in gauche conformations along the chain. With the addition of 15 mol per cent G_{M1} to the DPPC bilayer, the pretransition temperature determined by the I_{1130}/I_{1100} marker remains at 36 °C, whereas the temperature determined by the I_{2880}/I_{2940} ratio drops to 33.5 °C (Figure 11). In addition, in comparing the pretransition characteristics for both the DPPC and DPPC-G_{M1} multilayers, we note that the *intramolecular disorder, probed by the I_{1130}/I_{1000} ratios, is approximately the same* for both systems at the pretransition temperature (1.30 and 1.35, respectively), while the *laterial lattice disorder, reflected by the I_{2880}/I_{2940} ratios, is substantially increased* in the bilayer system containing the ganglioside (Figure 11). Although it appears that G_{M1} thus induces somewhat more rotational isomers in the acyl chains in the 21 °C to 30 °C temperature range by comparison to the pure DPPC liposomes, the intramolecular order of the bilayer is definitely increased from about 32 °C to 48 °C (Figure 11). In summary, these results are striking in that G_{M1} enhances the lateral or *intermolecular disorder* of the DPPC bilayer throughout the gel state while, at the same time, increasing the *intramolecular order* of the lipid chains

cause altered ion fluxes across the membrane and that these actions would be a cell signal independent of cAMP.[4–7] The role of gangliosides in this process is at present circumstantial but is suggested as possible and specific by the work of Poss *et al.*[50] and Tosteson and Tosteson.[51] The former workers showed that black lipid films containing G_{T1} were specifically perturbed by TSH to cause a transmembrane ion flux. Films with G_{D1a} and

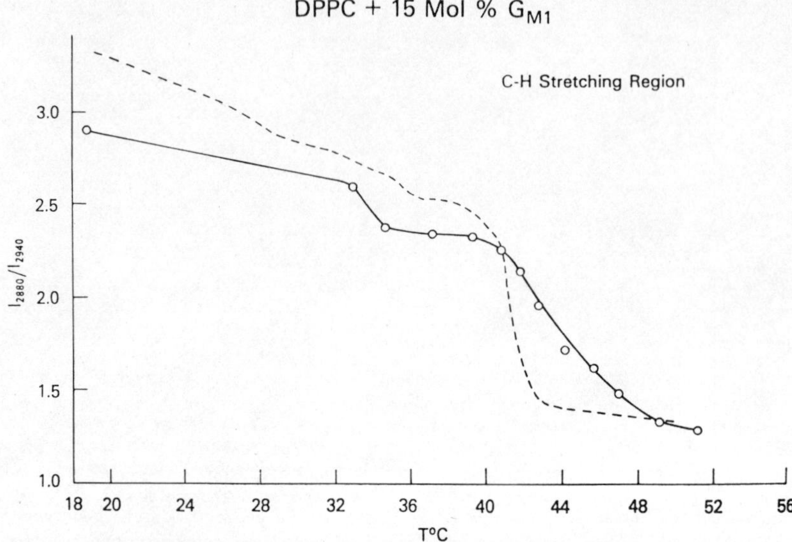

Figure 11. Temperature profile for DPPC-G_{M1} liposomes using the I_{2880}/I_{2940} peak height intensity ratio as a marker. The solid line represents the DPPC-G_{M1} system, while the dotted lines represent the pure DPPC bilayer assembly. The $2800 \, \text{cm}^{-1}$ to $3000 \, \text{cm}^{-1}$ area of the spectra defines the C–H stretching mode region. Thus, the $2880 \, \text{cm}^{-1}$ transition corresponds to the methylene chain CH_2 antisymmetric stretching modes, while the $2940 \, \text{cm}^{-1}$ mode reflects, in part, the symmetric C–H stretching mode of the terminal methyl group of the acyl chain in Fermi resonance with a CH_3 deformation overtone. In indicating both an increase and broadening of the transition temperature, the data clearly imply that G_{M1} is incorporated within the phospholipid bilayer and that the ganglioside and phospholipid exist as miscible components

G_{M1} were not similarly responsive. The latter workers reported similar results for G_{M1}—containing films and cholera toxin. The results of both Poss et al.[50] and Tosteson and Tosteson[51] also emphasize the specific conformational impact of the 'correct' ganglioside on TSH or cholera toxin and the dramatic impact of this 'correct conformational perturbation' on the lipid bilayer.

Figure 12. (*Continued*)

an overall, disordered state for the DPPC lipid matrix, as is usually found in the Raman spectra of phospholipid bilayer assemblies. Thus the spectrum for the liposomes with the added ganglioside and cholera toxin (panel E) represents a *mixture of DPPC lipid molecules of differing degrees of order*. Specifically the spectrum appears to consist of a small concentration of lipid molecules characteristic of the disorder normally present in the 39.5 °C to 44 °C temperature range superimposed upon the low temperature spectrum of DPPC

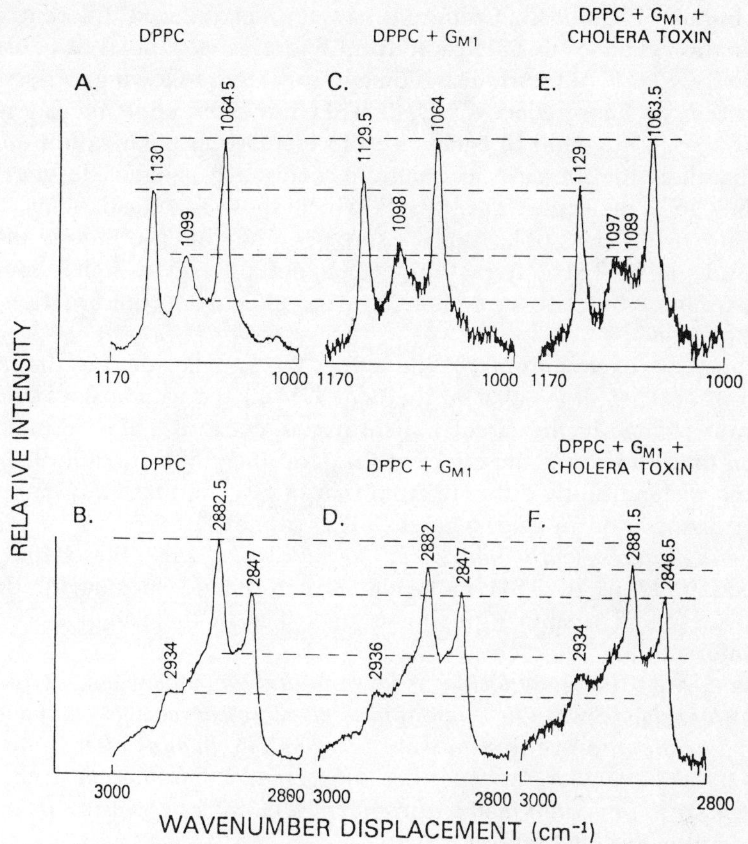

Figure 12. Comparison of the Raman spectra in the $1000\,\text{cm}^{-1}$ to $1170\,\text{cm}^{-1}$ C–C stretching region and in the $2800\,\text{cm}^{-1}$ to $3000\,\text{cm}^{-1}$ C–H stretching region for DPPC multilayers (panels A and B), for DPPC multilayers containing G_{M1} (panels C and D), and for DPPC multilayers containing G_{M1} and perturbed by cholera toxin (panels E and F). Spectra were obtained at $18\,^{\circ}\text{C}$ using a Spex Ramalog 6 spectrometer. Signal averaged spectra of DPPC+G_{M1} are the results of 11 scans, while 265 scans were required for the DPPC+G_{M1}+cholera toxin system. Spectra in the $1000\,\text{cm}^{-1}$ to $1170\,\text{cm}^{-1}$ region are normalized to the $1064\,\text{cm}^{-1}$ transition, while spectra in the $2800\,\text{cm}^{-1}$ to $3000\,\text{cm}^{-1}$ interval are normalized to the $2850\,\text{cm}^{-1}$ feature. Panel E (+ cholera toxin) compared to panel C (no cholera toxin) shows that the $1129\,\text{cm}^{-1}$ transition decreases in intensity on adding the toxin. Also, a redistribution of intensity occurs in the $1098\,\text{cm}^{-1}$ transition in which there is a slight shift to lower frequency of about $1\,\text{cm}^{-1}$ and a filling in of the trough from $1090\,\text{cm}^{-1}$ to $1070\,\text{cm}^{-1}$. Although the vibrational transitions cannot be completely resolved (panel E represents the signal averaging of 265 scans compared to a signal averaged spectrum for panel C of 11 scans), a second spectral transition is noted at $1089\,\text{cm}^{-1}$. The $1089\,\text{cm}^{-1}$ transition, assigned to gauche rotamers within the acyl chain, specifically characterizes disorder in the hydrophobic portion of the lipid bilayer. The important spectra aspect to note is that the pattern in panel E does not simply reflect

The initial indication that different gangliosides induced different specific conformation changes in TSH came from fluorescence studies of native TSH and DNS–TSH.[44] Support also comes from the following observations. Polarization of fluorescence of DNS–TSH increases upon the addition of G_{M1}, G_{D1a} or G_{D1b}, and in each case the increase in polarization quantitatively parallels the increase in quantum yield each ganglioside effected on the DNS–TSH molecule. The results which show a coincident increase in quantum yield and polarization support the interpretation that the ganglioside–DNS–TSH interaction results not only in a higher molecular weight complex but also in a change in the molecular conformation of the TSH molecule.[44]

Further, an examination of the effect of ganglioside on the circular dichroism of TSH shows that in the near UV all the gangliosides effected a significant change in the circular dichroism spectra of TSH.[44] Finally, acid titration data show that the effect of G_{D1b} on the conformation of the TSH molecule is significantly different from that of G_{D1a} in that G_{D1b} can prevent subunit dissociation in acid, whereas G_{D1a} cannot.[44]

The gangliosides clearly induce a conformational change; this change can be profound in terms of TSH molecular structure and specific for different gangliosides; and a similar phenomenon exists in the ganglioside–cholera toxin interaction.

In sum, then, the ganglioside or low affinity component can be argued to serve as a bridge between the high affinity glycoprotein receptor component and elements of the message transmission process. The ligand, after being 'emulsified' by the ganglioside and after the correct conformational change, is brought next to important membrane enzymes in order to transmit its information and to perturb the bilayer.

Finally, it is important to note that the ganglioside cannot be replaced by any acidic phospholipid such as phosphatidylinositol.[30] Thus, phosphatidylinositol has totally different properties of interacting with TSH and does not serve as a binding agent on the surface of liposomes. Phosphatidylinositol in particular is also not enriched on the outer surface of the cell.

5   DEFINITION OF THE RECEPTOR REVISITED

With the above data and models in mind, it is now important to re-examine the several questions raised earlier concerning the definition of a receptor.

5.1   A receptor must be target tissue-specific

This concept evolved from early studies defining a receptor by its response. Thus, TSH-stimulated adenylate cyclase activity and cAMP increases were measured in the thyroid and in thyroid membrane prepara-

tions, but not in liver, brain, muscle, adrenal gland, pancreas or even target
sites for LH and hCG. When binding was the assay of receptor recognition
and when receptor solubilization and purification became the issue, this rule
was immediately violated.

Thus when TSH binding was used to directly measure the recognition
function of the receptor, particularly using non-physiologic conditions, binding to fat cells, mouse L cells, brain membranes and even thyroid fibroblasts
could be measured. This binding was usually lower than to the thyroid
measured per mg membrane protein and by the old criteria could be
considered non-specific and irrelevant. In fact, however, functional responses to TSH can be measured in each of the non-thyroid tissues mentioned. Thus, fat cells do have a TSH-responsive adenylate cyclase.[52] In
mouse L cells, the TSH interaction results in inhibition of the antiviral effect
of interferon.[36] In thyroid fibroblasts, TSH appears to increase thymidine
uptake and cause the release of a growth factor or factors which are
important in establishing thyroid epithelial cell cultures (F. S. Ambesi-
Impiombato and H. G. Coon, manuscript in preparation). Even in neural
tissue, TSH can be shown to undergo retrograde transport as does tetanus
toxin.[53]

In sum, the binding has uncovered a 'receptor' site on non-target tissues
which may be coupled to very different responses in these tissues than the
functional response we have historically associated with thyroid action.
These 'receptors' are not at all 'non-specific' and may even be single
components of the recognition site in the target tissue save that they are
uncoupled to the classic response. Their study can be useful in understanding receptor structure and function.

Loss of target tissue specificity as a rigid requirement of receptor identification is not restricted to the TSH receptor. Thus, in the insulin field, white
blood cells (monocytes) bind with all the specificity of a target cell but with
no functional response. Yet the entire internalization route for insulin can
be studied in these cells.

In short, the binding assay has defined receptors in non-target tissues and
as a result we have defined new functional responses. The non-target tissues
may be like cell mutants with one component of a complex path missing,
thereby allowing better study of other systems. The target tissue specificity
rule must become relative rather than absolute and judged in a framework
of other criteria.

5.2 A receptor must have the same specificity as in its physiologic state

Again, it is evident that this criteria evolved from receptor assays measuring a response function. In the case of the TSH receptor on the thyroid, it
responded to TSH but not LH or hCG.

When binding assays were used to measure, characterize and isolate a solubilized receptor, the criteria immediately broke down. Thus, for example, hCG could inhibit TSH binding to the thyroid membranes at a low level[4–7,19,28,54] and, when looked for, a cyclase response could even be measured.[54] Further, the ability of hCG to inhibit TSH binding to solubilized preparations of the TSH receptor glycoprotein component was increased 10-fold by comparison even to a membrane preparation.[28,29]

Again, the loss in ligand specificity as an absolute requirement of a receptor is intuitive in the models above. If the recognition response process can be considered a series of enzymes in a linear response-generating pathway, each step can contribute to specificity. Further, each step amplifies the previous one. Thus, 95 per cent absolute specificity in the first binding step coupled with lesser (even 50 per cent) specificity in each of two subsequent transmission steps becomes not an additive specificity but one which is enormously multiplied.

This can even be seen if one compares the binding response of a defined component *in vitro* to its *in vitro* response using some measure of receptor function. For example, hCG can inhibit TSH binding to liposomes containing the glycoprotein component of the TSH receptor (Table 1).[31] This inhibition has the properties of a competitive inhibition process. hCG does not, however, cause 6-carboxyfluorescein release from liposomes containing these receptor components but does inhibit TSH-induced release (Figure 6).[31] In short, even if one simply assays the ability of the receptor-ligand complex to perturb lipid organization in a bilayer structure, one can see specificity not evident in the binding assay which measures the same component.

Specificity of inhibition can also be a problem if the membrane preparation or solubilization procedures alter receptor structure or uncover low affinity components not evident in the intact tissue. In this regard, thyroid cells which have no functioning receptor in their intact state may have a TSH-induced cyclase response after a membrane preparation is made.[55] Also, the ganglioside, measurable readily as a low affinity component in membrane preparations by using low salt binding assays,[19,22,44] can serve as an enormous reservoir of low affinity receptor sites which distort displacement studies (with cold ligand) used to measure specificity. This does not negate the ganglioside role, nor does it negate the specificity of the high affinity glycoprotein receptor component measured in direct binding assays. It requires, however, a sum of evidence to be obtained to place each in perspective aside from ligand specificity.

Once again, the chemistry of the system must be understood, and the rule must be taken in a relative sense. It is a rule which reflects, in the final analysis, a property which must be accounted for by the sum of the individual components to be identified, but not necessarily by each component.

5.3 The receptor must be high affinity

This argument reflects the starting point of receptor studies. For most ligands *in vivo* (hormones, for example), the concentration is very low, often below 1×10^{-9} M or 1×10^{-12} M. The initial response to this problem was that the receptor must be of high affinity to recognize such low concentrations of ligand.

Before accepting this as dogma, it is useful to review the history of work in the neurotransmitter field. In almost all cases, a ligand with the *highest binding affinity* is the best *antagonist*, whereas the best *agonist* is a ligand with a *low binding affinity*. If one gives equal credence to the possibility that a receptor with a high affinity for the ligand is not necessarily the best to effect a functional response, one can see an inherent conflict needing revision of the rule.

One way out of this dilemma is elegantly detailed in theoretical studies by A. P. Minton (see Chapter 2 in this book). Thus a high affinity and low affinity site working in concert should be definable: the high affinity for initial recognition, and the low affinity as the function coupler. Further, given the nature of a membrane, a low affinity site measured using 'solution' assays may be very relevant if there is a preconcentration step in the infinitely thin plane of the bilayer; i.e. low affinity becomes meaningless in this micro-environment.

Further, it should be recognized that in the simplest binding format (ligand + receptor $\rightleftarrows$ ligand − receptor − complex), binding under certain circumstances may be dominated by receptor capacity, not receptor affinity. Further, in a whole tissue, receptor capacity can be enormous, whereas it is small on each individual cell. This raises the point that 'apparent' high affinity may be a property of tissue organization rather than an inherent chemical property of a membrane or solubilized receptor in a test tube. The problem again is not to negate this rule but interpret it in a total picture. Experiments must define all components which interact with a ligand and which are associated with a defect where receptor malfunction has been an associated corollary.

In the case of the TSH receptor, the ganglioside exhibits low affinity and has lower specificity *in vitro* than anticipated from a 'physiologically relevant' receptor. Alone, this is true; in concert with a high affinity glycoprotein binding component, it may be an irrelevant concern. The necessity is to define the potential role by *in vitro* experimentation and then return to the *in vivo* state to see if the predictive role exists.

The low affinity component is not a dilemma solely of the TSH field. For example, the entire coated pit internalization process has been associated with the presence, in coated pits, of high affinity receptors.[56] It is clear from several studies (high density lipoprotein receptors, α-macroglobulin receptors, the asialoglycoprotein receptor) that internalization is not associated with the loss of high affinity receptors. The question then becomes, to what are these ligands binding when internalized.

Observations by I. H. Pastan and M. C. Willingham (manuscript in preparation) which define a low affinity component may be critical in resolving this question. Thus, bacitracin is one of a series of trans-glutaminase inhibitors which blocks α_2-macroglobulin internalization by a coated pit mechanism. Bacitracin in a binding experiment *blocks high affinity but not low affinity* binding of α_2-macroglobulin. If the low affinity component is acting as the intracellular 'functional receptor', this could contribute to data wherein membrane preparations exhibit 10-fold or more binding than cells in studies of the asialoglycoprotein receptor.[57] This could also account for the observation that the apparent high affinity external receptor pool may not interchange with the internal receptor pool in some experiments.[58] Thus, the external receptor can be a complex of low and high affinity components; the internal pool only low affinity. Physiologic function requires both.

5.4 *In vitro* conditions must be reflective of the physiologic milieu to have a true measure of a physiologic receptor

This argument exists in each new field when chemists start attacking a system. Each physiologic process has as its base a chemical process. In turn, each chemical process is optimized *in vitro* according to conditions which favour and reflect the nature of the chemical interactions—electrostatic or hydrophobic. *In vitro*, however, these interactions can be important either in the reaction itself or in the integrity and stability of the reactants in a subcellular preparation with a new microenvironment. As long as end response is the measurement of a receptor, the integrity of the system is less likely to survive non-physiologic conditions, i.e. conditions which may be extremely beneficial to a single component chemical process. When only the recognition process is measured, as in a binding assay, non-physiological optima almost become inherent in the assay.

A case in point is the physiologic temperature optimum. As an enzyme, adenylate cyclase activity has a positive temperature coefficient and increases its activity significantly at 37 °C. Thus it is no surprise that TSH-stimulated adenylate cyclase activity is better at 37 °C than at 0–4 °C.

In contrast, a recognition process involving a binding reaction can be detrimentally affected. Thus increased temperature could affect the off reaction more than the on reaction and actually decrease binding. This would be less likely if the receptor–ligand interaction were multivalent, since in an equilibrium situation not all interactions would be expected to be lost simultaneously. Binding may thus best be measured at 0–4 °C where processing reactions affected by enzymes are not presumed operative. For an enzymologist, low temperature assays are a normal procedure to distinguish a binding from a catalytic property. In a lipid bilayer, 0–4 °C assays would also emphasize hydrophobic interactions. Since these must be an important property of the ligand–receptor recognition process in a lipid bilayer environment, binding studies at 0–4 °C would again be better than those at 37 °C.

Requirements to use physiologic conditions of salt and pH in *in vitro* assays are also difficult to rationalize as an absolute criteria for measurements of a physiologically relevant receptor. The basis for this conclusion is evident in a study wherein the properties of trypsin were studied when free in solution or attached to a semipermeable surface.[59] Salt and pH optima shifted dramatically. The rationale expressed was that the microenvironment of the solid phase-associated enzyme was very different from the solution phase and that optimal conditions were a balance of the two. If one compares membranes in an aqueous suspension in a test tube with the closely packed state of a solid tissue, the same analogy holds.

What is the true microenvironment? Is the pH and salt concentration of the plasma or its immediate extracellular compartment relevant to a milieu in the shade of a forest of glycoproteins whose leaves are filled with negatively-charged sialic acid residues and whose floor is constantly subjected to floods or sudden withdrawals of protons and ions leaked from or into an interior compartment?

In sum, conditions of pH, salt and temperature may be 'physiologic' in an *in vitro* receptor assay, but this could be fortuitous rather than indicative of physiologic relevance. Further, it can be restrictive to examine chemical properties only in physiologic conditions. Had this been done in the past, numerous physiologically relevant enzymes would be undiscovered and *in vitro* protein synthesis experiments might still be missing tRNA supplements of the 'pH 5.0 supernatant'.

Clearly, however, non-physiologic assay conditions impose a new problem to workers in the area. Any component so defined will require alternative proof of relevance. In the case of the TSH receptor, critical early experiments showed that losses in TSH binding measured in a non-physiologic assay milieu were still able to be readily correlated with losses in TSH-stimulable adenylate cyclase activity[27] and membrane potential in more a physiologic or functional milieu.[8] Reconstitution experiments[31,34] and antibody definition experiments[32,33] are more direct evidence which will be necessary.

Before leaving this point, it is worth mentioning a corollary problem. If a binding assay is used, particularly one using conditions which enhance electrostatic interactions, interactions with plastic and glass surfaces may also be noted and compared. Again, it is easy to argue that this is a reflection of a 'non-specific' interaction; however, it is equally possible that the interaction with glass reflects an intrinsic physicochemical reaction which is highly amplified on the membrane. It is key to remember that an enzyme does not create a new reaction—it merely catalyses an existent one. The biologic membrane does not create new physicochemical interactions by comparison to glass and plastic—it merely incorporates them into a different planar surface and uses them as part of a catalytic or amplified process of cell signalling. Talc binds insulin, but talc is not present in a fat cell membrane. That this membrane has properties of talc is not impossible and in no way implies that the membrane binding assay is irrelevant. The assay is a means to an end; that end requires definitive evidence from alternative experiments. The burden of proof will, however, require again that the basic physiochemical interactions be characterized and explained, and shown to be a part of the amplication process.

5.5 Which component has the information: the ligand or receptor?

In many receptor fields, the ligand was believed to provide the major input of information to the message signalling process. In the case of interferon, species specificity of interferons was related to the function of the tissue and receptor rather than the ligand. In both cases the view was influenced by the technology.

The epinephrin and insulin molecules were well defined, derivatives existed or could be made, and functional correlates were easy to make. The receptor, in contrast, was an unknown structure, poorly susceptible to defined chemical modification experiments. In the interferon field, interferon was so difficult to obtain pure, it was easier to study biologic responses of the cell and confer properties to the receptor, not the ligand. Clearly, the situation is in the middle, as both carry key information.

The problem in large hormones or ligands such as TSH is to define 'active sites' and synthesize them. Recent work adapting the computer sequence analyses[60] to predictive sequence synthesis[45] suggests ligand structure will be more readily examined. In the case of the TSH α subunit, a sequence resembling oxytocin and vasopressin has been defined, a related sequence has been synthesized, and the peptide has been shown to stimulate adenylate cyclase in membrane preparations. The mechanism is now under investigation using other derivatives but allows an approach to active site chemistry. In the parathyroid hormone field, this is already far advanced.

In an equal but opposite sense, the identification of the ganglioside as a potential receptor component has led to experiments which indicated that the TSH receptor must have information to cause a critical conformation change in the ligand; that this is part of the recognition process; and that, in turn, a shift in receptor conformation is a key of the message transmission sequence.

5.6　Receptor definition revisited

In the context of these problems, it is evident that no single criteria qualifies a binding component to be a relevant component of the physiologic receptor, but neither does any single rule eliminate such a component. In the final analysis, an enzymologic approach is necessary. Purification will be followed by reconstitution. It will be necessary to develop antibodies to purified components, or use monoclonal antibodies to single determinants to resolve whether a purified or identified component is a valid part of the receptor. ('Monoclonal Receptor Antibodies' are discussed in Chapter 3 of this book.) Until then, evidence may rest on circumstantial associations of binding and function, such as correlations of abnormal function in association with lost binding.

Hopefully, chemically- and physiologically-oriented research will ultimately meet in a unified receptor concept. In the interim, the physiologist will hopefully recognize the chemist's problem and hope that this approach will provide new tools to probe physiologic function rather than be exploring blind alleys.

6　RECEPTOR AND THE BILAYER

Elsewhere in this book (Chapter 4 by S. M. Aloj), the role of phospholipids in modulating receptor expression will be discussed in detail. It may, however, be pertinent to emphasize that the state of the bilayer and the interaction of one cell upon its neighbour may have profound influence on receptor expression. As noted above, this relationship can be lost in a membrane preparation.

In the case of the TSH receptor, this issue has been approached in two separate ways in recent studies. Thus, Pekonnen and Weintraub[61] showed that exposing human thyroid membranes to 37 °C at pH 5.0 in high salt concentrations (1 M ammonium sulphate) results in a major increase in receptor number. The receptor could be assayed in higher salt concentrations presumed relevant to a physiologic milieu and retained a high degree of specificity in terms of inhibition by unlabelled TSH and other ligands. Their data suggested this could not result from uncovering sites with bound TSH and that this was related to the glycoprotein receptor component. The last conclusion is evident from studies showing trypsin sensitivity of TSH binding.

In a separate approach,[38] cell–cell interactions are argued to cause a major increase in receptor affinity and a shift in *in vitro* optima. Thus, ^{125}I-TSH binding to bovine thyroid membranes can be enhanced if concanavalin A (at concentrations as low as $5\,\mu\mathrm{g\,ml^{-1}}$) is used to pretreat the membranes or pretreat liposomes containing the glycoprotein component of the TSH receptor. Increased ^{125}I-TSH binding is not the consequence of an interaction between the hormone and plant lectin. Thus, concanavalin A and TSH do not form a complex under assay conditions as determined by gel filtration chromatography, and ^{125}I-TSH binding to pretreated membrane or liposome preparations is neither prevented nor reversed by 0.1 M α-methyl mannoside. Increased ^{125}I-TSH binding is affected only by the tetrameric form of concanavalin A, although both tetrameric and non-tetrameric preparations of concanavalin A appear to interact with the same membrane components, and neither interacts with the TSH receptor. Increased TSH binding by tetramer concanavalin A pretreated membranes is not associated with any increase in cholera toxin interactions with these same membranes.

Increased ^{125}I-TSH binding to membranes pretreated with tetramer concanavalin A can be both prevented and reversed by unlabelled TSH but not by unrelated hormones such as growth hormone, glucagon or prolactin. hCG inhibits poorly; cholera toxin does not inhibit, whereas it is an inhibitor with normal membranes. By comparison to untreated thyroid membranes, ^{125}I-TSH binding to tetramer concanavalin A pretreated membranes exhibits more physiologic optima; i.e. *in vitro* binding is readily measured at pH 7.5, 37 °C and in 0.1 M sodium chloride. The increased binding and shifted condition optima appear to result from a major increase in affinity of the glycoprotein receptor component (Figures 7 and 8 above).

Plant lectins are known to react with specific carbohydrate moieties on the glycoprotein or glycolipid components of cells, i.e. moieties which are important in regulating membrane recognition processes.[62] Concanavalin A is a particularly interesting lectin in this respect. Its action on cells and membranes has been linked to effects on the fluid state of the membrane or to the physical state of the membrane lipids.[63,64] This, in turn, causes changes in the function of other receptors on the surface of cells or liposomes.[65,66] In addition, dimeric and tetrameric forms of the lectin exhibit similar binding specificity, yet have very different effects in terms of cell agglutination and capping phenomena. Since TSH interactions with the glycoprotein receptor component are already known to be controlled by alterations in the physical state of the membrane[30,67,68] and receptor expression can be modified by changing the concentration of membranes used in an *in vitro* binding assay, i.e. by altering membrane–membrane interactions,[68] these results emphasize that the orientation and influence of the lipid bilayer is not neutral.

In this respect, it is important to emphasize that the presumed mechanism[38] is that the concanavalin A interacts with a site or sites on a membrane distinct from the TSH receptor. Further, it is presumed that binding *per se* is not the 'inducer' but rather that the 'agglutination' of these glycoproteins on the membrane surface restructures the lipid domain of the TSH receptor. Clearly, this is speculative, but this type of approach can be adapted to examine questions of receptor function *in situ,* i.e. to examine the effects of other agents with known effects on membrane structure.

7 THE RECEPTOR AND REGULATION OF MESSAGE TRANSMISSION

In several chapters in this book, receptor coupling to ionic fluxes will be dealt with in great detail. (See Chapter 8 by E. F. Grollman and Chapter 10 by D. L. Gill as examples.) Similarly, articles will deal with regulation of kinase activities by Ca^{2+}- and cAMP-mediated processes. (See Chapter 12 by J. D. Gardner, for example.)

Accordingly, attention will be devoted here to the regulation that might result from coupling of the receptor recognition and receptor transmission process. The reader is also referred to Chapter 13 by D. M. F. Cooper and C. Londos.

7.1 Receptor recognition, transmission of message and feedback regulation

As noted earlier, the original description of the TSH receptor used the end response, adenylate cyclase stimulation, as the assay, and the receptor was viewed as a complex with recognition and functional elements. With the recognition that adenylate cyclase was really a complex of multiple components[11] and that the recognition aspect was the primary province of different molecules, the question of receptor (recognition) coupling to the cyclase (transmission) action was a key problem. This is intensified by the knowledge that the receptor and cyclase can be in distinct membrane locales.

In the TSH receptor situation, the cholera toxin model was a starting reference point.[4–7,12,13,15,16,39,40] Thus, as noted above, TSH was shown to stimulate in a dose-dependent and hormone-specific manner thyroid membrane ADP ribosyltransferase activity.[15,16] The result of the action was to increase the incorporation of $[^{32}P]$ADP ribose into several membrane components, including an approximately 40,000 molecular weight component also ADP ribosylated by cholera toxin and believed to be the G regulatory subunit of the adenylate cyclase complex.

The labelling of the membrane components was, however, notable in two respects. First, the process appeared to precede in time the labelling of supernatant components, which migrated with TSH or the α subunit of TSH. Second, although the 40,000 molecular weight membrane component appeared more heavily labelled than several other membrane components in the presence of TSH, the basal and TSH-stimulated ADP ribosyltransferase activity appear to be ADP ribosylating the same membrane components. Whether this reflects a residual 'stable' TSH stimulatory activity in the membrane preparation or correlated with the 'basal' adenylate cyclase activity of membrane preparations was unclear.

As noted above, ADP ribosylation of the G regulatory component of the adenylate cyclase complex is believed to be the means by which cholera toxin stimulates adenylate cyclase activity.[12,13] It is this belief, coupled with the known sequence and receptor analogies between cholera toxin and TSH,[4-7] which readily led to the speculation that TSH regulation of thyroid membrane ADP ribosyltransferase activity may be important in the TSH regulation of adenylate cyclase activity. The ability of β-TSH to inhibit both ADP ribosyltransferase activity and adenylate cyclase activity supported this view as did an NAD effect on the TSH-stimulated cyclase in these membranes.

Functioning thyroid cells in culture can be grown by including TSH in the media.[27,55,69,70] These cells have a high cyclic AMP level, concentrate iodide and synthesize thyroglobulin. They are, however, desensitized[71] to further additions of TSH; i.e. more TSH does not increase any of the above functional parameters.[27,55,69,70] Non-functioning thyroid cells have a low cyclic AMP level; do not concentrate iodide; and do not synthesize thyroglobulin;[27,55,69,70] this phenomenon appears to relate to a lack of functionally exposed membrane receptors on these cells[32] until a membrane preparation is made.[55]

Membranes from functioning thyroid cells have been shown to have *low* ADP ribosyltransferase but *high* NAD glycohydrolase activities;[15,16] NAD glycohydrolase activity can be viewed as the half reaction of ADP ribosyltransferase wherein water is substituted for an acceptor protein. In exact opposition to these results, membranes from non-functioning thyroid cells have a *high* ADP ribosyltransferase but a *low* NAD glycohydrolase activity. Both membrane preparations have an enzyme activity capable of removing AMP from monoADP ribosylated acceptor (Figure 4).

Evidence also exists which shows that the different ADP ribosyltransferase activities in membranes from functioning and non-functioning thyroid cells really reflects a low level of available membrane acceptor in the functioning cell preparation and a high level in the non-functioning cell preparation. In short, the difference is not in the enzyme but in the acceptor. It is suggested that the 'desensitized' state of functioning thyroid cells could thus be accounted for by correlating the following: a high level of ADP ribosylated acceptor; a high cyclic AMP level; a stable functioning state; and a lack of responsiveness to additional amounts of TSH because of non-available free acceptor.

Integrating these results, the speculative model of receptor regulation of adenylate cyclase activity was proposed (Figure 4 above). There exist in thyroid membranes two enzymes: E_1 (ADP ribosyltransferase) catalyses the ADP ribosylation of the G regulatory protein and E_2 (the enzyme removing

AMP from the monoADP ribosylated acceptor) catalyses an initial step in the reversal of this process. TSH, by increasing the activity of E_1 or decreasing the activity of E_2, causes a net increase in activated ADP ribosylated G protein. In the absence of an acceptor substrate, i.e. the G protein, the ADP ribosyltransferase activity catalyses the conversion of NAD to free ADP ribose and nicotinamide and/or ADP ribosylates TSH itself. Thus, functioning thyroid cells in culture, chronically stimulated with TSH, might have a fully ADP ribosylated G regulatory protein; a stable activated adenylate cyclase activity; and a desensitized state with TSH undergoing ADP ribosylation or NAD undergoing hydrolysis. The 'turn-off' or desensitization to TSH is, in essence, an inherent regulatory property of the receptor coupling to the message transmission complex.

It is pertinent to point out the following in regard to this phenomenon. A preliminary report has already indicated that nicotinamide can desensitize thyroid cells chronically exposed to TSH.[72] It is also important to recognize that ADP ribosylated TSH may also be an intermediate in the cyclase activation process, rather than simply a product of the desensitization reaction. Thus it is known that cholera toxin causes 'auto' ribosylation; the 'auto' ADP ribosylation affects the α subunit of cholera toxin, i.e. the counterpart of the TSH α subunit; this ADP ribosylated α subunit is bioactive in the cyclase assay.[13] It is further notable that a stable stimulation of adenylate cyclase activity in membranes exposed to TSH and then washed free of ligand (i) is similar to the stable activation state effected by cholera toxin and (ii) has been potentially ascribed to a stable molecular derivative of TSH.[73]

Independent of these considerations, however, it is clear that a complex regulation system involving receptor, cyclase and other membrane enzymes is emerging. Already evidence is suggestive that GTP and ATP can inhibit the TSH-stimulated ADP ribosylation reaction.[6,15,16] Since these are activators of the cyclase independent of the receptor, this may be a means of insuring or regulating recognition of a transmitted response.

The peptide on the α subunit of TSH[45] can stimulate ADP ribosyltransferase activity of the thyroid membrane; however, this action is very related to the GTP and fluoride effect.[74] Again, a regulation point centred at the interaction of receptor and G regulatory subunit is raised.

Whether all of this is correct is the subject for intense study and will no doubt be actively debated in the field. *The key message for the reader, however, is that the regulation or turn-off of the receptor signal must be considered.* The recognition of these intermediate activities affords us new means of testing agents and models to approach an understanding of reality.

An alternative regulation scheme could be mentioned, i.e. receptor internalization and ligand/receptor destruction or coupling to internal protein synthetic sites. The scope of this chapter and book limited discussion of this aspect; however, the reader is referred to reference 75, where consideration of relationships between toxins and hormones dealt in detail with this problem.

7.2   Regulation by end products

Before finishing this discussion, some reference to recent observations of DeGroot and his coworkers is in order.[76] These workers showed that thyroglobulin could inhibit TSH binding to thyroid membranes.[76] Subsequent reports[77-79] have confirmed this and shown (i) that the thyroglobulin interaction with the membrane involves a binding site involving both GlcNAc residues on the B carbohydrate moiety of thyroglobulin and tyrosine residues of thyroglobulin and (ii) that the interaction may reflect the interaction of thyroglobulin with components of the iodination process. Although very speculative, it is easy to see that this *in vitro* observation can lead to the following regulatory proposal.

Thyroglobulin degradation or conversion to T_3 and T_4 is a TSH-regulated event. If thyroglobulin iodination and conversion to T_3/T_4 were, in turn, to regulate TSH receptor expression, a feedback loop could exist. New experiments will validate or disprove this idea. Nevertheless, the end result will be a new input to our picture of a dynamic-regulated receptor *in vivo* by comparison to the *in vitro* 'artifact' we must study.

8   SUMMARY

The purpose of this chapter is not to present the final or even correct model of TSH receptor structure and function. Rather, the current speculative model presented is used to open the door to a more broad view of the receptor problem and controversy as it has evolved today. Questions of how we define a receptor are clearly very much in flux and much more difficult than initially considered when a chemical approach is taken. Numerous binding components will be described and their relevance to the physiologic state will be debated. Some will be clearly erroneous in concept—yet the very debate and data will open new ideas and approaches other than repetitive membrane binding or response measurements. The remainder of this book will explore numerous other aspects of receptor structure, regulation and function. The reader may be disturbed by the complexity and extrapolations of data and the weakness of the models. The reader should, however, remember that the receptor is the key link of the cell to its environment. The complexities of this linkage are evident in our continued concern with knowledge of the mechanisms our bodily senses utilize. The controversy that will exist is evident in the arguments we have today over the agents in our environment which affect us and the mechanisms of these effects. It is hoped that this chapter and book will provide both the desire and some reference to follow and review the data in all receptor fields as they emerge in the next several years.

REFERENCES

1. Robbins, J., Rall, J. E., and Gordon, P. (1974). In *The Thyroid and Iodine Metabolism in Duncan's Diseases of Metabolism* (Bondy, P. K. and Rosenberg, L. E., eds), pp. 1009–1104, W. B. Saunders Co., Philadelphia, Pa.
2. Shoma, B., Liao, T. H., Howard, S. M., and Pierce, J. G. (1971). *J. Biol. Chem.*, **246**, 833–849.
3. McKenzie, J. M., and Zakarija, M. (1977). *Recent Prog. Horm. Res.*, **33**, 29–53.
4. Kohn, L. D. (1978). In *Receptors and Recognition* (Cuatrecasas, P. and Greaves, M. F., eds.) Vol. 5, Series A, pp. 134–212, Chapman and Hall, London.
5. Kohn, L. D., Consiglio, E., De Wolf, M. J. S., Grollman, E. F., Ledley, F. D., Lee, G., and Morris, N. P. (1980). In *Structure and Function of Gangliosides* (Svennerholm, L., Mandel, P., Dreyfus, H., and Urban, P.-F., eds.), pp. 487–504, Plenum Press, New York. *Adv. Exp. Med. Biol.*, **125**, 487–504.
6. De Wolf, M. J. S., Yavin, E., Yavin, Z., Consiglio, E., Vitti, P., Shifrin, S., Epstein, M., Gill, D. L., Grollman, E. F., Lee, G., and Kohn, L. D. (1980). In *Radioimmunoassay of Hormones, Proteins and Enzymes* (Albertini, E., ed.) pp. 3–12, Elsevier North-Holland, Amsterdam.
7. Kohn, L. D., Consiglio, E., Aloj, S. M., Beguinot, F., De Wolf, M. J. S., Yavin, E., Yavin, Z., Meldolesi, M. F., Shifrin, S., Gill, D. L., Vitti, P., Lee, G., Valente, W. A., and Grollman, E. F. (1981). In *International Cell Biology 1980–1981* (Schweiger, H. G., ed.) pp. 696–706, Lange and Springer, Berlin.
8. Grollman, E. F., Lee, G., Ambesi-Impiombato, F. D., Meldolesi, M. F., Aloj, A. M., Coon, H. G., Kaback, H. R., and Kohn, L. D. (1978). *Proc. Natl. Acad. Sci. U.S.A.*, **74**, 2352–2356.
9. Wang, J. H., and Waisman, D. M. (1979). *Curr. Top. Cell. Regul.*, **15**, 47–107.
10. Klee, C. B., Crouch, T. H., and Richman, P. G. (1980). *Annu. Rev. Biochem.*, **49**, 489–516.
11. Ross, E. M., and Gilman, A. G. (1980). *Annu. Rev. Biochem.*, **49**, 533–564.
12. Gill, D. M. (1977). In *Advances in Cyclic Nucleotide Research* (Greengard, P., and Robison, G. A., eds.) Vol. 8, pp. 85–118, Raven Press, New York.
13. Moss, J., and Vaughan, M. (1979). *Annu. Rev. Biochem.*, **48**, 581–600.
14. Moss, J., Ross, P. S., Agosto, G., Birken, S., Canfield, R. E., and Vaughan, M. (1978). *Endocrinology*, **102**, 415–419.
15. De Wolf, M. J. S., Vitti, P., Ambesi-Impiombato, F, S., and Kohn, L. D. (1981). *J. Biol. Chem.*, **256**, in press.
16. Vitti, P., De Wolf, M. J. S., Acquaviva, A. M., Epstein, M., and Kohn, L. D. (1981). *Proc. Natl. Acad. Sci. U.S.A.*, **78**, in press.
17. Wolf, J., and Jones, A. B. (1971). *J. Biol. Chem.*, **246**, 3939–3947.
18. Wolf, J., Winand, R. J., and Kohn, L. D. (1974). *Proc. Natl. Acad. Sci. U.S.A.*, **71**, 3460–3464.
19. Amir, S. M., Carraway, T. F., Jr., Kohn, L. D., and Winand, R. J. (1972). *J. Biol. Chem.*, **248**, 4092–4100.
20. Lissitzky, T., Fayet, G., Nerrier, B., Hennen, G., and Jacquet, P. (1973). *FEBS Letts.*, **29**, 20–24.
21. Moore, W. V., and Wolf, J. (1974). *J. Biol. Chem.*, **249**, 6255–6263.

22. Tate, R. L., Schwartz, H. I., Holmes, J. M., Winand, R. J., and Kohn, L. D. (1975). *J. Biol. Chem.*, **250,** 6509–6515.

23. Verrier, B., Planells, R., and Lissitzky, S. (1977). *Eur. J. Biochem.*, **74,** 243–252.

24. Azukizawa, M., Kurtzman, G., Pekary, E., and Hershman, J. M. (1977). *Endocrinology*, **101,** 1880–1889.

25. Manley, S. W., Bourke, J. R., and Hawker, R. (1974). *J. Endocrinol.*, **63,** 437–448.

26. Powell-Jones, C. H. J., Thomas, C. G., Jr., and Nayfeh, S. N. (1979). *Proc. Natl. Acad. Sci. U.S.A.*, **76,** 705–709.

27. Winand, R. J., and Kohn, L. D. (1975). *J. Biol. Chem.*, **250,** 6534–6540.

28. Tate, R. L., Holmes, J. M., Kohn, L. D., and Winand, R. J. (1975). *J. Biol. Chem.*, **250,** 6527–6533.

29. Tate, R. L., Winand, R. J., and Kohn, L. D. (1976). In *Thyroid Research* (Robbins, J., and Braverman, L. E., eds.) pp. 57–60, American Elsevier Publishing Co., New York.

30. Aloj, S. M., Lee, G., Grollman, E. F., Beguinot, F., Consiglio, E., and Kohn, L. D. (1979). *J. Biol. Chem.*, **254,** 9040–9049.

31. Beguinot, F., Formisano, S., Consiglio, E., Kohn, L. D., and Aloj, S. M. (1981). *J. Biol. Chem.*, **256,** manuscript submitted.

32. Yavin, E., Yavin, Z., Schneider, M. D., and Kohn, L. D. (1981). *Proc. Natl. Acad. Sci. U.S.A.*, **78,** 3180–3184.

33. Yavin, E., Yavin, Z., Schneider, M. D., and Kohn, L. D. (1981). In *Monoclonal Antibodies in Endocrine Research* (Fellows, R. E., and Eisenbarth, G., eds). Raven Press, New York, 53–67.

34. Maheshwari, R. K., Lazo, P. S., Friedman, R. M., and Kohn, L. D. (1980). In *Interferon: Properties and Clinical Uses* (Khan, A., Hill, N. O., and Dorn, G. L., eds.) pp. 386–397, Leland Fikes Foundation Press of Wadley Institutes of Molecular Medicine, Dallas, Texas.

35. Besancon, F., Ankel, H., and Basu, S. (1976). *Nature*, **249,** 576–578.

36. Kohn, L. D., Friedman, R. M., Holmes, J. M., and Lee, G. (1976). *Proc. Natl. Acad. Sci. U.S.A.*, **73,** 3695–3699.

37. Grollman, E. F., Lee, G., Ramos, S., Lazo, P. S., Kaback, H. R., Friedman, R. M., and Kohn, L. D. (1978). *Cancer Res.*, **38,** 4172–4185.

38. Tombaccini, D., Shifrin, S., De Wolf, M. J. S., Lee, G., Valente, W., Yavin, E., Epstein, M., and Kohn, L. D. (1981). *J. Biol. Chem.*, **256,** submitted.

39. Bennett, V., and Cuatrecasas, P. (1977). In *The Specificity and Action of Animal, Bacterial, and Plant Toxins* (Cuatrecasas, P., and Greaves, M. F., eds.), pp. 3–66, Chapman and Hall, London.

40. Holmgren, J. (1978). In *Bacterial Toxins and Cell Membranes* (Jeljaszewick, J., and Wadström, T., eds.), pp. 333–366, Academic Press, New York.

41. Mullin, B. R., Pacuszka, T., Lee, G., Kohn, L. D., Brady, R. O., and Fishman, P. H. (1978). *Science*, **199,** 77–79.

42. Meldolesi, M. F., Fishman, P. H., Aloj, S. M., Kohn, L. D., and Brady, R. O. (1976). *Proc. Natl. Acad. Sci. U.S.A.*, **73,** 4060–4064.

43. Aloj, S. M., Kohn, L. D., Lee, G., and Meldolesi, M. F. (1977). *Biochem. Biophys. Res. Commun.*, **74,** 1053–1059.

44. Aloj, S. M., Lee, G., Consiglio, E., Formisano, S., Minton, A. P., and Kohn, L. D. (1979). *J. Biol. Chem.*, **254,** 9030–9039.

45. Epstein, M., Ross, E. M., De Wolf, M. J. S., Fridkin, M., and Kohn, L. D. (1980). In *Novel ADP-Ribosylation of Regulatory Enzymes and Proteins* (Sugimura, T., and Smulson, M., eds.), pp. 369–379, Elsevier/North-Holland, New York.

46. Lee, G., Aloj, S. M., Brady, R. O., and Kohn, L. D. (1976). *Biochem. Biophys. Res. Commun.*, **73,** 1976–1984.
47. Jaques, L. W., Brown, E. B., Barrett, J. M., Brey, W. S., Jr., and Weltner, W., Jr. (1977). *J. Biol. Chem.*, **252,** 4533–4538.
48. Formisano, S., Lee, G., Johnson, M. L., Aloj, S. M., and Edelhoch, H. (1979). *Biochemistry*, **18,** 1119–1124.
49. De Wolf, M. J. S., Fridkin, M., and Kohn, L. D. (1981). *J. Biol. Chem.*, **256,** 5489–5496.
50. Poss, A., De Leers, M., and Ruysschaert, J. M. (1978). *FEBS Letts.*, **86,** 160–162.
51. Tosteson, M. T., and Tosteson, D. C. (1978). *Nature*, **275,** 142–145.
52. Mullin, B. R., Lee, G., Ledley, F. D., Winand, R. J., and Kohn, L. D. (1976). *Biochem. Biophys. Res. Commun.*, **69,** 55–62.
53. Lee, G., Grollman, E. F., Dyer, S., Beguinot, F., Kohn, L. D., Habig, W. H., and Hardegree, M. C. (1979). *J. Biol. Chem.*, **254,** 3826–3832.
54. Carayon, P., Lefort, G., and Nisuka, B. C. (1980). *Endocrinology*, **106,** 1907–1916.
55. Wadeleux, P. A., Etienne-Decerf, J., Winand, R. J., and Kohn, L. D. (1978). *Endocrinology*, **102,** 889–902.
56. Goldstein, J. L., and Brown, M. S. (1977). *Annu. Rev. Biochem.*, **46,** 897–930.
57. Neufeld, E., and Ashwell, G. (1980). In *The Biochemistry of Glycoproteins and Proteoglycans* (Lennarz, W., ed.), pp. 241–266, Plenum Press, New York.
58. Stockert, R. J., Howard, D. J., Morell, A. G., and Scheinberg, I. T. (1980). *J. Biol. Chem.*, **255,** 9028–9029.
59. Goldman, R., Kedem, O., and Katchalski, E. (1968). *Biochemistry*, **7,** 4518–4524.
60. Dayhoff, M. O. (1973). *Atlas of Protein Sequence and Structure*, National Biomedical Research Foundation, Washington, D.C.
61. Pekonen, F., and Weintraub, B. D. (1980). *J. Biol. Chem.*, **255,** 8121–8127.
62. Gesner, B. M., and Ginsburg, V. (1964). *Proc. Natl. Acad. Sci. U.S.A.*, **52,** 750–755.
63. Singer, S. J., and Nicholson, G. L. (1972). *Science*, **175,** 720–731.
64. Horowitz, A. F., Holten, M. E., and Burger, M. M. (1974). *Proc. Natl. Acad. Sci. U.S.A.*, **71,** 3115–3119.
65. Vahara, I., and Edelman, G. (1972). *Proc. Natl. Acad. Sci. U.S.A.*, **69,** 608–612.
66. Vahara, I., and Edelman, G. (1975). *Proc. Natl. Acad. Sci. U.S.A.*, **72,** 1579–1583.
67. Lee, G., Consiglio, E., Habig, W. H., Dyer, S., Hardegree, M. C., and Kohn, L. D. (1978). *Biochem. Biophys. Res. Commun.*, **83,** 313–320.
68. Morris, N. P., Consiglio, E., Kohn, L. D., Habig, W. H., Hardegree, M. C., and Helting, T. B. (1980). *J. Biol. Chem.*, **255,** 6071–6076.
69. Ambesi-Impiombato, F. S., Parks, L. A. M., and Coon, H. G. (1980). *Proc. Natl. Acad. Sci. U.S.A.*, **77,** 3455–3459.
70. Lissitzky, S., Fayet, G., Giraud, A., Verrier, B., and Torresani, J. (1971). *Eur. J. Biochem.*, **24,** 88–99.
71. Rapoport, B. (1976). *Endocrinology*, **98,** 1189–1197.
72. Filetti, S., Takai, N. A., and Rapoport, B. (1980). *Abstracts of the American Thyroid Society, 56th Meeting*, San Diego, California, November 5–8, 1980, p. 12.
73. Ross, P. S., Moss, J., and Vaughan, M. (1979). *Endocrinology*, **104,** 1036–1040.
74. De Wolf, M. J. S., Epstein, M., Kohn, L. D., and Fridkin, M. (1981). *J. Biol. Chem.*, **256,** 5481–5488.

75. Middlebrook, J., and Kohn, L. D., eds. (1981). *Receptor Mediated Binding and Internalization of Toxins and Hormones,* Academic Press, New York.
76. Hashizuma, K., Fenzi, G., and De Groot, L. J. (1978). *J. Clin. Endocrinol. Metab.,* **46,** 679–689.
77. Consiglio, E., Salvatore, G., Rall, J. E., and Kohn, L. D. (1979). *J. Biol. Chem.,* **254,** 5065–5672.
78. Consiglio, E., Shifrin, S., Rall, J. E., Salvatore, G., and Kohn, L. D. (1981). *J. Biol. Chem.,* **256,** in press.
79. Shifrin, S. and Kohn, L. D. (1981). *J. Biol. Chem.,* **256,** in press.

Hormone Receptors
Edited by L. D. Kohn
© 1982, John Wiley & Sons, Ltd

2 Steady-state relations between hormone binding and elicited response: quantitative mechanistic models

A. P. Minton

1 INTRODUCTION

When hormone is added to a suspension of target cells, a variety of measurable responses may be elicited, which shall be termed primary, secondary, ..., in order of their appearance in time. In the most general case, one cannot discern, *a priori*, whether secondary and subsequent responses are directly elicited by hormone or by an intermediate substance (such as cyclic AMP), the production of which is hormone-mediated. We shall therefore restrict our attention to the elicitation of the primary response, a process which may be assumed to take place in the vicinity of the hormone binding site(s) in the plasma membrane.

The purpose of this chapter is to review several models for the elicitation of the primary response by hormone, with emphasis upon steady-state relations between hormone binding and the magnitude of elicited response. Different types of steady-state relations reported in the literature are classified, and the ability of each model to account for the various classes of relations examined. In the concluding section, measurements suggested as alternative methods of discriminating between the various models are discussed.

2 DEFINITIONS AND CLASSIFICATION OF EXPERIMENTAL OBSERVATIONS

Before attempting to classify the types of relations between hormone binding and response elicitation which have been reported in the literature, several often-used terms and concepts will be defined.

43

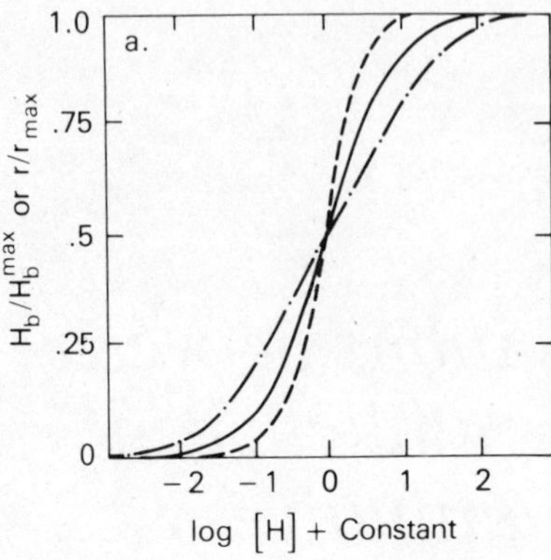 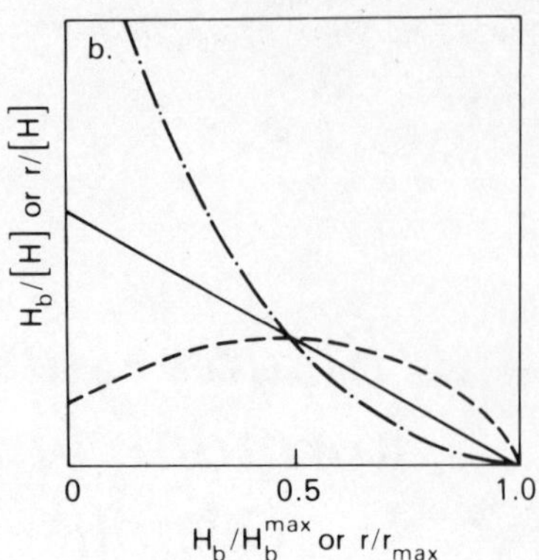

Figure 1. Titration plot (a) and Scatchard plot (b) of three phenomenological types of binding or response isotherms. Reference (Langmuir) isotherm, ————; apparent multiple site class isotherm, – · – · –; apparent cooperative isotherm, – – – –

A *binding isotherm* is a mathematical or graphical representation of the dependence, at constant temperature, of the amount of hormone bound to surface sites at equilibrium (or steady-state) upon the concentration of free hormone or other ligand. A *response isotherm* is a mathematical or graphical representation of the dependence of the level of steady-state response upon the concentration of free hormone. We shall describe the various types of isotherms reported in the literature on the basis of the appearance of a titration or Scatchard plot of the data.

As a reference, we select the isotherm generated by the simplest possible hormone binding reaction: the equilibrium between monovalent hormone and a set of n equivalent, independent surface sites

$$H + S \rightleftharpoons HS \tag{2.1}$$

with an equilibrium association constant K. The equilibrium average amount of hormone bound to surface sites is given by

$$H_b = n \frac{K[H]}{1 + K[H]} \tag{2.2}$$

where [H] is the concentration of free hormone. We shall refer to equation (2.2) as the *reference isotherm*. The titration plot of the reference isotherm (Figure 1a, solid curve) is sigmoid, symmetrical about the half-saturation concentration of hormone, $[H]_{50}$, with two log units in free hormone concentration separating the 9 per cent and 91 per cent levels of site saturation. The Scatchard plot of the reference isotherm (Figure 1b, solid curve) is a straight line.

Two other general types of isotherms frequently observed are plotted in Figures 1a and 1b. We shall refer to curves resembling the short dashed curves in Figures 1a and 1b as *apparent cooperative* isotherms. Less than two log units in free hormone concentration separate the 9 per cent and 91 per cent levels of site saturation in a titration plot, and the Scatchard plot is convex upward. We shall refer to curves resembling the long dashed curves in Figures 1a and 1b as *apparent multisite* isotherms. More than two log units in free hormone concentration separate the 9 per cent and 91 per cent levels of site saturation in a titration plot, and the Scatchard plot is concave upward. Both the apparent cooperative and apparent multi-site isotherms may be generated by a variety of mechanisms.[1] The names used to describe these curves are purely descriptive and do not imply foreknowledge of the underlying mechanism.

The concentration of free hormone required to half-saturate a binding isotherm will be denoted by $[H]_{50}^B$; that required to half-saturate a response isotherm will be denoted by $[H]_{50}^R$.

Many studies have been published on the response of cells and tissues as a function of hormone concentration. Similarly, many studies have been published on the binding of hormone to cells, tissues and membranes as a function of varying hormone concentration. However, relatively few studies have been published in which the steady-state levels of hormone binding

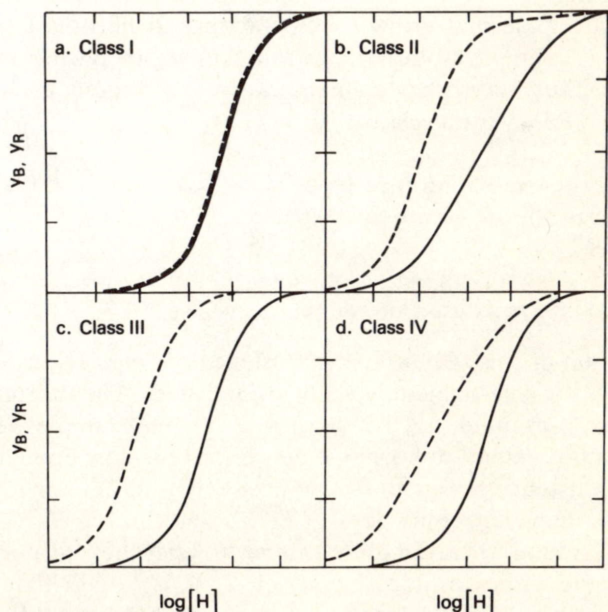

Figure 2. Schematic depiction of four different classes of relations between hormone binding isotherms (————) and response isotherms (– – – –)

and elicited response have been measured on parallel samples under identical or similar conditions. The results of these studies may be classified as follows.

Class I (Fig. 2a)
Binding isotherm—reference type.
Response isotherm—reference type.
$[H]_{50}^{B} \simeq [H]_{50}^{R}$
Example: Insulin acting on rat thymocytes to stimulate uptake of α-amino isobutyric acid.[2]

Class II (Fig. 2b)
Binding isotherm—multi-site type.
Response isotherm—reference type.
$[H]_{50}^{B} > [H]_{50}^{R}$
Examples: Insulin acting on fat cells to stimulate glucose transport/oxidation.[3,4] Glucagon acting on rat liver plasma membranes to stimulate activation of adenylate cyclase in the presence of GTP.[5]

Class III (Fig. 2c)
Binding isotherm—reference type (??).
Response isotherm—reference type.
$[H]_{50}^{B} > [H]_{50}^{R}$
Examples: Epidermal growth factor acting on fibroblasts to stimulate α-amino isobutyric acid intake.[6] Epinephrine acting on turkey erythrocyte membranes to stimulate activation of adenylate cyclase.[7]

Class IV (Fig. 2d)
Binding isotherm—reference type.
Response isotherm—multi-site type.
$[H]_{50}^{B} > [H]_{50}^{R}$
Example: Vasopressin acting on renal medullary membranes to stimulate activation of adenylate cyclase.[8,9]

In the opinion of this reviewer, the existence of Class III, as distinct from Class II, has not been unequivocally established. The hormone binding measurements presented in the studies cited as examples of Class III behaviour did not extend to hormone concentrations low enough to rule out the possible existence of up to 20 per cent of total sites with apparently higher affinity than the remainder.

There is no reason to expect all hormones to elict their respective primary responses in the same fashion. Nonetheless, it is hoped that the various classes of binding/response relations listed above represent differing expressions of common underlying principles. Thus we seek molecular models which can utilize a small set of testable mechanistic assumptions to generate as many different classes of observed behaviour as possible.

3 THE ELEMENTARY EFFECTOR

The elicitation of cellular response by hormone must involve at least one step in which a rate process is regulated by the binding of a ligand. That molecular species, presumably an enzyme, which modulates the rate process is conventionally referred to as *effector*, in distinction to the hormone-binding species conventionally referred to as *receptor*. In the present section we explore the properties of the simplest possible effector: an enzyme, E, with a single binding site for substrate, S, and a single regulatory ligand, L. To further simplify the analysis, it will be assumed that in the absence of regulatory ligand, the rate of conversion of substrate to produce P is zero. The reaction scheme characterizing this elementary effector is as follows:

$$
\begin{aligned}
\text{(a)} \qquad & L+E \underset{k_{-1}}{\overset{k_1}{\rightleftharpoons}} LE \\[2ex]
\text{(b)} \qquad & L+ES \underset{k_{-2}}{\overset{k_2}{\rightleftharpoons}} LES \\[2ex]
\text{(c)} \qquad & E+S \underset{k_{-3}}{\overset{k_3}{\rightleftharpoons}} ES \\[2ex]
\text{(d)} \qquad & LE+S \underset{k_{-4}}{\overset{k_4}{\rightleftharpoons}} LES \\[2ex]
\text{(e)} \qquad & LES \xrightarrow{k_5} LE+P
\end{aligned}
\qquad (3.1)
$$

When ligand L is present, substrate S is continuously and irreversibly converted into product P *via* reaction (3.1e), and the system is not at equilibrium. The steady-state concentrations of the species E, LE, ES and LES are obtained as functions of [L], the concentration of free ligand; [S], the concentration of free substrate; and E_T, the total concentration of enzyme, by applying the steady–state condition (i.e. zero rate of change of intermediate concentrations) to each of the rate equations derived from the mechanism shown in equation (3,1). Given the steady-state concentrations of LE and LES, the amount of bound ligand, L_b, and the reaction rate, r, may be calculated as follows:

$$L_b = [LE]+[LES] \qquad (3.2)$$

$$r = k_5[LES] \qquad (3.3)$$

In certain cases, the elementary effector model yields particularly simple analytical expressions for L_b and r as explicit functions of [L]. These cases are described below.

(a) *Quasi-equilibrium:* When k_5 is much smaller than all other rate constants, the rate at which LES irreversibly dissociates to form product will be small relative to the rate at which equilibrium between E, ES, LE and

LES is attained. Equilibrium methods may then be used to calculate, to a very good approximation, the steady-state concentrations of these four intermediates as functions of [L], [S] and E_T.

(b) *Ligand binding prerequisite to substrate binding:* Reactions (3.1b) and (3.1c) do not take place.

(c) *Substrate binding prerequisite to ligand binding:* Reactions (3.1a) and (3.1d) do not take place. LE formed in reaction (3.1e) dissociates instantaneously to L and E; hence reaction (3.1e) may be rewritten

$$\text{LES} \xrightarrow{k_5} \text{L} + \text{E} + \text{P}$$

It will be shown elsewhere (A. P. Minton, manuscript in preparation) that in all three of these cases, the steady-state values of L_b and r are given by expressions of the following type:

$$L_b = K_1[L]/(1 + K_2[L]) \tag{3.4}$$

$$r = K_3 L_b \tag{3.5}$$

where the constants K_i are functions of the k_i, [S], and E_T, which differ in each of the three cases. The form of equation (3.4) is identical to that of the reference isotherm.

In the most general case, the dependence of L_b and r upon [L] do not take the form of equations (3.4) and (3.5). In Figures 3(a), 3(b) and 3(c), titration plots are presented which are generated by the elementary effector model,

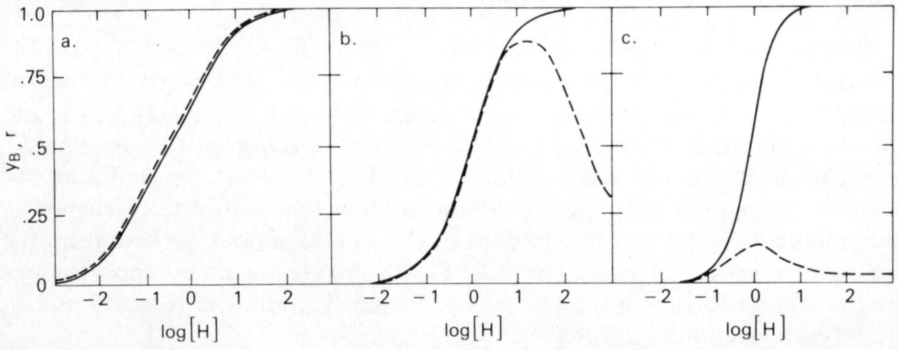

Figure 3. Binding (————) and response (– – – –) isotherms generated by the elementary effector model with the following sets of parameter values:

	k_1	k_{-1}	k_2	k_{-2}	k_3	k_{-3}	k_4	k_{-4}	k_5
(a)	0.01	0.1	10	1	0.01	0.001	100	0.01	10
(b)	1	1	1	1	100	1	0.1	0.001	1
(c)	1000	1	0.1	0.01	100	0.01	0.01	0.0001	1

Response is given in arbitrary units which are scaled so that response is equal to binding at low levels of binding

given the parameter values shown in the figure legend. The binding isotherm in Figure 3(a) is of the multi-site type, and the response is proportional to L at all concentrations of L. The binding isotherm in Figure 3(b) is of the reference type, and the response is proportional to L_b up to a site occupancy of about 0.8. However, with further increase in L_b, the response reaches a maximum at $L_b \approx 0.9\, L_b^{max}$ and then declines. The binding isotherm in Figure 3(c) is of the cooperative type, and the response is proportional to L_b only at very low levels of L_b. The response reaches a maximum at $L_b \approx 0.5\, L_b^{max}$ and declines as L_b increases further. Response isotherms of the type seen in Figures 3(b) and 3(c) are of interest because they may account for the observation that some ligands can act as an agonist at low concentrations and as an antagonist at higher concentrations[10] and the observation that certain hormones and hormone derivatives appear to elicit a diminished steady-state activation of adenylate cyclase at the highest hormone concentrations examined.[12]

The simplest possible model for response elicitation by hormone consists of an elementary effector whose regulatory ligand is hormone itself. All of the results presented above apply to this *unified receptor-effector model*. Under conditions such that hormone binding and response isotherms are given by equations (3.4) and (3.5), respectively, this model can account for Class I behaviour. In subsequent sections it will be shown that more complex models in which the receptor function is assigned to species other than effector can account for the other classes of behaviour as well.

4 THE FLOATING RECEPTOR MODEL

For all classes of hormone dependency relations other than Class I, $[H]_{50}^B > [H]_{50}^R$. The unified receptor–effector model cannot generate this type of behavior. One simple way to account for the difference between $[H]_{50}^B$ and $[H]_{50}^R$ is to postulate a second class of surface binding sites for hormone, in addition to receptor–effector, which is unrelated to hormone function. If the number of non-functional sites is much greater than the number of receptor–effector sites, and if the affinity of non-functional sites for hormone is less than that of receptor–effector sites, then $[H]_{50}^B$ will be greater than $[H]_{50}^R$. However, it is generally observed that the rank order of $[H]_{50}^R$ among a given series of related hormones and/or hormone derivatives, as established by parallel dose-response experiments, is the same as the rank order of $[H]_{50}^B$. This result implies that the bulk of specific binding sites are related to hormone function and therefore invalidates the hypothesis of a large excess of non-functional hormone binding sites.

The floating receptor model emerged as the first attempt to deal in a quantitative fashion with the problem posed above. Similar versions were independently proposed at about the same time by DeHaën[13] and Jacobs and Cuatrecasas.[14] According to this model, receptor and effector functions are assigned to distinct molecular species in the cell membrane. The stoichiometric ratio of effector units to receptor units is not constrained to be unity and may be considerably less than unity. In order for effector to function as a catalyst, thereby activating the primary response, receptor must be bound. In other words, the regulatory ligand for effector in this model is receptor. Activation of effector is regulated by hormone because the association constant for binding of receptor to effector is greater when hormone is bound to receptor than when receptor is hormone-free.

The model may be most simply described in molecular terms using a modification of the equilibrium formalism introduced by Jacobs and Cuatrecasas:[14]

$$\left.\begin{array}{ll} \text{(a)} & \quad H + R \overset{K_1}{\rightleftharpoons} HR \\[2ex] \text{(b)} & \quad R + E \overset{K_2}{\rightleftharpoons} RE \\[2ex] \text{(c)} & \quad HR + E \overset{\alpha K_2}{\rightleftharpoons} HRE \end{array}\right\} \tag{4.1}$$

where $\alpha > 1$. The constant α characterizes the degree to which binding of hormone to receptor enhances the interaction of receptor with effector. The magnitude of steady-state response is assumed to be proportional to the total concentration of receptor–effector complex, i.e. the sum of [RE] and [HRE]. Since K_2 is non-zero, some RE will be present even in the absence of hormone. Thus the magnitude of *elicited* response is defined to be the difference between the levels of steady-state response in the presence and absence of hormone.

Some of the properties of the floating receptor model have been described by DeHaën[13] and Jacobs and Cuatrecasas.[14] The main features of hormone binding and response isotherms generated by this model are summarized below and illustrated in Figures 4(a)–4(d).

1. Elicited response isotherms are of the reference type.

2. When the ratio of total effector to total receptor, $E_T/R_T \approx 1$, then the binding isotherm is of the reference type and $[H]_{50}^B \approx [H]_{50}^R$ (Figure 4(a)).

3. When $E_T/R_T < 1$ and $\alpha K_2 < K_1$, the binding isotherm is of the reference type and $[H]_{50}^B \approx [H]_{50}^R$ (Figure 4(b)).

4. When $E_T/R_T < 1$ and $\alpha K_2 > K_1$, the binding isotherm is of the apparent two-site type and $[H]_{50}^R < [H]_{50}^B$ (Figure 4(c)). An apparent exception to this occurs when $E_T/R_T = X$, where $X \leqslant 0.05$. In this case, the binding isotherm will appear to be of the reference type unless close attention is paid to binding at occupancy values less than X, where deviations from the reference isotherm may be observed (Figure 4(d)).

The floating receptor model is seen to be capable of generating Class I, Class II, and apparent Class III behaviour. It will be shown subsequently

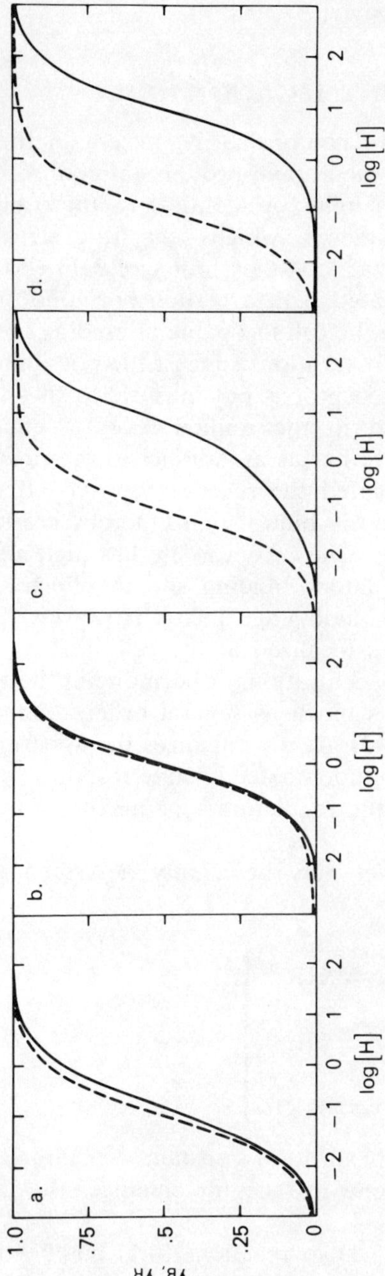

Figure 4. Binding (———) and response (– – –) isotherms generated by the floating receptor model with $K_1 = 1$, $K_2 = 0.001$, and the following additional parameter values:

	α	E_T/R_T
(a)	10^5	1.0
(b)	10^2	0.2
(c)	10^5	0.2
(d)	10^5	0.02

y_r is the scaled response, defined as $[HRE]/E_T$

that a slightly generalized version of the floating receptor model is capable of generating Class IV behaviour as well.

5 THE BIVALENT LIGAND MODEL

There exist substances which are not ordinarily present in the cellular environment, such as lectins, anti-membrane-protein antibodies and some drugs, which can elicit responses similar or identical to those elicited by certain hormones.[10,15] These substances, which bear little structural resemblance to the hormones whose actions they mimic, can in certain cases elicit the 'normal' response associated with a certain hormone even after chemical modification has rendered the cell incapable of binding or responding to the hormone.[15,16,17] These observations suggest the possibility that at least in some cases the hormone receptor is not intrinsic to the process of response elicitation, as is implied in the unified receptor–effector and floating receptor models, but acts rather as an adjunct to the process.

The above considerations stimulated the recent proposal[18] that certain types of hormones may simultaneously bind to, and thereby crosslink, two independent classes of surface sites. The receptor site has high affinity and specificity for hormone. The regulatory binding site of effector has low affinity for free hormone and low binding specificity. However, when hormone binds to receptor it becomes localized at the cell surface; the thermodynamic activity, or effective concentration, of hormone in the vicinity of effector is thereby magnified by as much as several orders of magnitude. Thus binding of hormone to receptor greatly enhances the apparent affinity of effector for hormone. The magnitude of steady-state response is assumed proportional to the occupancy of the regulatory binding site of effector by hormone (or other ligand).

The basic bivalent ligand model may be simply described using the following equilibrium formalism:

$$
\left.
\begin{aligned}
\text{(a)} \qquad & H + R \underset{}{\overset{K_1}{\rightleftharpoons}} HR \\[2em]
\text{(b)} \qquad & H + E \underset{}{\overset{K_2}{\rightleftharpoons}} HE \\[2em]
\text{(c)} \qquad & HR + E \underset{}{\overset{\alpha K_2}{\rightleftharpoons}} HRE
\end{aligned}
\right\} \qquad (5.1)
$$

where $K_2 \ll K_1$ and $\alpha \gg 1$. The localization constant α characterizes the degree to which binding of hormone to receptor enhances the interaction between hormone and effector.

The resemblance of reactions (5.1) to reactions (4.1) has been noted.[18] Under conditions such that the species on the right hand side of reactions (4.1b) and (5.1b) are not formed in appreciable quantity, the floating receptor and basic bivalent ligand models become mathematically—but not conceptually—equivalent. Hence the types of binding and response isotherms which may be generated by the floating receptor model (Figures 4(a)–4(d)) can, for the most part, also be generated by the bivalent ligand model, and vice versa. For this reason, the two models cannot be reliably distinguished on the basis of experimentally measured hormone binding and response isotherms alone. The kinds of experiments which would enable such a distinction to be made will be considered in the concluding section of this chapter.

Like the floating receptor model, the basic bivalent ligand model is capable of generating Class I, Class II, and apparent Class III behaviour, and may be generalized to allow the generation of Class IV behaviour as well.

6 GENERALIZATIONS OF THE FLOATING RECEPTOR AND BIVALENT LIGAND MODELS

Class IV behavior is characterized by a reference-type binding isotherm and multi-site response isotherm, with $[H]_{50}^{R} < [H]_{50}^{B}$. None of the models presented above are capable of accounting for such behaviour. In the present section we shall describe two generalizations, each of which may be applied to either the floating receptor or bivalent ligand models. Either of these generalizations permits both the floating receptor and bivalent ligand models to simulate Class IV behaviour.

For the purpose of demonstrating the effect of each generalization, we shall neglect the binding of receptor to effector in the absence of hormone (in the context of the floating receptor model) and the binding of hormone to effector in the absence of receptor (in the context of the bivalent ligand model). In this limit the two models become formally identical and may be represented by a common set of equilibria, called the *two-step model:*[1]

$$
\left.
\begin{array}{ll}
\text{(a)} & \quad H + R \underset{}{\overset{K_{R}}{\rightleftharpoons}} HR \\[2em]
\text{(b)} & \quad HR + E \underset{}{\overset{K_{E}}{\rightleftharpoons}} HRE
\end{array}
\right\}
\qquad (6.1)
$$

Generalized effector. In the floating receptor and bivalent ligand models, it was assumed that the binding of HR to E could be represented by a single equilibrium association constant, and that the magnitude of steady-state response was proportional to the steady-state concentration of HRE. In section 3, it was shown that the binding of regulatory ligand to an elementary effector may not always be describable by a single apparent equilibrium constant. In particular, under certain circumstances the elementary effector can exhibit binding and response isotherms of the multi-site type (Figure 2(a)).

Let us postulate that the regulatory ligand L appearing in reaction scheme (3.1) is the complex HR. If K_R represents the equilibrium association constant for binding of H to R, and if the ratio of total effector, E_T, to total receptor, R_T, is quite small, then

$$[L] \equiv [HR] \simeq R_T \frac{K_R[H]}{1 + K_R[H]} \qquad (6.2)$$

Under these conditions the binding isotherm will approximate the reference isotherm. The steady-state solution to the rate equations governing the reaction scheme (3.1), together with equation (6.2), may then be obtained as a function of hormone concentration. Binding and response isotherms calculated in this manner for one set of parameters are plotted in Figure 5.

Heterogeneous effector. In the absence of information to the contrary, one cannot rule out the possibility that effector is present, not as a single chemical species, but as a mixture of closely related species, whose affinity for the hormone-receptor complex may be subject to some variability. Let us consider the simplest possible example of a heterogeneous effector, i.e. one composed of two classes of affinity for the hormone-receptor complex. The generalized equilibrium relations corresponding to equation (6.1) are

$$
\begin{aligned}
\text{(a)} \qquad & H + R \underset{}{\overset{K_R}{\rightleftharpoons}} HR \\[1em]
\text{(b)} \qquad & HR + E^{(1)} \underset{}{\overset{K_E^{(1)}}{\rightleftharpoons}} HRE^{(1)} \\[1em]
\text{(c)} \qquad & HR + E^{(2)} \underset{}{\overset{K_E^{(2)}}{\rightleftharpoons}} HRE^{(2)}
\end{aligned}
\right\} \qquad (6.3)
$$

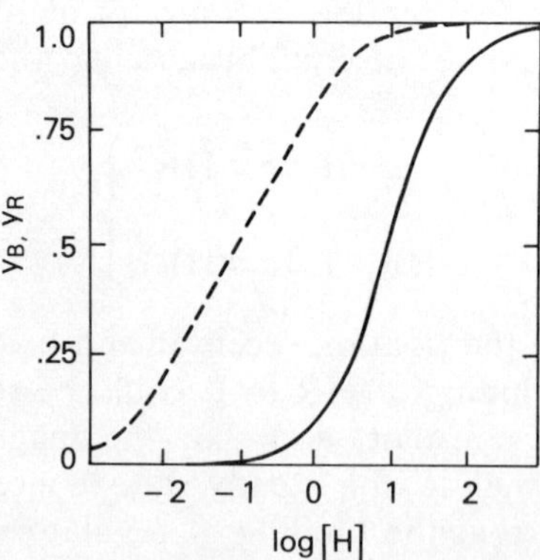

Figure 5. Binding (———) and response (– – – –) isotherms generated by the generalized effector model, using the following parameter values: $K_R = 0.1$, $E_T/R_T = 0.033$, $k_1 = 0.01$, $k_{-1} = 0.1$, $k_2 = 10$, $k_{-2} = 1$, $k_3 = 0.01$, $k_{-3} = 0.001$, $k_4 = 100$, $k_{-4} = 0.01$, $k_5 = 10$

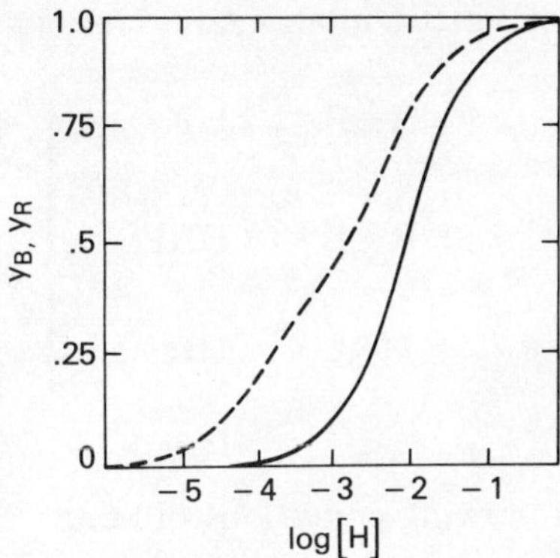

Figure 6. Binding (———) and response (––––) isotherms generated by the heterogeneous effector model, using the following parameter values: $K_R = 100$, $E_T^{(1)}/R_T = 0.003$, $E_T^{(2)}/R_T = 0.01$, $K_E^{(1)} = 100$, $K_E^{(2)} = 10$

The magnitude of steady-state response is assumed proportional to the sum of the concentrations of $HRE^{(1)}$ and $HRE^{(2)}$. If it is further assumed that the total concentration of both classes of effector is small relative to the concentration of receptor, then the binding isotherm will approximate the reference isotherm. Binding and response isotherms calculated using equation (6.3) for one set of parameters are plotted in Figure 6.

Since both generalized effector and heterogeneous effector models can simulate Class IV behaviour, experimental data other than steady-state binding and response isotherms are required to distinguish between them.

7　THE COLLISION COUPLING MODEL

It has been found that when turkey erythrocyte membranes are incubated with mixtures of a hormone agonist (1-epinephrine) and an antagonist (propanolol), the hormone-stimulated adenylate cyclase activity decreases as the concentration of propanolol increases, in a manner which may be interpreted as normal competitive inhibition. However, when membranes are incubated in the presence of hormone alone, the subsequent addition of a large excess of antagonist does not result in an inhibition of enzyme activity (Sevilla *et al.*, 1976). These findings led to the proposal that under the conditions of the experiment, activation was due to the quasi-irreversible formation of an activated form of cyclase, which could remain active even in the absence of bound agonist.

In order to account for these and related observations, Tolkovsky and Levitski (1978) proposed a model in which elicited response is assumed to be proportional to the amount of an activated form of effector, E^*, which is associated with neither hormone nor receptor. This model may be described by the following set of kinetic relations:

$$
\begin{array}{ll}
\text{(a)} & H + R \underset{k_{-1}}{\overset{k_1}{\rightleftharpoons}} HR \\
\\
\text{(b)} & HR + E \underset{k_{-2}}{\overset{k_2}{\rightleftharpoons}} HRE \\
\\
\text{(c)} & HRE \xrightarrow{k_3} HR + E^* \\
\\
\text{(d)} & E^* \xrightarrow{k_4} E
\end{array}
\right\} \quad (7.1)
$$

At steady-state, the following conditions obtain:

$$
\begin{array}{ll}
\text{(a)} & [HR] = K_1[H][R] \\
\text{(b)} & [HRE] = K_2[HR][E] \\
\text{(c)} & [E^*] = K_3[HRE]
\end{array}
\right\} \quad (7.2)
$$

where $K_1 \equiv k_1/k_{-1}$, $K_2 \equiv k_2/(k_{-2}+k_3)$, and $K_3 \equiv k_3/k_4$.

Depending upon the values of K_1, K_2 and K_3, and E_T/R_T, the collision coupling model is capable of generating Class I, Class II and Class III behaviour (Figures 7(a)–7(c)). This model may not be reliably distinguished from the floating receptor or bivalent ligand models on the basis of experimental steady-state hormone binding or response isotherms alone. The types of experiments required to make such a distinction will be discussed in the concluding section of this chapter.

The collision coupling model may be considered to be a generalization of the floating receptor or bivalent ligand models, in which the HR complex

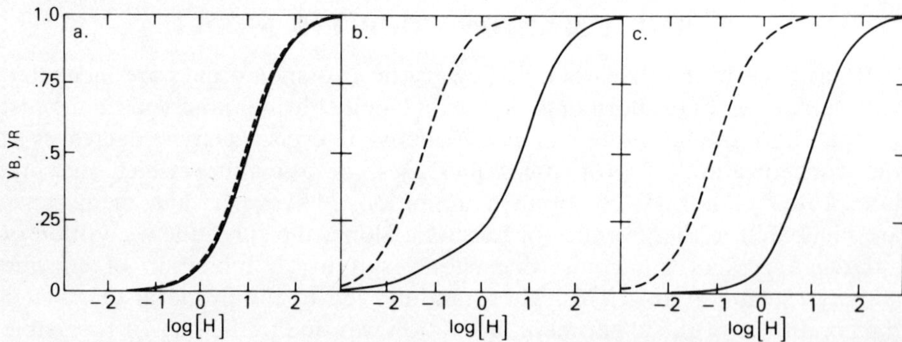

Figure 7. Binding (———) and response (– – – –) isotherms generated by the collision coupling model using $K_1 = 0.1$ and the following additional parameter values:

	K_2	K_3	E_T/R_T
(a)	0.001	100	0.2
(b)	100	1	0.3
(c)	1	100	0.3

may potentially activate more than one effector unit. In this respect, it is conceptually similar to that of the random-hit matrix model proposed by Bergman and Hechter.[19] However, unlike the collision coupling model, the random-hit matrix model is formulated in terms of a large number of *ad hoc* rules of computation, rather than chemical rate processes. Moreover, some of these rules bear no obvious analogy to known chemical rate processes. For this reason, and because of the complexity of calculations in the random-hit matrix model, a direct comparison between the predictions of this model and the 'chemical' models reviewed here is beyond the scope of the present chapter.

8 SYMMETRIC TWO-MODULATOR MODEL

One class of hormone-elicited responses, the production of cyclic AMP, is known to require the presence of a second ligand, guanosine triphosphate (GTP), or an analog thereof. The role of GTP in the mechanism of activation of adenylate cyclase has been the subject of intensive study (see References 20 and 21 and References therein). Rodbell[20] has proposed the following extension of the two-step model to account for the effects of both hormone H and nucleotide triphosphate G upon cyclase activity†

$$
\begin{aligned}
\text{(a)} && H + R &\xrightleftharpoons{K_R} HR \\[6pt]
\text{(b)} && G + R &\xrightleftharpoons{K_G} GR \\[6pt]
\text{(c)} && H + RG &\xrightleftharpoons{\beta K_R} HRG \\[6pt]
\text{(d)} && HR + G &\xrightleftharpoons{\beta K_G} HRG \\[6pt]
\text{(e)} && HRG + E &\xrightleftharpoons{K_E} HRGE
\end{aligned}
\qquad (8.1)
$$

The competition factor β indicates the degree to which binding of one ligand (either H or G) to R modifies the affinity of R for the second ligand. The rate of cyclic AMP production is presumed to be proportional to the concentration of the 'activated' effector, HRGE. We shall call this model the symmetric two-modulator model, since from a formal (but not conceptual) standpoint, the species H and G appearing in the above reaction scheme are interchangeable. Inspection of reaction scheme (8.1) reveals that in the presence of saturating concentrations of G, all receptor will be present in the

---

† Rodbell (1980) conceives of the species denoted here by R as a complex of a hormone binding protein and a nucleotide binding protein. As no function-related dissociation of the complex is postulated, we treat it as a single functional species.

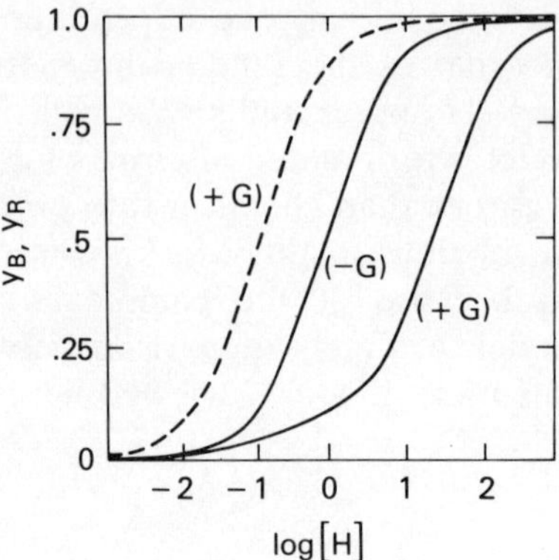

Figure 8. Binding (———) and response (− − − −) isotherms generated by the symmetric and asymmetric two-modulator models in the absence of G $(-G)$ and in the presence of a saturating concentration of G $(+G)$, calculated using the following parameter values: $K_R = 1$, $K_E = 300$, $\beta = 0.03$, $E_T/R_T = 0.1$. No response is generated by these models in the absence of G

form of RG, and this model reduces to the simple two-step model given by reaction scheme (6.1).

If the competition factor β is less than one, then addition of G will lower the overall affinity of receptor for hormone. However, the binding of G to R also makes possible the coupling of the hormone-receptor complex to effector, a process which results in enhancement of the affinity of some fraction of hormone binding sites for hormone (see section 4). Thus the symmetric two-modulator model is capable of simultaneously generating a GTP-induced right-shift in the overall hormone binding isotherm together with a GTP-induced appearance of an apparent high-affinity subset of hormone sites associated with cyclase activation (Figure 8). Such behaviour has been observed in the glucagon-stimulated production of cyclic AMP in rat liver membranes.[5]

Reaction scheme (8.1) may be readily understood in the context of the floating receptor hypothesis. It may also arise from the bivalent ligand model if it is assumed that the binding of G to R results both in a change of affinity of R for the binding of H and in a substantial increase in the localization factor α (equation 5.1(c)). This assumption is unattractive as the physical basis for a postulated GTP-induced increase in the localization factor is not apparent.

9 ASYMMETRIC TWO-MODULATOR MODEL

Lad *et al.*[22] reported a series of observations which are presented as evidence for two distinct sites of action for GTP in the hormone-stimulated production of cyclic AMP. On the basis of these observations, it was proposed that one site of GTP action resided on receptor and another on effector. While some of the observations reported by Lad *et al.*[22] may be readily accommodated by the symmetric two-modulator model described in the preceding section, others may be more naturally interpreted if, like Lad *et al.*,[22] we postulate that binding sites for GTP exist on both receptor and effector. The following reaction scheme represents one possible extension of the two-step model in this direction:

$$
\begin{aligned}
\text{(a)} \qquad & H + R \underset{}{\overset{K_R}{\rightleftharpoons}} HR \\[6pt]
\text{(b)} \qquad & R + G \underset{}{\overset{K_{RG}}{\rightleftharpoons}} HRG \\[6pt]
\text{(c)} \qquad & H + RG \underset{}{\overset{\beta K_R}{\rightleftharpoons}} HRG \\[6pt]
\text{(d)} \qquad & HR + G \underset{}{\overset{\beta K_{RG}}{\rightleftharpoons}} HRG \\[6pt]
\text{(e)} \qquad & E + G \underset{}{\overset{K_{EG}}{\rightleftharpoons}} EG \\[6pt]
\text{(f)} \qquad & HR + EG \underset{}{\overset{K_E}{\rightleftharpoons}} HREG \\[6pt]
\text{(g)} \qquad & HRG + EG \underset{}{\overset{K_E}{\rightleftharpoons}} HREG_2
\end{aligned}
\qquad (9.1)
$$

The steady-state response (rate of cyclic AMP production) is assumed to be proportional to the sum of the activated species HREG and $HREG_2$. Since H and G are not interchangeable in this scheme, we refer to it as the asymmetric two-modulator model.

Like the symmetric model, the symmetric model can generate a GTP-induced right shift in the overall hormone binding isotherm, together with a GTP-induced appearance of an apparent high-affinity subset of hormone sites associated with response elicitation (Figure 8 above). The difference between the asymmetric and symmetric models becomes apparent only at intermediate concentrations of G. As an illustration of this difference, let us consider the different effects of GTP and the GTP analog guanyl-5'-yl imidodiphosphate [Gpp(NH)p] upon the binding of hormone and upon elicited response at fixed hormone concentration, as reported by Lad *et al.*[22] Both GTP and Gpp(NH)p decrease hormone binding and increase elicited response. Both substances appear to be equally potent with respect to their effect upon response elicitation, but GTP appears to be a much more potent modulator of hormone binding than Gpp(NH)p, having an effect upon the level of hormone binding at concentrations which are 10–100 times lower than the concentrations of Gpp(NH)p required to have an equivalent effect. This difference may be readily accounted for in the context of the asymmetric two-modulator model if it is assumed that GTP has an affinity for R

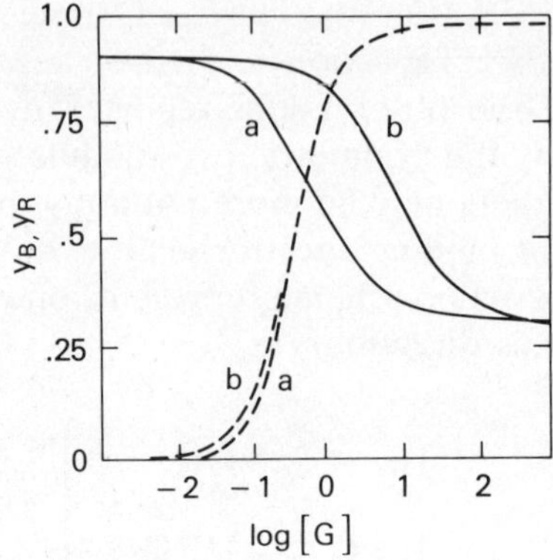

Figure 9. Binding (———) and response (– – – –) at constant hormone concentration, calculated as a function of $[G]$ using the asymmetric two-modulator two-step model with $K_E = 300$, $K_{EG} = 1$, $\beta = 0.03$, $E_T/R_T = 0.1$, $[H] = 10$. (a) $K_G = 10$; (b) $K_G = 1$

which is 10–100 times as great as that of Gpp(NH)p, while the two substances have approximately the same affinity for E (Figure 9). This assumption is reasonable, as the structurally distinct binding sites for G on R and E need not discriminate equally between GTP and Gpp(NH)p. In contrast, no simple explanation of the difference between the effects of GTP and Gpp(NH)p is afforded by the symmetric two-modulator model.

10   DISCUSSION

Six models for steady-state relations between hormone binding and response elicitation have been reviewed in the preceding sections. Five of these models are capable of simulating all of the classes of experimental behaviour delineated in section 2, and two of them are additionally capable of describing some of the combined effects of hormone and a second modulating substance upon elicited response. These models are by no means the only models which may be capable of accommodating the presently available data, but they are probably the simplest and most readily analysed. The question naturally arises whether, given the comparable abilities of these models to account for the steady-state data, there exist different types of measurements which permit some (or perhaps even all) of these models to be ruled out as possible descriptions of the mechanism of response elicitation in a particular hormone-cell system. In this section we shall consider possible means of discriminating between these models.

10.1   Kinetic measurements

Tolkovsky and Levitski[7] proposed that measurements of the rate of appearance of response (presumed proportional to the concentration of activated effector) as a function of hormone concentration can serve to discriminate between alternate mechanistic models relating hormone binding to response elicitation. Models containing as many individual reaction steps as those discussed in this chapter give rise to rate equations containing numerous adjustable parameters. In the reviewer's experience, such rate equations can be made to fit almost any observed kinetic result by judicious manipulation of parameter values. The problem of discrimination is particularly acute when one is attempting to compare models which are very similar. A concrete illustration of this difficulty follows.

Tolkovsky and Levitski[7] reported that upon addition of 1-epinephrine to turkey erythrocyte membranes in the presence of saturating amounts of Gpp(NH)p, the appearance of adenylate cyclase activity followed a first-order rate law of the form

$$\ln \frac{E^*(\infty) - E^*(t)}{E^*(\infty)} = -\lambda t \tag{10.1}$$

where $E^*(t)$ is the concentration of the active form of effector (cyclase) at time t, $E^*(\infty)$ is the concentration in the long-time (steady-state) limit, and λ is a rate constant which is a function of hormone concentration. Tolkovsky and Levitski[7] claimed that such behaviour is consistent with predictions of the collision coupling model but inconsistent with predictions of the floating receptor model, hence ruling out the latter model as a valid description of the system under study. Under closer inspection, this claim does not appear to be substantiated.

Tolkovsky and Levitski[7] reported that a 'maximal' response may be elicited in their system by a concentration of hormone which corresponds to a receptor occupancy of about one per cent. In the context of the floating receptor model, this result implies that $E_T < 0.01 R_T$. Under these conditions the floating receptor model predicts that the hormone binding isotherm will be of a reference type, as reported by Tolkovsky and Levitski.[7]

According to the two-step model, which is a simplification of both the floating receptor and bivalent ligand models, the appearance of the catalytically active species HRE will be given by the rate equation

$$\frac{d[HRE]}{dt} = k_E[HR][E] - k_{-E}[HRE] \tag{10.2}$$

where k_E and k_{-E} are, respectively, the forward and reverse rate constants for reaction (6.1b). If we assume that reaction (6.1a) equilibrates essentially instantaneously with respect to the time scale over which the evolution of HRE is studied, then equation (10.2) may be rewritten

$$\frac{d[HRE]}{dt} = k_E[HR]_{eq}[E] - k_{-E}[HRE] \tag{10.3}$$

where $[HR]_{eq}$ is the equilibrium concentration of HR, given by

$$[HR]_{eq} \cong R_T K_R [H]/(1 + K_R[H]) \tag{10.4}$$

and hence constant (to a very good approximation) for fixed hormone concentration. Under these conditions it may be readily shown that the rate of appearance of HRE will be given by Equation (10.1) with $E^* = HRE$ and

$$\lambda = k_E [HR]_{eq} + k_{-E} \tag{10.5}$$

Thus it may be seen that the kinetic measurements of Tolkovsky and Levitski[7] cannot rule out the floating receptor (or bivalent ligand) model as a valid description of the system studied.

10.2   Effect of total receptor blocking upon response elicitability

According to the bivalent ligand model [reaction scheme (5.1)] a hormone can, in principle, elicit a response of normal magnitude without binding to receptor, if the concentration of hormone is sufficiently high (of order $1/K_2$ or larger). Thus one can envisage an experiment in which the hormone binding site is rendered incapable of binding hormone. One possible means for doing so would to covalently crosslink antagonist to the receptor. Alternatively, the hormone binding site might be non-covalently blocked by carrying out the experiment on cells or membranes which have been preincubated with limiting high concentrations of antagonist. The elicitation of a normal response by a higher than normal concentration of hormone under such conditions would be consistent with the bivalent ligand model but not with a single-site (floating receptor or collision coupling) model.

10.3   Direct demonstration of the presence of two distinct hormone binding species in the cell membrane

In certain instances it may be possible to solubilize and isolate hormone-binding constituents of the cell membrane. The bivalent ligand model predicts that one component of the cell membrane (receptor) should bind hormone strongly, a second component (effector) should bind hormone weakly, and that the combination of the two should bind hormone more strongly than either separately.[18] Such behaviour appears to be exhibited by insulin-binding components of liver cell membranes.[23] An even stronger test of the bivalent ligand hypothesis would be to demonstrate that while hormone can bind to both the putative receptor and putative effector species, an antagonist can bind only to the putative receptor.

10.4   Persistence of effector activation

The collision coupling model differs from the other models presented here in that it postulates the existence of an activated state of effector (E^*) which need not be bound to the hormone-receptor complex. It has been suggested[7] that when Gpp(NH)p is used in place of GTP, the rate of reaction (7.1d) becomes so low that the formation of E* is effectively irreversible within the duration of a conventional experiment. If this is the case, then it should be possible, in principle, to stimulate a certain level of cyclic AMP production by addition of hormone, remove hormone and/or receptor rigorously from the system, and observe either total persistence or a very slow decline in the level of cyclic AMP production following removal of hormone and/or receptor.

Since epinephrine-stimulated cyclase activity in turkey erythrocyte membranes persists long after addition of a large excess of the epinephrine antagonist propanolol,[12] some investigators (for example Tolkovsky and Levitski,[7] and Citri and Schramm[21]) have concluded that the activation of cyclase requires only a transient interaction with the hormone-receptor complex. However, an alternative explanation of this observation is possible. According to the floating receptor and bivalent ligand models, the formation of the HRE complex from HR and E is highly favored at equilibrium. Thus if the kinetics of *association* to form HRE are such as to be easily observed over the time course of an experiment requiring minutes or tens of minutes,[7] then it is quite likely that the *dissociation* of the HRE complex is too slow to proceed significantly over the same period of time. Since the intrinsic rate of hormone dissociation is not altered by antagonist, the addition of a large excess of antagonist following equilibration with hormone does not assure the removal of hormone from high affinity sites associated with response elicitation within the duration of a conventional experiment.

In order to unequivocally establish that effector is activated by a transient rather than continuous interaction with the hormone–receptor complex, it will be necessary to devise some means to chemically dissociate the HRE complex, to separate each of the components, to quantitatively characterize the effectiveness of separation, and to measure the activity of effector in the absence of hormone and receptor.

10.5   Simultaneous effect of two modulators upon modulator binding and elicited response

In section 9 it was pointed out that data on the different effects of GTP and Gpp(NH)p on hormone binding and elicited response could be more easily interpreted in the context of the asymmetric two-modulator model than in the context of the symmetric two-modulator model. This finding suggests that the measurement of the entire hormone binding and response isotherms at several different GTP [or Gpp(NH)p] concentrations in the intermediate concentration range could provide a firm basis for discriminating between some two-modulator models.

10.6 Conclusion

Several models have been shown to be consistent with a variety of steady-state data on the relationship between hormone binding and response elicitation, the simplest of which are the floating receptor and bivalent ligand models. Unless modulators other than hormone (such as GTP) are subject to controlled variation, steady-state and kinetic measurements on intact hormone-cell or hormone-membrane systems are unlikely to provide unequivocal discrimination between the various possible model descriptions. It appears that, at least in principle, physical separation of the molecular species involved, and measurement of their individual binding and/or catalytic properties, offers the simplest means of discriminating between the validity of the various models as descriptions of the elicitation of primary response in a particular target cell by a particular hormone.

REFERENCES

1. Boeynaems, J. M., and Dumont, J. E. (1975). Quantitative analysis of the binding of ligands to their receptors. *J. Cyclic Nucleotide Res.*, **1**, 123–142.
2. Goldfine, I. D., Gardner, J. D., and Neville, D. M. (1972). Insulin action in isolated rat thymocytes. *J. Biol. Chem.*, **247**, 6919–6926.
3. Cuatrecasas, P. (1971). Insulin-receptor interactions in adipose tissue cells: direct measurement and properties. *Proc. Natl. Acad. Sci. U.S.A.*, **68**, 1264–1268.
4. Kahn, C. R., Freychet, P., Roth, J., and Neville, D. M. (1974). Quantitative aspects of the insulin-receptor interaction in liver plasma membranes. *J. Biol. Chem.*, **249**, 2249–2257.
5. Lin, M. C., Nicosia, S., Lad, P. M., and Rodbell, M. (1977). Effects of GTP upon binding of [^{3}H]glucagon to receptors in rat hepatic plasma membranes. *J. Biol. Chem.*, **252**, 2790–2792.
6. Hollenberg, M. D., and Cuatrecasas, P. (1975). Insulin and epidermal growth factor. *J. Biol. Chem.*, **250**, 3845–3853.
7. Tolkovsky, A. M., and Levitski, A. (1978). Mode of coupling between the β-adrenergic receptor and adenylate cyclase in turkey erythrocytes. *Biochemistry*, **17**, 3795–3810.
8. Bockaert, J., Roy, C., Rajerison, R., and Jard, S. (1973). Specific binding of [^{3}H]lysine-vasopressin to pig kidney plasma membranes. *J. Biol. Chem.*, **248**, 5922–5931.
9. Hechter, O., Tarada, S., Nakahara, T., Flouret, G., and Bergman, R. N. (1978). Relationship between hormonal occupancy of neurohypophyseal hormone receptor sites and adenylate cyclase activation. *J. Biol. Chem.*, **253**, 3219–3229.
10. Cuatrecasas, P., and Hollenberg, M. D. (1976). Membrane receptors and hormone action. *Adv. Protein Chem.*, **30**, 251–451.
11. Roy, C., Barth, T., and Jard, S. (1975). Vasopressin-sensitive kidney adenylate cyclase. *J. Biol. Chem.*, **250**, 3149–3156.
12. Sevilla, N., Steer, M. L., and Levitski, A. (1976). Synergistic activation of adenylate cyclase by guanylyl imidophosphate and epinephrine. *Biochemistry*, **15**, 3493–3499.

13. DeHaën, C. (1976). The non-stoichiometric floating receptor model for hormone-sensitive adenylyl cyclase. *J. Theor. Biol.*, **58**, 383–400.
14. Jacobs, S., and Cuatrecasas, P. (1976). The mobile receptor hypothesis and 'cooperativity' of hormone binding. *Biochim. Biophys. Acta*, **433**, 482–495.
15. Kahn, C. R., Baird, K., Flier, J., and Jarrett, D. B. (1977). Effects of antibodies to the insulin receptor on isolated adipocytes. *J. Clin. Invest.*, **60**, 1094–1106.
16. Kahn, C. R., Baird, K., Jarrett, D. B., and Flier, J. S. (1978). Direct demonstration that receptor crosslinking or aggregation is important in insulin action. *Proc. Natl. Acad. Sci. U.S.A.*, **75**, 4209–4213.
17. Pillion, D. F., Grantham, J. R., and Czech, M. P. (1979). Biological properties of antibodies against rat adipocyte intrinsic membrane proteins. *J. Biol. Chem.*, **254**, 3211–3220.
18. Minton, A. P. (1981). The bivalent ligand hypothesis: a quantitative model for hormone action. *Mol. Pharmacol.*, **19**, 1–14.
19. Bergman, R. N., and Hechter, O. (1978). A random-hit matrix model for coupling in a hormone-sensitive adenylate cyclase system. *J. Biol. Chem.*, **253**, 3238–3250.
20. Rodbell, M. (1980). The role of hormone receptors and GTP-regulatory proteins in membrane transduction. *Nature*, **284**, 17–22.
21. Citri, Y., and Schramm, M. (1980). Resolution, reconstitution, and kinetics of the primary action of a hormone receptor. *Nature*, **287**, 297–300.
22. Lad, P. L., Welton, A. F., and Rodbell, M. (1977). Evidence for distinct guanine nucleotide sides in the regulation of the glucagon receptor and of adenylate cyclase activity. *J. Biol. Chem.*, **252**, 5942–5946.
23. Maturo, J. M., III and Hollenberg, M. D. (1978). Insulin receptor: interaction with nonreceptor glycoprotein from rat liver membranes. *Proc. Natl. Acad. Sci. U.S.A.*, **75**, 3070–3074.

3 Perspectives for monoclonal antibodies in receptor research

E. Yavin

1 INTRODUCTION

It is now well established that many hormones and neurotransmitters are stored in subcellular structures and, upon release, interact by first binding to specific receptors on target tissues to elicit physiological responses. Following binding, certain signals are triggered which cause either activation of enzymes regulating the cyclic nucleotide intracellular pool, and/or introduce perturbations of ion transport systems across the membrane by mechanisms which are still poorly understood. Finally, the signal is coupled to mechanisms that promote or inhibit a physiological response such as cell proliferation, motility, secretion, electrical excitability, and other cellular post-translational events. The uniqueness of the physiological response is determined at the binding step by virtue of the ligand and the appropriate receptor on the target cell rather than by the subsequent steps which follow their interaction.[1] This was elegantly demonstrated in the case of the adenylate cyclase;[2,3] it is possible to confer on cells that possess an adenylate cyclase system, new hormonal responses by inserting a receptor into their cell membrane. Membrane fusion results in functional coupling between preexisting components.

Investigators from a variety of disciplines have established the conceptual and technical framework for analysis of receptor structure-function relationships. In the following section I will address primarily the potential use of immunological approaches for the dissection of the molecular structure, function and turnover of receptor complexes. Using selected examples from both the endocrine and the nervous system I shall discuss the advantages and the disadvantages of antibodies and the potential use of monoclonal antibodies particularly for studying receptor-associated autoimmune disease.

2 SOME BASIC PROPERTIES OF CELL
SURFACE RECEPTORS

Current work suggests that cell receptors are molecular complexes which conceptually consist of at least two functionally and possibly structurally different components; a binding site which provides the specific recognition milieu for the ligand and an additional site, the role of which is to transduce the signal to a secondary complex as it is illustrated schematically in Figure 1. The secondary complex could assume the form of either a specific enzyme, an ionophore or a structural protein, or possibly be any combination of the three. The question whether the receptor complex is an integral part or an allosteric regulator of the secondary complex, and whether the latter is affected directly or indirectly, by the receptor is presently unknown. The fact that both complexes are integral parts of the cell plasma membrane postulates an active role for the lipid microenvironment in signal processing and message amplification. Studies in this direction (for a detailed account see Aloj, this volume) have indicated that a wide variety of lipids may modulate the expression of cell surface antigens. Resolution of the molecular determinants which participate in the sequelae of events initiated at the binding site and terminated with the cellular response is thus essential for an understanding of the molecular basis of receptor function.

Early approaches used to identify receptor sites were based on the ability of biologically active compounds to interact with a target tissue and elicit a physiological response. Although in most cases the compound of preference was the purified natural ligand, synthetic compounds displaying agonistic or antagonistic properties have also been widely utilized. Pharmacological studies using these compounds contributed a great deal in determining apparent affinity constants, binding site multiplicity, and receptor density distribution. The pharmacological approach also provided insights into the steric configuration of the receptor complex, thus establishing a rational basis for drug therapy. This strategy was used, for example, in characterizing the cholinergic and adrenergic receptors of the neutrotransmitter apparatus. In the endocrine system much of the impetus for investigating the binding characteristics of various polypeptide hormones was provided with the development of assay systems employing appropriate target cells such as

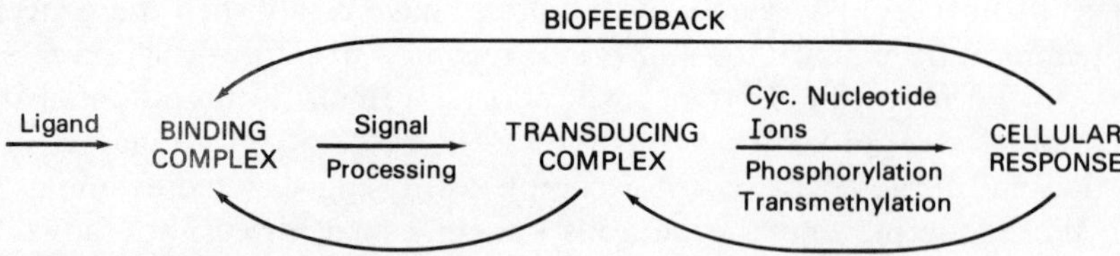

Figure 1. Ligand-elicited cellular response

hepatocytes, adipocytes and endocrine end-organ cells along with the introduction of ^{125}I-labelled peptide hormones as radio-ligands. These techniques prompted a large number of studies aimed at elucidating the interaction of polypeptide hormones and target tissues. Although rigorous examination of the labelled probe and suitable controls to account for non-specific binding were required in each case to ensure specificity, a great deal of information on the kinetic and physical properties of these hormone-receptor interactions were obtained.

The relatively large number of putative receptors identified by these techniques has prompted parallel efforts aimed at the purification of the components involved. Currently, however, there are only very few receptor complexes which have been purified to homogeneity. The most serious problem encountered in purifying a peptide hormone receptor has been its presence in exceedingly low quantities in the target tissue. To purify, for example, the insulin receptor to homogeneity would require 10^3 to 10^4 fold more material than that required for the purification of the nicotinic acetylcholine receptor obtained from the Torpedo marine ray or the Electrophorus fish electric organs. This largely explains the more advanced state of receptor purification from these latter sources.

Second, for the most part, receptors are integral membrane components and therefore solubilization is a prerequisite for any subsequent purification. A broad spectrum of detergents has been employed to achieve effective and mild extraction. However, solubilization of many receptors into an artificial, lipid-free environment has increased their instability and also altered significantly characteristic properties.

A great deal of effort has been devoted to integrate solubilized and partially purified receptors into appropriate artificial membranes prepared either from lipid mixtures extracted from the original tissue or comprising of pure synthetic lipids. Although this approach has been successfully applied for studying several membrane proteins (for review see Reference 4), it was only recently that successful reconstitution was achieved for the relatively long purified receptor for acetylcholine.[5,6] In most cases, the multiplicity of components necessary for the receptor function and their correct assembly into artificial membranes in a functional configuration, is probably one of the major obstacles in obtaining a successful reconstitution. The unequivocal demonstration that a reconstituted acetylcholine receptor exhibited the properties of the native receptor, should have great advantage for the understanding structure–function relationship of the receptor complex. Another approach for studying structure–function relationship is the use of well defined antibodies directed to specific determinants on the receptor complex. These probes may not only provide means for affinity purification of receptor components but also enable studies on receptor function as will be emphasized below.

3 RECEPTOR-ASSOCIATED DISORDERS

3.1 Anti-thyroid reactive antibodies and Graves' disease

Historically, the immunological approach to studying receptor-associated disorders of the thyroid dates back to early observations by Adams and Purves,[7] who first proposed the involvement of anti-thyroid antibodies in patients with Graves' disease. Later it was noted that the antibody activity which resembled the thyrotropin (TSH) hormone in its overall effect on thyroid stimulation resided in a heterogeneous population of immunoglobulins.[8] Unlike the native hormone, the thyroid stimulating autoantibody (TSab) has a prolonged time course of action. It was further suggested that the enhanced cAMP production elicited by TSab as well as TSH, both in thyroid slices[9] and thyroid membrane preparation,[10] is indicative of coupling between the cyclase and the TSH receptor.

Immunologic efforts to characterize further the structure and function of the glycoprotein component of the TSH receptor (the primary high affinity binding site) were unfortunately limited, as antibodies made in rabbits to the glycopeptide fragment of the receptor did not readily block TSH binding to cells or membranes[11] using current assay procedures, despite the ability of these antisera to precipitate solubilized receptor activity.[12,13] One element of this problem was the sensitivity of TSH binding assays to normal sera and gamma globulin preparations.[14,15,16,17] Second, the rabbit antibodies, appeared to be directed at molecular determinants which did not affect binding, much the same as described for antibodies to solubilized acetylcholine[18] and insulin receptors.[19,20] Third, the glycoprotein or glycopeptide receptor fragment, as the result of solubilization and trypsinization, did not have receptor determinants in the same configuration as on the intact cell or the native membrane[21,22].

Despite these problems immunologic probes remained extremely attractive. In particular they enable one to investigate the question whether in the polyclonal autoimmune response, antibody molecules capable of inhibiting TSH binding are identical to those which stimulate TSH-dependent adenylate cyclase activity. Thus the question of whether antibodies inhibiting TSH binding were directed at the binding determinant of the TSH receptor, and whether reactivity against this determinant alone was responsible for stimulating cyclase activity, has been an actively debated issue which might be resolved with important clinical consequences by the experimental production of antibodies, each reactive with only one antigenic domain of the TSH receptor.

3.2 Anti-insulin receptor antibodies and insulin-resistant diabetes

The existence of a diabetic syndrome associated with autoantibodies to the insulin receptor was first reported by Flier *et al.*[23] Binding of insulin to monocytes was inhibited by a polyclonal gamma globulin isolated from sera of some diabetic patients resistant to conventional doses of exogenous insulin. The F(ab')$_2$ and F(ab') fragments were also inhibitory.[24] The antibodies inhibited insulin binding to adipocytes, hepatocytes and erythrocytes obtained from various sources such as human, rat and avian species. Unlike monovalent F(ab') fragments which antagonize both binding and activity,[24] bivalent receptor antibodies have complete agonist activity in their ability to stimulate glucose incorporation. Similarly, the IgG or F(ab')$_2$ fragment isolated from the serum of a patient with insulin-resistant diabetes were as effective as insulin, in stimulating rat hepatocyte glycogenesis.[25]

3.3 Anti-acetylcholine receptor antibodies and myasthenia gravis

The possible involvement in the ethiology of myasthenia gravis of circulating antibodies which prevent the action of acetylcholine, at the neuromuscular junction disease in man has been suggested by Simpson.[26] More direct clues to the involvement of the acetylcholine receptor were offered by Fambrough *et al.*[27] and by Patrick and Lindstrom[28] who produced conditions analogous to myasthenia patients in rabbits injected with a preparation of acetylcholine receptor. Additionally, antibodies directed against the acetylcholine receptor have subsequently been detected in patients with myasthenia gravis.[29]

The precise mechanism by which antibodies to the receptor cause impairment of neuromuscular conduction and produce the syndrome of muscle weakness and fatigability is still unknown. The observation that no single antigenic determinant on the receptor complex is recognized by all myasthenic sera most likely suggests that there are several classes of antireceptor antibodies, each presumably with a different affinity and possibly directed at a different antigenic site. Either acting as single components or in concert, these antibodies seem to lead to a loss of the acetylcholine receptor. Thus the basic lesion of the acetylcholine receptor, both in the experimental myasthenic model as well as in the myasthenic patient, is attributed to an enhanced receptor degradation[30,31,32] as a result of intermolecular crosslinking of the receptors by the antibodies. In addition, complement-mediated-destruction or prolonged occupancy of the active site of the receptor by the antibody may also be implicated in the ethiology of the disease.

3.4 Conclusion

The abnormal ligand–receptor interactions under each of the human disease states described above, is the consequence of organ-specific immunity, with circulating antibodies which specifically bind cell surface membrane receptors. The means by which circulating anti-receptor antibodies could

POSSIBLE MECHANISMS OF INHIBITION OF
BINDING BY ANTIBODY

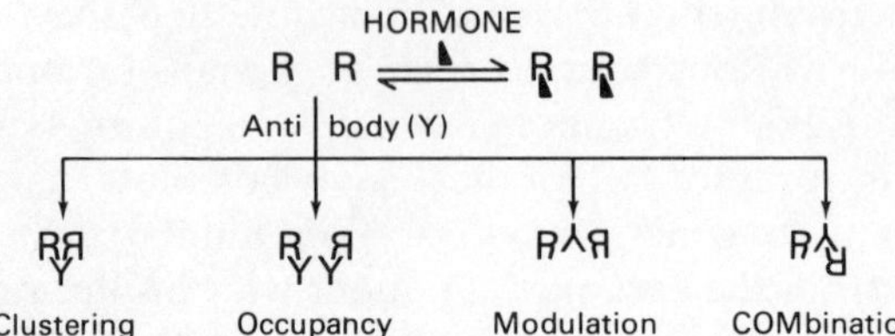

Figure 2. Possible mechanisms of inhibition of binding by antibody

influence receptor sites are illustrated in Figure 2. First, the antibody could competitively occupy a portion of the active site, thus preventing ligand binding. Second, it could modulate the receptor by acting indirectly at an allosteric site by shifting the receptor, for example, from a high to a low affinity state. Third, the antibody may crosslink receptors to form microaggregates. By doing so, it may render receptors to enhanced endocytic processes which could result in a decrease in the actual number as it has been suggested for the acetylcholine receptor.[33]

4   RECEPTORS AS IMMUNOGENS

4.1   Polyclonal antisera to cell surface determinants

Proteins and glycoproteins are key components present in the majority of receptor complexes studied so far. As part of the plasma membrane repertoire, they usually exist in small quantities and most frequently are unstable when membrane integrity is disrupted. A possible approach for identification, characterization and molecular dissection of these ubiquitous receptor components is to take advantage of their antigenic properties by producing antibodies of appropriate specificity and affinity.

Antibodies are almost uniquely suitable probes for studying receptor structure-function relationships. One major advantage is the possibility of generating antibodies directed against essentially any number of antigenic determinants on the receptor. By coupling the appropriate tag to antibodies, visualization of receptors can be followed by immunofluorescence or by high power electron microscopy. Receptor mobility in living cells can be visualized and quantified with image intensified camera, using fluorescent photobleaching recovery techniques.[34,35]

The acetylcholine receptor is presumably the best characterized example for utilization of antibodies in receptor research. Although the probes available for this receptor, which include toxins, analogues and affinity labelling reagents, are perhaps more abundant and versatile than any other receptor system studied, the immunological approach has made significant contributions.[36] This is in part because antibodies, unlike analogues and toxins, can provide important clues as to the role of sites other than the active site of the receptor complex.

Until recently, the efficiency in producing antibodies against discrete cellular antigens was relatively limited. When an immunogen is injected into a laboratory animal, the immune response elicited usually consists of a large variety of antibodies directed against multiple determinants often on the same molecule. Thus, the conventional polyclonal antibodies against cell surface antigens are usually complex and often themselves require extensive purification procedures.

4.2 Monoclonal antibodies to cell surface determinants

Somatic cell hybridization technique has recently dramatically changed concepts on the utilization of the immunological approach for studying surface antigens at the cellular, subcellular and molecular level. The technique involves fusion of spleen cells obtained from immunized mice with a plasma cell tumour (myeloma) adapted to tissue culture. The myeloma renders the spleen cells 'immortal' in culture and is a permissive vehicle for the expression of immunoglobulins encoded by the parental spleen cell genome. Cloning of these hybrid cell lines results in the production of a set of homogeneous antibodies each reactive with a single antigenic determinant. In a most elegant study, Milstein and his associates[37,38] were able to develop continuous hybrid cell lines which can produce antibodies of defined specificity against a number of cell antigens. The protocol used for selecting the appropriate clone is in principle similar to that shown in Figure 3. Only the hybrid plasmacytoma cell lines (hybridomas) survive in selection medium containing hypoxanthine–aminopterine–thymidine, whereas, parental myeloma cells which lack hypoxanthine–guanine–ribosyl–transferase, and parental spleen cells, do not. The culture medium from each colony is screened for secreted antibodies reacting against the injected antigen, by a radioimmunoassay employing the hybridoma antibody and a second, labelled antibody against mouse immunoglobulin (phase 1).

After expansion of cells in culture, the secreted antibodies are screened by a more specific assay (phase II) to isolate those antibody-secreting hybridoma cells of particular interest. For most receptor antigens, suitable antibody-radioligand competition assays can be established as helpful criteria (see below). In addition, if the antigen in concern expresses a biological activity, i.e. enzymatic activity, transport, agglutination, the ability of the antibody to stimulate or inhibit such an activity, can be used in

PROTOCOL FOR MONOCLONAL ANTIBODIES PRODUCTION

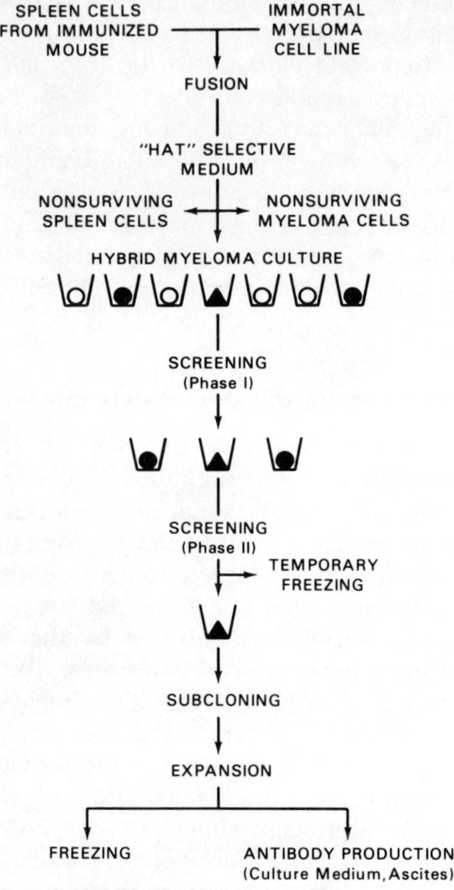

Figure 3. Protocol for monoclonal antibody production

principle as a screening method. Given the impurity of most receptor complexes, monoclonal antibodies may prove to be an excellent tool for receptor identification and purification. Once the desired colonies are selected, the cells are subcloned to obtain pure clonal lines which can then be maintained in culture or injected into suitable animals to generate the specific monoclonal antibody. Thus large quantities of antibodies with high reproducibility can be synthesized.

Production of a monoclonal antibody directed against a specific determinant on a discrete receptor site raises several experimental problems. The first and perhaps the most acute relates to the fact that the number of cell surface receptors is usually small in comparison to many other cellular antigens. Partial purification of the receptor is therefore desirable. A second problem pertains to the level of the immune response elicited by a weak antigen; a concern for any immune approach. Special procedures may be required to select out a specific antibody-secreting hybridoma amidst those secreting non-specific immunoglobulins. Third, the criteria of specificity for some of the monoclonal antibodies, by the more refined screening procedures (i.e. binding displacement, bioactivity interference), may be overlooked if the antigenic site has no immediate effect on the receptor site. This concern can be experimentally tested by mixing a number of monoclonal antibodies in the hope to generate the desired biological effect. This approach may be utilized also when receptor purification is attempted by either immunoprecipitation or by immunoaffinity chromatography.

4.3   Monoclonal antibodies to the thyrotropin receptor

Evidence now exists to suggest that the receptor for the thyrotropin hormone consists of a two component system: a glycoprotein and a ganglioside.[21,22] While the glycoprotein component is the primary high affinity binding site, the ganglioside may participate in the modulation of the hormone-receptor interaction and may also affect signal transduction. Our laboratory has recently used the hybridoma technique to produce monoclonal antibodies[39] to study the structure and function of the TSH-receptor and to resolve questions concerning whether antibodies which inhibit TSH binding are also thyroid activators.

Partially purified thyrotropin receptor preparations from bovine thyroid tissue were injected into Balb/C mice and spleen cells from immunized

COATING SEQUENCES FOR SOLID SURFACE
RADIOASSAY

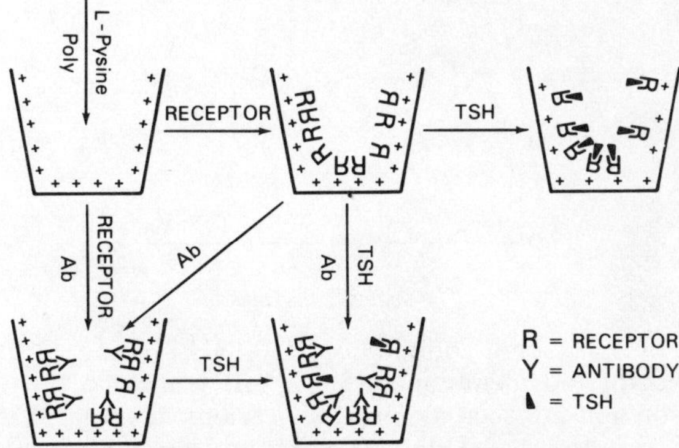

Figure 4. Coating sequences for solid surface radioassay

Table 1. Effect of antibody on ^{125}I-TSH binding to human membranes, or to the glycoprotein component of the bovine TSH receptor

	TSH receptor preparation		
	Human (Graves') thyroid membranes	Glycoprotein component (from bovine thyroid)	
		Free	In liposomes
Control	10.0	3.1	15.3
Monoclonal antibody	1.9	1.0	3.1

Aliquots of 0.1 ml of either human thyroid membranes (50 μg/ml) or lithium diiodosalicylate (LIS) solubilized bovine thyroid membrane preparations (40 μg/ml) were added to 96 well-plates. After four hours at 4° the unattached membranes were removed and wells rinsed twice with buffer before addition of ^{125}I-TSH (55,000 c.p.m./well). Monoclonal antibody raised in ascites fluid (at a dilution of 1:100) was added along with the labeled hormone. Incubation was carried for one hour at 4°. (For details, see Yavin *et al.*, 1980). Values are expressed as c.p.m. $\times 10^{-3}$.

animals fused with P3-NS1/1-Ag4-1 myeloma cells using polyethylene glycol.

In essence, following the protocol depicted in Figure 2 above, we have isolated five clones which secrete antibodies whose binding to thyroid membranes is specifically inhibited by unlabeled thyrotropin (10^{-6} M). In addition, we have developed a solid surface radioligand assay capable of

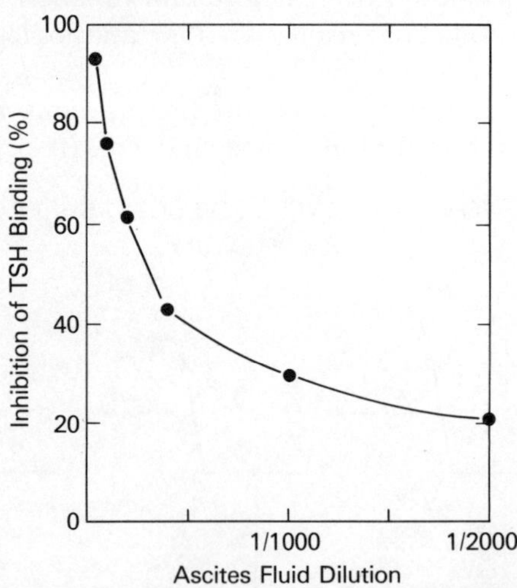

Figure 5. Effect of hybridoma antibody on ^{125}I-TSH binding. Coating of microtiter wells with bovine thyroid membranes (50 μg/ml) was done for 16 hours at 4 °C. After removing the non-attached membranes, various dilutions of the ascites fluid in 20 mM Tris pH 6.7 and ^{125}I-TSH (40,000 c.p.m./well) were added. The values expressed as per cent inhibition of TSH bound represent averages of 2–3 wells

measuring ^{125}I-TSH binding to thyroid membrane and to assess the effect of the monoclonal antibody on this reaction. Figure 4 represents a schematic illustration of the coating sequences employed for the binding assay. Microtiter well plates precoated with poly-L-lysine are incubated with either soluble (lithium diiodosalicylate extract) or membrane bound TSH receptor preparations for 4 to 24 hours. Unadsorbed membranes are removed and antibodies at desired dilutions are added for specified times. Alternatively, antibodies can be added concurrently with the receptor preparations. The unadsorbed antibodies are washed off with a buffer solution containing bovine serum albumin and ^{125}I-labelled TSH is added to the wells. TSH binding can also be carried out in the presence of the antibody. After reaching a steady state at 4° the cells are rinsed and the radioactivity associated with the cells is measured in a gamma counter. Using this technique we could demonstrate that these antibodies prevented ^{125}I-TSH binding to bovine or human (Graves' disease) thyroid membrane preparations (Table 1). They were capable of competitively blocking TSH binding to bovine thyroid membranes as illustrated in Figure 5. Furthermore, TSH

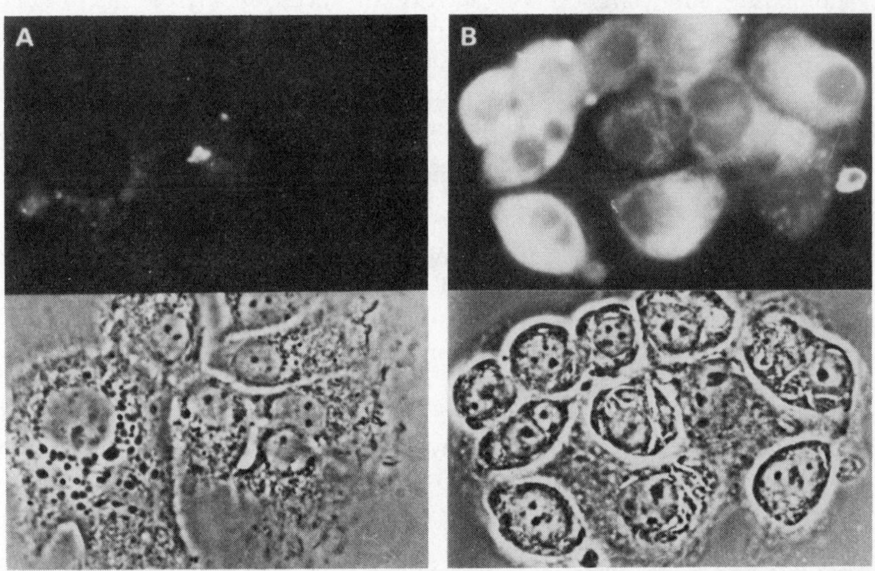

Figure 6. Indirect immunofluorescence labelling of thyroid cells by the anti-TSH receptor monoclonal antibody. Rat thyroid cells[42] grown on poly-L-lysine precoated glass coverslips for 3–4 days were rinsed in phosphate buffered saline and fixed with 1 per cent glutaraldehyde for 30 minutes. After rinsing monoclonal antibody (1:10 dilution) was layered over the cells in the presence (a) or absence (b) of 1 mg/ml TSH for 1 hour at room temperature. The solution was removed and after washing, cells were exposed to fluorescent conjugated (FITC) goat antimouse immunoglobulin and after washing, cells were visualized under fluorescent optics

Table 2. Effect of monoclonal antibodies to TSH receptor on the adenylate cyclase activity of thyroid membranes

Assay addition[a]	Adenylate cyclase activity picomoles cAMP/15 min/mg membrane protein
None (basal)	106
+TSH $(1 \times 10^{-7}$ M)	388
+T-11E8	104
+T-13D11	105
+T-18A3	110

[a] Antisera were tested as 0–45 per cent ammonium sulphate fractions of culture media at 10–12 mg/ml. No changes in adenylate cyclase activity were evident at 0.1 or 0.01 mg/ml protein.

efficiently blocked the monoclonal antibody interaction with rat thyroid cells in culture as evident by the indirect immunofluorescence technique (Figure 6). On the other hand, none of the antibodies tested so far had the capability of stimulating adenylate cyclase activity in slices or thyroid membrane preparations (Table 2). This suggests that autoantibodies in sera of patients with Graves' disease which inhibit binding to thyroid membranes and which are considered as potential antibodies to the thyrotopin receptor, need not be thyroid adenylate cyclase stimulators.

4.4 Prospects for future studies

Monoclonal antibodies of defined specificity have prompted many studies aimed at analysing both soluble and cell surface antigens which are associated with tumorigenicity, differentiation and both viral and bacterial infections. Aside from being better immunodiagnostic tools, the potential biomedical application of hybridoma products has been already demonstrated for example in the immunopurification of interferon. It should not be long before monoclonal antibodies directed against circulating, antireceptor autoantibodies could be used for direct therapy.

The advantages of using monoclonal antibodies as probes for studying receptor structure-function relationships are forseen in three major aspects; (1) to introduce molecular probes for studying mechanism of ligand-receptor interaction at sites other than the binding site of the receptor complex; (2) to study the dynamics of receptor internalization, synthesis and degradation irrespective of the ligand and localize receptors when other probes are not available; and finally (3) to purify receptors by affinity chromatography. Preliminary reports on the use of monoclonal antibodies in receptor research are very promising.[41,42] Recent demonstration of myasthenia gravis-like syndrome in rats injected with monoclonal antibodies against the acetylcholine receptor has already suggested to Richman *et al.*,[43] that binding to a single antigenic determinant may elicit the entire autoimmune syndrome. Whether the same mechanism is applicable to other receptor-associated autoimmune diseases remains to be investigated. From our own studies with the TSH receptor, it seems that more than one antibody may be necessary to block TSH binding and also elicit the adenylate cyclase response.

5 ACKNOWLEDGEMENT

I thank my colleagues M. Schneider, Z. Yavin and L. Kohn for sharing a great deal of laboratory effort and discussion time on some of the experiments described in this paper and to B. Armstrong for her patience in preparing the manuscript.

REFERENCES

1. DeHaën, C. (1976). The nonstoichiometric floating receptor model for hormone-sensitive adenylyl cyclase. *J. Theor. Biol.*, **58**, 383–400.
2. Orly, J., and Schramm, M. (1976). Coupling of catecholamine receptor from one cell with adenylate cyclase from another cell by cell fusion. *Proc. Natl. Acad. Sci. U.S.A.*, **73**, 4410–4414.
3. Citri, Y., and Schramm, M. (1980). Resolution, reconstitution and kinetics of the primary action of a hormone receptor. *Nature*, **287**, 297–300.
4. Miller C., and Racker, E. (1979). Reconstitution of membrane transport functions. in *The Receptors*, ed. O'Brien R. D., (Plenum Publishing, New York), Vol. 1, pp. 1–31.
5. Nelson, N., Anholt, R., Lindstrom J., and Montal, M. (1980). Reconstitution of purified acetylcholine receptors with functional ion channels in planar lipid bilayers. *Proc. Natl. Acad. Sci. U.S.A.*, **77**, 3057–3061.
6. Schindler, H., and Quast, V. (1980). Functional acetylcholine receptor from *Torpedo marmorata* in planar membranes. *Proc. Natl. Acad. Sci. U.S.A.*, **77**, 3052–3056.
7. Adams, D. D., and Purves, H. D. (1956). Abnormal response in the assay of thyrotropin. *Proc. Univ. Otago Med. Sch.*, **34**, 11–12.
8. Smith, B. R., Munro, D. S., and Dorrington, K. J. (1969). The distribution of the long-acting thyroid stimulator among γ-G immunoglobulins. *Biochim. Biophys. Acta*, **188**, 89–100.
9. Kendall-Taylor, P. (1973). Effects of long-acting thyroid stimulator (LATS) and LATS protector on human thyroid adenylcyclase activity. *Br. Med. J.*, **3**, 72–75.
10. Mukhtar, E. D., Smith, B. R., Pyle, G. A., Hall, R., and Vice, P. (1975). Relation of thyroid-stimulating immunoglobulin to thyroid function and effect of surgery, radioiodine and antithyroid drugs. *Lancet*, **1**, 713–715.
11. Kohn, L. D., *et al.* (1981). Unpublished observations.
12. Tate, R. L., Holmes, J. M., Kohn, L. D., and Winand, R. J. (1975). Characteristics of a solubilized thyrotropin receptor from bovine thyroid plasma membranes. *J. Biol. Chem.*, **250**, 6527–6533.
13. Tate, R. L., Winand, R. J., and Kohn, L. D. (1976). Solubilization and partial purification of the thyrotropin receptor; in *Thyroid Research*, eds. Robbins, J.,

and Braverman, L. *Exc. Med. Int. Cong. Ser.* (Elsevier/North-Holland, Amsterdam) No. 378, pp. 57–60.

14. Amir, S. M., Carraway, T. F., Jr., Kohn, L. D., and Winand, R. J. (1972). Thyrotropin binding to thyroid plasma membranes. *J. Biol. Chem.*, **248**, 4092–4100.

15. Fenzi, G., Macchia, E., Bartalena, L., Mazzanti, F., Baschieri, L., and De Groot, L. J. (1978). Radioreceptor assay of TSH; its use to detect thyroid-stimulating immunoglobulins. *J. Endocrinol. Invest.*, **1**, 17–24.

16. Hall, R., Smith, B. R., and Mukhtar, E. D. (1975). Thyroid stimulators in health and disease. *Clin Endocrinol.*, **4**, 213–230.

17. McKenzie, J. M., and Zakarija, M. (1977). LATS in Graves' disease. *Recent Prog. Horm. Res.*, **33**, 29–53.

18. Stanley, E. F., and Drachman, D. B. (1978). Effect of myasthenic immunoglobulin on acetylcholine receptors of intact mammalian neuromuscular junction. *Science*, **200**, 1285–1287.

19. Jacobs, S., Chang, K., and Cuatrecasas, P. (1978). Antibodies to purified insulin receptor have insulin-like activity. *Science*, **200**, 1283–1284.

20. Pillion, D. J., and Czech, M. P. (1978). Antibodies against intrinsic adypocyte plasma membrane proteins activate D-glucose transport independent of interaction with insulin binding sites. *J. Biol. Chem.*, **253**, 3761–3764.

21. Kohn, L. D. (1978). Relationship in the structure and function of receptors for glycoprotein hormones, bacterial toxins, and interferon. In *Receptors and Recognition*, eds. Cuatrecasas, P., and Greaves, M. F. (Chapman and Hall, London), Series A, Vol. 5, pp. 134–212.

22. Kohn, L. D., Consiglio, E., De Wolf, M. J. S., Grollman, E. F., Ledley, F. D., Lee, G., and Morris, N. P. (1980). Thyrotropin receptors and gangliosides; in *Structure and Function of Gangliosides*, eds. Svennerholm, L., Mandel, P., Dreyfus, H., and Urban, P.-F. *Adv. Exp. Med. Biol.* (Plenum Press, New York), **125**, 487–504.

23. Flier, J. S., Kahn, C. R., Roth, J., and Bar, R. S. (1975). Antibodies that impair insulin receptor binding in an unusual diabetic syndrome with severe insulin resistance. *Science*, **190**, 63–65.

24. Kahn, C. R., Baird, K. L., Jarrett, D. B., and Flier, J. S. (1978). Direct demonstration that receptor crosslinking or aggregation is important in insulin action. *Proc. Natl. Acad. Sci. U.S.A.*, **75**, 4209–4213.

25. Baldwin, D., Jr., Terris, S., and Steiner D. F. (1980). Characterization of insulin-like actions of anti-insulin receptor antibodies. *J. Biol. Chem.*, **255**, 4028–4034.

26. Simpson, J. A. (1960). Myasthenia gravis: A new hypothesis. *Scott. Med. J.*, **5**, 419–436.

27. Fambrough, D. M., Drachman, D. B., and Satyamurti, S. (1973). Neuromuscular junction in myasthenia gravis: Decreased acetylcholine receptors. *Science*, **182**, 293–295.

28. Patrick, J., and Lindstrom, J. (1973). Autoimmune response to acetylcholine receptor. *Science*, **180**, 871–872.

29. Lennon, V. (1975). Humoral factors in myasthenia gravis. *Nature (Lond)*., **258**, 11–12.

30. Appel, S. H., Anwyl, R., McAdams, M. W., and Elias, S. (1977). Accelerated degradation of acetylcholine receptor from cultured rat myotubes with myasthenia gravis sera and globulins. *Proc. Natl. Acad. Sci. U.S.A.*, **74**, 2130–2134.

31. Heinemann, S., Bevan, S., Kullberg, R., Lindstrom, J., and Rice, J. (1977).

Modulation of acetylcholine receptor by antibody against the receptor. *Proc. Natl. Acad. Sci. U.S.A.*, **74,** 3090–3094.

32. Drachman, D. B., Angus, C. W., Adams, R. N., and Kao, I. (1978a). Effect of myasthenic patients' immunoglobulin on acetylcholine receptor turnover; selectivity of degradation process. *Proc. Natl. Acad. Sci. U.S.A.*, **75,** 3422–3426.

33. Drachman, D. B., Angus, C. W., Adams, R. N., Michelson, J. D., and Hoffman, G. J. (1978b) Myasthenic antibodies cross-link acetylcholine receptors to accelerate degradation. *N. Engl. J. Med.*, **298,** 1116–1122.

34. Koppel, D. E., Axelrod, D., Schlessinger, J., Elson, E. L., and Webb, W. W. (1976). Dynamics of fluorescence marker concentration as a probe of mobility. *Biophys. J.*, **16,** 1315–1329.

35. Schlessinger, J., Schechter, Y., Willingham, M. C., and Pastan, I. (1978). Quantitative determination of the lateral diffusion coefficients of the hormone-receptor complexes of insulin and epidermal growth factor on the plasma membrane of cultured fibroblasts. *Proc. Natl. Acad. Sci. U.S.A.*, **75,** 5353–5357.

36. Lindstrom, J. (1979). Antibodies to the acetylcholine receptor molecule and its component polypeptide chains. In *Adv. Cytopharmacol.* eds. Ceccareli, B., and Clementi, F., (Raven Press, New York), Vol. 3, pp. 245–253.

37. Köhler, G., and Milstein, C. (1975). Continuous cultures of fused cells secreting antibody of predefined specificity. *Nature (Lond).*, **256,** 495–497.

38. Galfre, G., Howe, S. C., Milstein, C., Butcher, G. W., and Howard, J. C. (1977). Antibodies to major histocompatibility antigens produced by hybrid cell lines. *Nature (Lond).*, **266,** 550–552.

39. Yavin, E., Yavin, Z., Schneider, M. D., and Kohn, L. D. (1981). Monoclonal antibodies to the thyrotropin receptor: implications for receptor structure and the action of autoantibodies in Graves' disease. *Proc. Natl. Acad. Sci. U.S.A.* **78,** 3180–3184.

40. Ambesi-Impiombato, F. S., Parks, L. A., and Coon, H. G. (1980). Culture of hormone-dependent functional epithelial cells from rat thyroids. *Proc. Natl. Acad. Sci. U.S.A.*, **77,** 3455–3459.

41. Greene, G. L., Fitch, F. W., and Jensen, E. V. (1980). Monoclonal antibodies to estrophilin: Probes for the study of estrogen receptors. *Proc. Natl. Acad. Sci. U.S.A.*, **77,** 157–161.

42. Tzartos, S. J., and Lindstrom, J. M. (1980). Monoclonal antibodies used to probe acetylcholine receptor structure: Localization of the main immunogenic region and detection of similarities between subunits. *Proc. Natl. Acad. Sci. U.S.A.*, **77,** 755–759.

43. Richman, D. P., Gomez, C. M., Berman, P. W., Burres, S. A., Fitch, F. W., and Arnason, B. G. W. (1980). Monoclonal anti-acetylcholine receptor antibodies can cause experimental myasthenia. *Nature*, **286,** 738–739.

4 Membrane lipids and modulation of hormone receptor expression†

S. M. Aloj

In the first volume of this series, in the article by D. A. Vessey and D. Zakim on regulation of membrane-bound enzymes, it was stated right at the onset that 'Because of their capacity to establish and maintain microenvironments, membranes form the base from which life emanates'. This is a fairly strong statement which, indeed, turns out to be most appropriate to introduce our discussion on the role of membrane components in the expression of receptor function, given the vital importance of membrane mediated biological effects of hormones, antibodies, drugs, etc. It should be pointed out that most of the experimentation and writing on the relevance of lipids, sugars, proteins, membrane fluidity, hydrophobicity, polarity, etc. on the regulation of membrane associated enzymes could profitably be applicable to the regulation and expression of membrane receptors. In fact receptors may resemble membrane-bound enzymes in their structural properties and in their relationships with the membrane lipid bilayer. It is also relevant, in this regard, that hormone receptors in plasma membranes are believed to be closely associated with enzymes, such as adenylate cyclase, and, although it is now accepted that the receptor and the enzyme are two distinct entities, their structural and functional requirements must be very similar since they are best expressed when the integrity of membrane composition is preserved.

Let us now define a receptor in terms of its structural and functional properties in order to establish the purpose and limits of this presentation.

A simple-minded view would define receptor as a component of the plasma-membrane capable of recognizing and binding, with high specificity and affinity, a ligand present in its close environment at a very low concentration (10^{-9}–10^{-12} M). Within these limits many receptor systems have been described for hormones, antigens, antibodies, drugs, toxins, lectins, neurotransmitters and plasma proteins. A very large number of

* Supported by N.I.H. grant 1 R01 AM 21689-02.

83

studies have been devoted to characterize such receptors and the general approach has been to analyse the ability of more or less homogeneous membrane preparations to bind ligands isotopically labelled to a high specific activity. Thus many putative receptors have been described in terms of their binding properties such as thermodynamic parameters, cooperative behaviour, sensitivity to temperature, pH, ions concentration, etc. A more physiological approach to hormone action should not only limit receptor function to recognition and binding of the ligand, but it should also involve coupling of the latter to a ligand-mediated biological response and an integral part of receptor function should be the ability to transmit a signal to the cell machinery. Within these wider limits receptor function becomes more complex and it is hard to imagine that the full expression of this function requires the involvement of only one type of molecules amongst the diverse components of the plasma membrane. Thus it appears that ligand binding and message transmission are the result of complex interactions of the major membrane components, i.e. proteins and lipids. The question then arises as to which is the role of the various components in the chain of events which is triggered by the interaction of the ligand with the cell membrane and eventually leads to a biologic response which is typical for each ligand.

The following discussion will analyse some of the pertinent data and speculations, which are currently available, to answer that question; with a special emphasis on the role of membrane lipids. However it will not take long to realize that, in spite of the many elegant studies on lipid involvement in receptor regulation, the fundamental aspects of the problem remain at the frontier of biochemical knowledge. These aspects will be stressed in order to stimulate in the reader's mind new ideas and new experimental approaches which will contribute to deepen our understanding of the biochemical mechanisms of cell regulation and, most important, will clarify the molecular basis of diseases which depend on derangement of such mechanisms.

1 PLASMA MEMBRANE COMPONENTS AND THEIR RELEVANCE TO HORMONE RECEPTOR INTERACTION AND MESSAGE TRANSMISSION

1.1 Lipids

Lipids account for a large proportion of the components of cell membranes. Their percentage of total dry weight varies from 25 per cent in mitochondrial membranes up to 80 per cent in myelin, the multilayer coat of plasma membranes of the Schwann's cells which lay adjacent to many neurons. There are many kinds of lipids, present in substantial amounts, among these the major classes are phospholipids, sphingomyelin, glycolipids and cholesterol. Lipid distribution varies widely between membranes of different origin making it difficult to generalize about membrane composition. However phospholipids are always present in large proportion since they constitute from 40 per cent to over 90 per cent of the total lipids.

Phospholipids are derived from either the three-carbon-alcohol glycerol or the more complex amino-alcohol sphingosine. The glycerol derivatives are also called phosphoglycerides. A phosphoglyceride is characterized by a glycerol backbone with two long chain, saturated or unsaturated, fatty acids (typically between 14 and 24 carbon atoms) and a phosphoric acid residue linked in an ester bond to an alcoholic hydroxyl group. The resulting compound is phosphatidic acid, the simplest phosphoglyceride, present only in trace amounts in plasma membranes but an important intermediate in phosphoglyceride biosynthesis. The predominant phosphoglycerides are derivatives of phosphatidic acid in which the phosphate group becomes esterified to the hydroxyl group of one of several alcohols. The most common phosphoglycerides are phosphatidylcholine, phosphatidylethanolamine, phosphatidylserine and phosphatidylinositol. The first two, together with sphingomyelin, are generally referred to as 'neutral' phospholipids, whereas phosphatidyl-serine, phosphatidylinositol and phosphatidic acid are considered acidic phospholipids. As we shall see later on this grouping of phospho-lipids in two classes is useful since sensitivities to phospholipases of different origins are different in the two classes so is the involvement in receptor function.

Membranes also contain glycolipids and cholesterol; the former, as their name implies, are sugar containing lipids which derive, like sphingomyelin, from sphingosine. Glycolipids differ from sphingomyelin in that the unit linked to the primary hydroxyl group of sphingosine is not phosphoryl-choline but one or more sugars. The simplest glyco-lipid is cerebroside containing only one sugar residue, either glucose or galactose. More complex glycolipids are the gangliosides in which there is a complex oligosaccharide chain containing sialic acid.

Because of the presence in their structure of long, non-polar, hydrocarbon chains and charged headgroups phospholipids and glycolipids are amphipatic molecules i.e. they contain both hydrophylic and hydrophobic moieties. It is obvious that, in an aqueous environment, their hydrophylic headgroups will have affinity for water molecules whereas their hydrocarbon chains will avoid water. Indeed the hydrophobic interactions are the major driving force in the formation and stability of the lipid bilayer.

1.2   Proteins

In the average proteins account for about one half of membrane dry weight and it appears that they dwell in a close contact with lipids. In many cases it has been difficult to isolate proteins from plasma membranes because of the tendency of these proteins to become denatured when separated from their lipid environment. This proves that the membrane bilayer plays an important role in maintaining the native structure of the imbedded protein. There are membrane proteins which can be extracted under relatively mild conditions such as high ionic strength (e.g. 1–2 M salt); in contrast separation of other proteins can only be accomplished by using detergents or organic solvents to destroy the lipid moiety of the membrane. These differences in dissociability have led to believe that the location of proteins in the membranes can be either peripheral or integral i.e. exposed to the aqueous environment or deeply embedded in the lipid matrix. In either cases interaction with lipids must be vital to maintain the protein in its functionally proper conformation.

It is now well substantiated that membrane proteins, as well as lipids, move around the membrane, and it is widely accepted that this dynamic character of plasma membranes plays a primary role in modulation of membrane receptors '*in vivo*'. The mobility of membrane proteins depends mainly on the fluidity of the lipid matrix. Thus the functional properties of receptors can be markedly affected by changes in microviscosity. Lipid protein interaction may not only affect the lateral (sidewise) and rotational (on their own axis) mobility of the proteins but also their degree of exposure to the outer surroundings. Increasing microviscosity may reflect a tighter lipid–lipid interaction with a decrease in lipid–protein interaction; as a result membrane proteins may be squeezed out and become more exposed to the external environment. Decreasing microviscosity would otherwise reflect stronger interaction between the phospholipid acyl chains and the hydrophobic part of the protein and result in a lesser exposure. It could then be speculated that membrane fluidity and protein exposure are inversely related. Thus cholesterol, by increasing microviscosity, promotes the exposure of membrane proteins; contrarywise cholesterol depletion increases membrane fluidity and causes shielding of membrane proteins.

With these considerations in mind several roles for lipids can be anticipated in the regulation of hormone receptor interactions. One possibility would be the ability to interact directly with the ligand i.e. provide binding sites for the hormone accessory or even integral to the receptor site. This possibility has been extensively investigated and a role for gangliosides as components of receptor for glycoprotein hormones, bacterial toxins and the antiviral agent interferon has been well substantiated and discussed in many publications, even in this series.[1–3] Alternatively, polar lipids may play a structural role essential for receptor conformation and stability, and provide the suitable environment in the plasma membrane to make the coupling process between receptor and intracellular effector possible and efficient.

This possibility is suggested by experiments in which it has been shown that, in several hormone responsive tissues, treatment with phospholipases alters the response to hormones and that the addition of specific phospholipids to solubilized preparations of adenylate cyclase, restores the hormonal response of this enzyme activity.

A third possible role of membrane lipids may involve these molecules as modulators of receptor expression and the message transmission process either by direct interaction with receptor components or by affecting the physical state of the membrane matrix, for instance by changing its microviscosity, and increasing or decreasing the exposure of membrane proteins. Clearly these possibilities are not mutually exclusive and it is quite plausible to anticipate a multifunctional role for membrane lipids in hormone receptor interaction and transmembrane signalling.

2 RECEPTORS ARE PROTEINS

To establish unequivocally the chemical nature of hormone receptors it is necessary to separate them from the membrane matrix. Many studies have pursued this goal and many soluble receptor preparations have been described. However, when analysed in an aqueous environment, the binding properties of these preparations become significantly different from those of the parent plasma membranes, therefore, at the present time, the statement can be made that no plasma membrane receptor has been obtained in a pure, homogeneous form in aqueous solution. Nonetheless there is strong evidence, both direct and indirect, that proteins or glycoproteins are key components of peptide hormone receptors. The direct evidence comes from studies on the composition of soluble receptor preparations and the indirect evidence comes from the many studies in which treatment of plasma membranes with proteolitic enzymes, such as trypsin, causes a significant loss of receptor activity. The concept that membrane receptors should be identified with a protein is now a well established fact. The enormous difficulties encountered in the various attempts to purify receptors raise however a couple of very important considerations. Firstly it should be emphasized that, within the complexity of the membrane structure, there might be many hormone binding components, with different binding properties and different stabilities toward the reagents used to fractionate the plasma membrane, and the identification of these binding components with the physiologic receptor is an extremely difficult task. Secondly receptor activity, as it is expressed in a living cell, is a complex function which is contributed for by several membrane components, widely different in their chemical nature, interacting with each other in a cooperative manner. It is most likely that this complex interaction is a cogent need which makes impossible to separate from the membrane context the component which selectively recognizes or discriminates a hormone in a specific binding reaction.

Two basic approaches can be followed to establish the role of membrane components in receptor function. As it has been already discussed, one is to evaluate the effects of specific degradation of a membrane component on hormone binding or hormonal response; the second is to supplement uniquely defective mutant plasma membranes with the missing component, and assess whether the reconstitution of the normal composition is associated with the normalization of abnormal or absent hormonal response. The latter would be an ideal system to establish unequivocally the function of membrane constituents. Alternatively, membrane components, with potential receptor function, could be incorporated into artificial lipid bilayers (liposomes) thus a suitable model system would be available for hormone binding studies and even studies of hormone sensitive membrane functions.

3 ROLE OF PHOSPHOLIPIDS

3.1 Historical

The involvement of phospholipids in the biologic activity of protein hormones became an obvious possibility when it was discovered that the binding of the hormone on the cell surface was a key reaction for hormonal action. In 1967 Macchia and Pastan[4] showed that treatment of dog thyroid slices with a crude enzyme preparation obtained from the growth medium of *Clostridium perfringens* made them insensitive to thyrotropin (TSH) stimulation, as measured by the increase of oxidation of glucose-1-^{14}C to ^{14}CO$_2$. The enzyme responsible for the alteration of TSH response was identified as a phospholipase C which appeared to act by perturbing the receptor site for TSH on the thyroid cell membrane since glucose oxidation in thyroid slices, treated with the same enzyme, proceeded normally in the absence of TSH, and could be enhanced from basal by acetylcholine, which was also known to increase glucose oxidation in thyroids.

There are many studies in which hormone binding kinetics have been analysed using plasma membranes previously treated with various phospholipases, in analogy to studies in which trypsin was used to prove the protein nature of hormone receptors. In both cases alteration of the binding reaction would indirectly prove the importance of the enzymatically degraded component in the receptor site.

Studies on the role of phospholipids in hormonal action have looked at the effects of specific degradations on ligand stimulable adenylate cyclase. Based on the work of Sutherland and the discovery of 3'-5'-cyclic adenosine monophosphate (cAMP)[5] many polypeptide hormones and catecholamine are believed to exert their action via the so called 'second messenger'. According to this theory the hormone binding reaction activates a membrane bound enzyme or effector system which, in turn, promotes the synthesis of an intracellular compound i.e. the messenger of hormone action within the cell. In many cases the membrane bound enzyme is the adenylate cyclase which promotes the synthesis of the cyclic nucleotide cAMP generally referred to as the 'second messenger'.[6] Thus the nature of hormone membrane interaction can also be investigated by studying adenylate cyclase activation. This will provide information on both the binding reaction and the ability of the membrane to perceive the hormonal message and generate the second messenger. It is now accepted that hormone receptors and adenylate cyclase are physically distinct; however they are integral part of the plasma membrane therefore the coupling of hormone binding to cyclase activation is likely to be dependent on the status of the plasma membrane.

3.2 Studies with phospholipases

Phospholipases are enzymes capable of cleaving phospholipids at different sites. They have become widely used in studies aimed at the investigation of the role of phospholipids in the structure of hormone receptors and hormone dependent effects in plasma membranes. The experimental approach has been very similar in the many studies reported in the literature. Thus either hormone binding parameters (hormone bound, number of binding sites, affinity, etc.) or hormone stimulation of adenylate cyclase have been measured on membranes or solubilized receptor preparations after treatment with phospholipases. Careful interpretation of such experiments requires knowledge of the sites of cleavage on the phospholipid molecule of the various phospholipases, since different enzymes give rise to entirely different degradation products. In Figure 1 are depicted the sites of action of

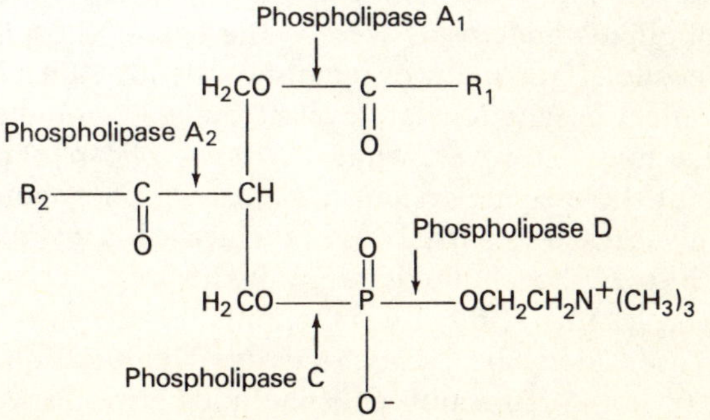

Phospholipase A_1

H_2CO ── C ── R_1

Phospholipase A_2

R_2 ── C ── CH

Phospholipase D

Phospholipase C

H_2CO ── P ── $OCH_2CH_2N^+(CH_3)_3$

O^-

Figure 1. Sites of action of phospholipases on phosphatidylcholine

the major groups of phospholipases. Phospholipase A_1 is specific for the ester bond adjacent to carbon 1, in the stereospecific numbering of the glycerol moiety, removing the fatty acid in position 1. Phospholipase A_2 removes the fatty acid from position 2. Cleavage of a fatty acyl residue from a phosphoglyceride generates a lysophosphoglyceride, for instance lyso-phosphatidylcholine (also known as lysolecithin). Lysophosphoglycerides are powerful sufactants which are present in tissue or cells in minute concentrations as intermediates in phospholipids metabolism. At high concentrations are toxic since they can solubilize plasma membranes. Phospholipase B is a mixture of phospholipases A_1 and A_2 capable of removing sequentially the fatty acids from phosphoglycerol, it also generates lysophosphoglycerides and fatty acids. Quite different are the products of phospholipases C and D which hydrolyse the ester bonds of phosphate with glycerol and the amino-alcohol respectively, removing the so called 'polar head' from the phospholipid molecule.

It should be obvious that the interpretation of receptor perturbation, as a consequence of treatment of plasma membranes with phospholipases A, C or D, may be entirely different since the end products of the hydrolysis of phospholipids caused by the three enzymes have very different properties. The powerful detergent lysophosphoglyceride and fatty acid, generated by phospholipases A, are likely to alter to a large extent the membrane structure therefore one can learn very little on the direct involvement of phospholipids in receptor function.

Of the many studies on the effects of phospholipases on membrane receptors and receptor coupling to adenylate cyclase activation, those of Rubalcava and Rodbell[7] on the glucagon sensitive adenylate cyclase system in liver plasma membrane, and those of Menon and coworkers[8,9] on the gonadotropin receptors on plasma membranes of corpus luteum appear more careful and extensive. In the former the effect of phospholipase C from two different sources was investigated. The two enzyme preparations, a crude one from *Clostridium perfringens* and a highly purified one from *Bacillus cereus*, showed different specificities in that the *Cl. perfringens* hydrolysed neutral phospholipids and the B. cereus the acidic phosphatidylserine and phosphatidylinositol. Treatment of membranes with either enzyme preparations did not affect basal adenylate cyclase activity, nor the enzyme activation by fluoride ions. However, while *B. cereus* phospholipase C abolished the sensitivity of the enzyme system to glucagon, phospholipase C from *Cl. perfringens* only caused a partial loss of glucagon sensitivity when 60 per cent of membrane phospholipids were hydrolized. These results clearly established a role for acidic phospholipids in the mechanism whereby glucagon increases adenylate cyclase activity. The question was then asked: at what stage of hormonal action phospholipids have a specific role and the three obvious possibilities were considered, i.e. (a) hormone binding, (b) transmembrane signalling, (c) structural role by providing the suitable environment for the receptor and the receptor–enzyme coupling to be operational.

The possibility that acidic phospholipids participate in the binding of glucagon was ruled out because no reduction of binding sites could be measured in phospholipase treated membranes. However a tenfold decrease in affinity of these sites was found. Interestingly, treatment of membranes with low concentration of *B. cereus* phospholipase C, i.e. hydrolysis of phosphatidylsereine and phosphatidylinositol, consistently showed increased binding of glucagon at concentrations of 10^{-8} and above suggesting the possibility that indeed phospholipids may also shield some binding sites, causing binding inhibition in special case such as, for instance, high hormone concentration. It was also pointed out that the affinity for intact liver membranes of des-His-glucagon, a competitive, inactive analogue of glucagon, missing the amino terminal histidine, was comparable to that of native glucagon for phospholipase C treated membranes and did not change after phospholipase treatment. Guanosine-triphosphate (GTP), which is required for glucagon action, did not increase the dissociation rate of glucagon bound to phospholipase treated membranes as it did for intact membranes. Putting all these data together a role for acidic phospholipids was postulated as regulators of the binding sites for the NH_2-terminal histidine residue of glucagon, and the nucleotide GTP both of which play an unique role in glucagon binding and hormone action on the cyclase system.

At variance with the glucagon-liver membrane system, treatment of corpus luteum membranes with phospholipases A and C caused a major decrease of specific binding of ^{125}I-choriogonadotropin (hCG) which was attributed to a decrease in the number of binding sites in the absence of changes in the affinity of the hormone receptor complex. Phospholipase D treatment did not affect the binding. It was noted that soluble phosphorylaminoalcohols, produced by phospholipase C, were completely released into the medium whereas phospholipase A end products remained associated with the membranes. The former and the insoluble diglycerides produced by phospholipase C were without effect on receptor activity while lysophosphoglycerides and fatty acids were strong inhibitors on both the membrane associated and solubilized receptor activity. Washing of the phospholipase A treated membranes with defatted serum albumin removed the phospholipid by-products leading to a complete restoration of the receptor properties. It was concluded that phospholipase dependent inhibition of choriogonadotropin binding to corpus luteum membranes was due to perturbation of phospholipid structure in the case of phospholipase C and by interference of phospholipids hydrolysis end-products in the case of phospholipase A. This study, while establishing unequivocally the relevance of phospholipids in receptor activity, failed to elucidate whether phospholipids are directly involved as receptor components or participate indirectly to receptor conformation and stability.

4 STUDIES ON RECONSTITUTED HORMONE RESPONSIVE SYSTEMS

While phospholipase experiments are important for establishing a role for phospholipids in hormonal action additional, and perhaps more convincing, evidence on this role comes from experiments in which addition of specific phospholipids restores normal responsiveness in membranes or solubilized receptor-adenylate cyclase preparations which have become insensitive to hormones as a consequence of either phospholipase or detergent treatment. Along this line several observations have been made of partial restoration of homonal response in systems reconstituted by the addition of specific phospholipids.

Levey[10,11] demonstrated that adenylate cyclase activity can be solubilized from cat myocardial homogenates using the non-ionic detergent Lubrol-PX (an ethylene oxide condensate of dodecanol). The enzyme activity could be increased by sodium fluoride but was unresponsive to the hormones that normally stimulate the membrane bound adenylate cyclase. However it was shown that the addition of specific phospholipids could selectively restore hormonal response. Thus in the presence of phosphatidylserine the solubilized enzyme became fully responsive to glucagon and histamine, but not to nor-epinephrine. Response to catecholamine was totally restored by the addition of phosphatidylinositol. This was a very elegant demonstration of specific phospholipids involvement in the expression of hormonal response; also the results of reconstitution experiments were consistent with the phospholipase C data of Rubalcava and Rodbell[7] showing the specific role of acidic phospholipids in glucagon action.

The specificity of phospholipid requirements supports their role of active transducers for hormonal signals. It seems, in fact, unlikely that phospholipids only contribute the suitable apolar environment to maintain the receptor–effector complex in its proper conformation since, if this were the case, apolar detergent micelles could provide a substitute for phospholipids and this does not seem to be the case, at least in the myocardial adenylate cyclase system.[10,11]

5 ACIDIC PHOSPHOLIPIDS AND THE RECEPTOR FOR THYROTROPIN: A NEW APPROACH

Conventional phospholipase studies indicated that the binding of TSH to thyroid plasma membranes was enhanced by treating membranes or thyroid slices with phospholipase A or phospholipase C[12,13] whereas similar treatment reduced TSH effects, such as stimulation of adenylate cyclase activity, iodide transport, phospholipid metabolism and, as it has been discussed already, glucose oxidation. Reconstitution experiments also showed that phospholipid supplements could partially restore TSH-stimulable adenylate cyclase activity of phospholipase treated membranes.[14,15] Thus the role of phospholipids in the activity of the receptor for TSH appears to be a complex one.

It has been proposed that at least two membrane components, a glycoprotein and a ganglioside, are integral part of the receptor for TSH and evidence has been presented which shows that both components are necessary for the transmission of the hormonal message. Glycolipids and phospholipids show important structural similarities and similar orientations within the membrane bilayer; thus the evidence that glycolipids have a role in the structure of the TSH receptor[1-3] prompted analogous experimental approaches to probe the role of phospholipids as components of the TSH receptor or modulators of its functional responsiveness. The involvement of gangliosides as components of the receptor for TSH was based on the following evidence: gangliosides interact with TSH, as shown by physical techniques, such as fluorescence spectroscopy and analytical ultracentrifugation, and inhibit the binding of TSH to plasma membranes. The efficacy of binding inhibition depends on the number and location of sialic acid residues on the ganglioside molecule. More direct evidence came from studies which showed that gangliosides containing liposomes (artificial model membranes made up of phospholipids and cholesterol), bind specifically ^{125}I-TSH, and that the amount of hormone bound correlates with the molar ratio gangliosides/phospholipids in the liposomes. The binding reaction resembled that of authentic plasma membranes in its affinity, and its temperature and salt sensitivity.[16]

In many respects the behaviour of acidic phospholipids resembles that of gangliosides in experiments in which their interaction with TSH was investigated by fluorescence and fluorescence polarization and in their ability to inhibit ^{125}I-TSH binding to thyroid plasma membranes.[17] Studies of the interaction of TSH with phospholipids have shown that phosphatidylinositol, and to a lesser extent, phosphatidylserine and phosphatidylethanolamine interact with TSH forming an adduct which does not bind to thyroid plasma membranes (Figures 2 and 3). At concentrations ranging between 10^{-6} and 10^{-4} M phosphatidylinositol is the most potent inhibitor of ^{125}I-TSH binding. This inhibition is comparable to that of the ganglioside, G_{D1b}, which has been posbulated to be a component of the receptor for TSH.

The combination of a negatively charged 'polar head' and the hydrophobic portion of the molecule was a cogent requirement for adduct formation with TSH and membrane binding inhibition. In fact, inositol, inositol monophosphate, serine and ethanolamine were devoid of inhibitory

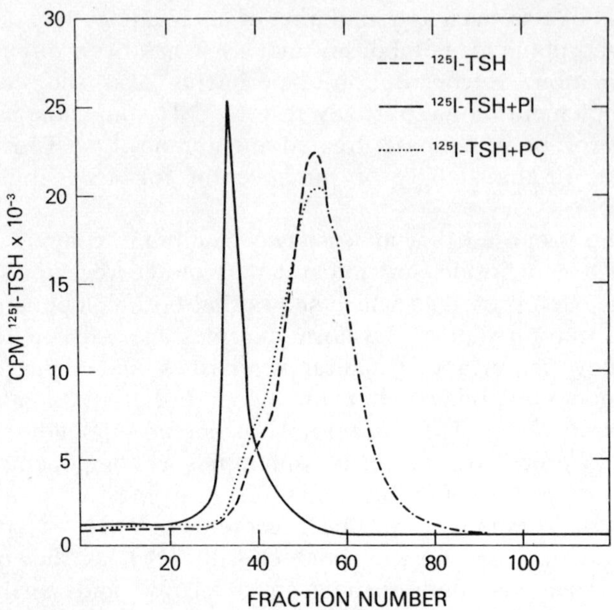

Figure 2. The effect of phosphatidylinositol and phosphatidylcholine on elution of ^{125}I-TSH from Sephadex G-100. Experimental details may be found in Reference 17

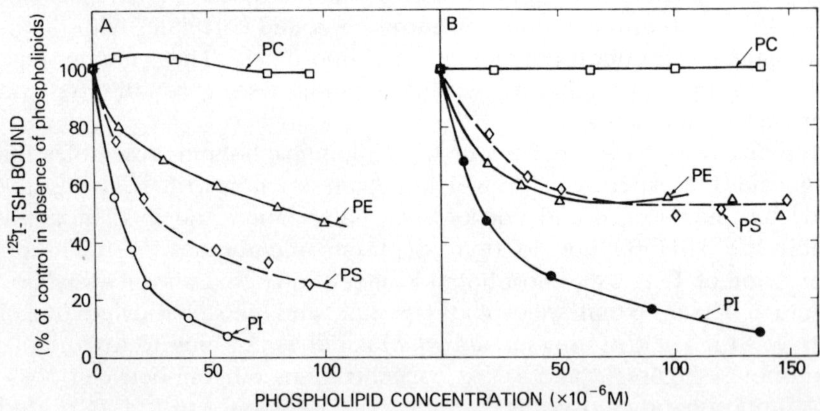

Figure 3. The effect of phospholipids on specific ^{125}I-TSH binding to bovine thyroid plasma membranes. In (a), phospholipids and ^{125}I-TSH were preincubated under assay contitions (see below) for 1 hour before the addition of membranes. In (b) the ^{125}I-TSH and membranes were preincubated for 1 hour before the addition of phospholipids. Incubations were for 1 hour at 0 °C in 0.025 M tris-acetate, pH 6.0, containing 0.6 per cent albumin; ^{125}I-TSH concentration was 2×10^{-9} M, membrane protein was 10 μg/assay[17]

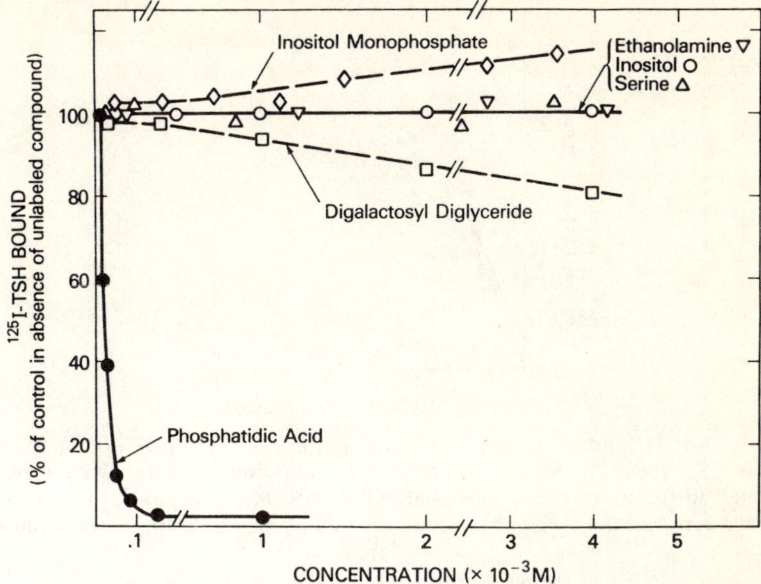

Figure 4. The effect of various substances on the binding of ^{125}I-TSH to bovine thyroid plasma membranes. Conditions are identical to those of Figure 3(a)[17]

activity; so was phosphatidylcholine; whereas phosphatidic acid was nearly as potent as phosphatidylinositol in its inhibitory effect (Figure 4). In spite of the overall similarities in the effects of gangliosides and phosphatidylinositol on the interaction with TSH and inhibition of TSH binding to thyroid membranes, there are important differences which strongly suggest that phosphatidylinositol is not a TSH receptor analogue. At variance with gangliosides, phosphatidylinositol interaction with TSH is not salt sensitive; it should be recalled that salt sensitivity of TSH gangliosides interaction resembles the salt sensitivity of TSH binding to plasma membranes. Moreover phosphatidylinositol does not bind TSH when it is incorporated into liposomes. An important feature of phosphatidylinositol is that it can prevent reconstitution of TSH receptor into liposomes by decreasing the activity of the glycoprotein component (Figure 5). The effect on the latter is exerted in two ways: (i) by decreasing its incorporation into the lipid matrix; and (ii) by inhibiting the binding activity of the fraction of glycoprotein which is incorporated. It is noteworthy that neither phosphatidylinositol, nor other phospholipids have any effect on the incorporation and the expression of receptor properties of gangliosides into liposomes as measured by TSH binding activity of reconstituted gangliosides-liposomes (Figure 6).

Putting all these observations together a role for phosphatidylinositol, emerges as a negative modulator of TSH receptor expression by either

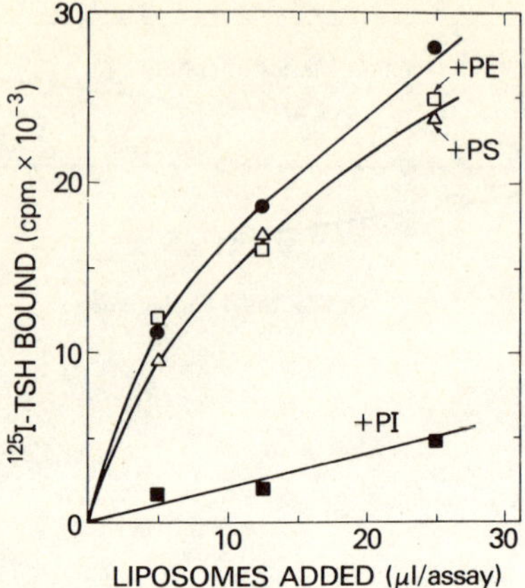

Figure 5. ^{125}I-TSH binding to liposomes containing the glycoprotein component of the thyroid TSH receptor (●) and to liposomes containing the same glycoprotein but were formed in the presence of phosphatidylinositol (■), phosphatidylserine (△), or phosphatidylethanolamine (□). Experimental details may be found in Reference 17

direct interaction with TSH or by its ability to control the incorporation of the glycoprotein component of the receptor into the membrane bilayer. It is important to point that studies on the functional roles of gangliosides and glycoprotein of TSH receptor suggest that the glycoprotein serves as the primary binding determinant.

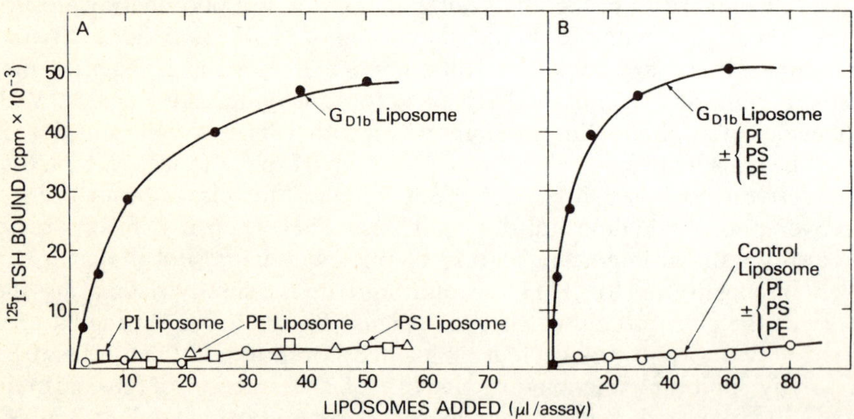

Figure 6. ^{125}I-TSH binding to phosphatidylcholine/cholesterol liposomes containing: (a) ganglioside G_{D1b} (●), phosphatidylinositol (PI), phosphatidylserine (PS), or phosphatidylethanolamine (PE). In (b), the binding of ^{125}I-TSH to phosphatidylcholine/cholesterol liposomes containing G_{D1b} as well as the noted phospholipids. Experimental details on the preparation of liposomes and the binding assay may be found in References 16 and 17

Interpretation of the phosphatidylinositol effects on TSH receptor activity can lead to some considerations on the potential role of phospholipids as modulators of receptor expression. At the present stage of experimental data these considerations remain largely speculative and much more work is clearly needed. It is however, worthwhile pursuing these ideas because of the following observations on hormonal effects on phosphatidylinositol turnover. This widespread phenomenon was firstly detected as an increased rate of ^{32}Pi incorporation into phosphatidylinositol, and also into phosphatidic acid, which occurs in tissues shortly after the application of a stimulus and persists for as long as the stimulus lasts. The common feature of these stimuli is that, in the majority of the cases, they exert their effects on their target tissues through the interaction with cell surface, most likely, with specific membrane receptors. Since the enhanced incorporation of ^{32}Pi and inositol was not accompanied by increased incorporation of glycerol[18] it was concluded that the process did not involve the enhancement of '*de novo*' synthesis of phosphatidylinositol. The response of thyroid to TSH, however, represents the only well-established exception since, in this tissue, the synthesis of the whole phosphatidylinositol molecule is enhanced as a consequence of stimulation. Even more important, the enhancement of phosphatidylinositol synthesis appears to be independent of the levels of cAMP since: (i) it occurs at concentrations of TSH which are to low for adenylate cyclase stimulation; (ii) it is not affected by theophilline and prostaglandin E_2 both of which are known to increase the levels of cAMP in the thyroid; (iii) it is not affected by dibutyryl-cAMP which mimics most of the effects of TSH.

With regard to the possible direct interaction of the glycoprotein component of TSH receptor with phosphatidylinositol it is pertinent to recall that membrane glycoprotein biosynthesis is initiated at the rough endoplasmic reticulum where phosphatidylinositol levels have been estimated to be the highest.[19]

The speculative implications of these observations may suggest that TSH effects on phosphatidylinositol metabolism and phosphatidylinositol effects on receptor expression represent a regulatory mechanism whereby the hormone controls the activity of its own receptor.

Consistent with this hypothesis are also some observations of the association of thyroid phospholipid abnormalities and altered thyroid response to TSH. In the experimental rat thyroid tumor 1–8 characterized by decreased TSH binding and defective TSH stimulation of adenylate cyclase, the binding abnormality may also be related to a 2–3 fold higher level of phosphatidylinositol in this tumour relative to normal rat thyroid.[20,21] On the opposite end hyperfunctioning thyroid membrane from patients with thyrotoxicosis or Graves' disease, have a deficient phospholipids content by comparison to normal membranes. This abnormality may be responsible for normal TSH binding activity despite gangliosides alterations in their membranes, and an increased sensitivity to TSH stimulation of adenylate cyclase.[17,22] This inverse relationship between phospholipid content and hormone binding is consistent with the data of Rubalcava and Rodbell, on glucagon binding to liver membranes discussed earlier in this review.

6 SUMMARY AND CONCLUSIONS

A wealth of experimental data has unequivocally established that the functional properties of plasma membrane receptors are the result of complex interactions of proteins, glycolipids and phospholipids. Different experimental approaches have been used to prove the role of membrane lipids in receptor activity. These have included the effects of selective degradation of phospholipids on hormone binding and hormone dependent functions; specific phospholipid supplement, in phospholipase treated membranes or in detergent solubilized adenylate cyclase preparations, has been, in some cases, associated with restoration of hormone responsiveness. Reconstitution experiments, in which the components of the receptor for thyrotropin have been incorporated into liposome model membranes have shown that phosphatidylinositol may be responsible for 'downregulation' of this receptor. The original question that was asked was: can we assign a specific role to each membrane component in the reaction which is initiated by the interaction of the hormone with its receptor site and eventually generates the so called 'hormonal effect'? Do we have now a definitive answer to that question? Obviously we don't! However, we have sufficient information to construct a fairly reasonable picture where we can place all the elements which are important in receptor ligand interaction and transmembrane signalling. Although largely speculative, being based on circumstantial evidence, this picture is helpful in directing future investigation. Do membrane lipids play an important role in this picture? The answer is undoubtedly yes! The evidence is overwhelming that phospholipid perturbation affects significantly the function of many hormone receptor systems. Although the possibility that phospholipids participate directly in the structure of the hormone binding site can be generally ruled out, it is fairly well established that hormone receptor interaction and transduction of the chemical signal generated by this interaction require the integrity of specific phospholipid complement. It is not clear yet how phospholipids modulate the structural and functional properties of the receptor site. One obvious consideration would take into account the amphipatic character of phospholipids and speculate that different regions of the receptor molecule may need a suitable environment to attain the thermodynamically favourable conformation for ligand recognition and signal transmission. Thus the charged polar heads could contribute to electrostatic interactions with charged groups of the receptor proteins and also contribute to the proper organization of the surrounding water. It is known that the interaction of polar groups with water, such as hydrogen bonding, bear greater relevance to protein stability in the presence of salts at concentrations high enough to mask electrostatic interactions. It is conceivable that the action of phospholipase C, by removing the phosphorylalcohol from phospholipids, may cause a major rearrangement of the interactions at the membrane surface. At the other end the diacylglyceride provides the appropriate environment for the non-polar residues to interact and contribute to stabilize the proteins.[23] These observations are substantiated by studies showing important conformational changes in proteins upon interaction with phospholipid and detergent micelles.[24,26]

The role of glycolipids has been discussed extensively elsewhere therefore has been largely neglected in this review. It is however important to recall their structural similarities with phospholipids which suggests that they might play similar roles in receptor structure. However their more complex oligosaccharide chains may involve them directly in the structure of the receptor site and play a role in ligand recognition and binding.

One final point should be made with regard to the role of cholesterol and non-polar lipids in general. This topic has been somewhat overlooked in hormone receptor studies. There are however experimental data which relate directly cholesterol levels with membranes microviscosity and exposure of membrane proteins. These studies also raise an interesting connection between membrane dynamics and functional features of cells.[27] With these observations in mind it is attractive to speculate that the altered growth behaviour of a strain of mutant chinese hamster ovary cells, defective in cholesterol, might be due to alterations in membrane fluidity and hormone receptor exposure.[28]

In conclusion I should like to stress which is, in my opinion, the important message that we should perceive out of the complexity of receptor structure and function. It is now evident, from studies on many differentiated cell lines grown under chemically defined conditions, that normal cell growth and differentiation is dependent upon the concerted actions of several hormones and growth factors. It is also evident that the availability of such factors has to be associated with the ability of the cells to make use of them, for the development and maintainance of biological functions. It is conceivable, therefore, that qualitative or quantitative alterations of membrane receptor components may be the underlying defect in many examples of abnormal cell growth and development. This possibility should always be considered as a potential pathogenetic factor especially in those disease states in which signs and symptoms of a particular hormone deficiency are associated with normal circulating levels of that hormone.

REFERENCES

1. Kohn, L. D. (1977). in *Horizons in Biochemistry and Biophysics*, Vol. 3, (Quagliariello, E., ed.), pp. 123–163, Addison-Wesley Publishing Co., Reading, Massachusetts.

2. Kohn, L. D., Lee, G. Grollman, E. F., Ledley, F. D., Mullin, B. R., Friedman, R. M., Meldolesi, M. F., and Aloj, S. M. (1978). In *Cell Surface carbohydrate Chemistry* (Harmon, R. E. ed.), pp. 103–133, Academic Press, New York.

3. Kohn, L. D. (1978). In *Receptors and Recognition* (Quatrecasas, P., and Greaves, M. F., eds.), Series A, vol. 5, pp. 134–212, Chapman and Hall, London.

4. Macchia, V., and Pastan, I. (1967). *J. Biol. Chem.*, **242**, 1864–1869.

5. Sutherland, E. W., and Rall, T. W. (1960). *Pharmacol. Rev.*, **12**, 265–299.

6. Robinson, G. A., Butcher, R. W., and Sutherland, E. W. (1968). *Ann. Rev. Biochem.*, **37**, 149–174.

7. Rubalcava, B., and Rodbell, M. (1973). *J. Biol. Chem.*, **248**, 3831–3837.

8. Azhar, S., and Menon, J. K. M. (1976). *J. Biol. Chem.*, **251**, 7398–7404.

9. Azhar, S. Hajra, A. K., and Menon, J. K. M. (1976). *J. Biol. Chem.*, **251**, 7405–7412.

10. Levey, G. S. *Biochem. Biophys. Res. Commun.* (1970). **38**, 86–92.

11. Levey, G. S. *J. Biol. Chem.* (1971). **246**, 7405–7410.

12. Moore, W. V., and Wolff, J. (1974). *J. Biol. Chem.*, **249**, 6255–6263.

13. Amir, S. M., Goldfine, I. D., and Ingbar, S. H. (1976). *J. Biol. Chem.*, **251**, 4693–4699.

14. Wolff, J. *Isr. J. Med. Sci.* (1972). **8**, 1856.

15. Yamashita, K., and Field, J. B. (1973). *Biochim. Biophys. Acta.*, **304**, 686–692.

16. Aloj, S. M., Kohn, L. D., Lee, G., and Meldolesi, M. F. (1977). *Biochem. Biophys. Res. Commun.*, **74**, 1053–1059.

17. Aloj, S. M., Lee, G., Grollman, E. F., Beguinot, F., Consiglio, E., and Kohn, L. D. (1979). *J. Biol. Chem.*, **254**, 9040–9049.

18. Hokin, M. R., and Hokin, L. E. (1953). *J. Biol. Chem.*, **203**, 967–977.

19. Michell, R. H. (1975). *Biochim. Biophys. Acta*, **415**, 81–147.

20. Macchia, V., Meldolesi, M. F., and Chiariello, M. (1972). *Endocrinology*, **90**, 1483–1491.

21. Mandato, V., Meldolesi, M. F., and Macchia, V. (1975). *Cancer Res.*, **35**, 3089–3093.

22. Lee, G. Grollman, E. F., Aloj, S. M., Kohn, L. D., and Winand, R. J. (1977). *Biochem. Biophys. Res. Commun.*, **77**, 139–146.

23. Edelhoch, H., and Osborne, J. C. Jr. (1976). *Advan. Protein Chem.*, **30**, 183–250.

24. Bornet, H., and Edelhoch, H. (1971). *J. Biol. Chem.*, **246**, 1785–1782.

25. Schneider, A. B., and Edelhoch, H. (1972). *J. Biol. Chem.*, **247**, 4986–4991.

26. Schneider, A. B., and Edelhoch, H. (1972). *J. Biol. Chem.*, **247**, 4992–4995.

27. Borochov, H., and Shinitzky, M. (1976). *Proc. Natl. Acad. Sci. U.S.A.*, **73**, 4526–4530.

28. Chang, T. Y., Telakowski, C., Vanden Heuvel, W., Alberts, A. W., and Vagelos, P. R. (1977). *Proc. Natl. Acad. Sci. U.S.A.* **74**, 832–836.

5 Biosynthesis and assembly of membrane proteins

S. Bonatti

1 INTRODUCTION

Some fundamental concepts of membrane structure and of the cellular protein synthetic machinery must be taken into consideration in any model of the biosynthesis and assembly of membrane proteins. These include:

(a) the existence of two very different classes of membrane proteins, peripheral and integral proteins. According to the classic work of Singer and Nicholson,[1] peripheral membrane proteins are characterized by their weak interaction with the membrane; they can be extracted from the membrane in a soluble, lipid-free form by mild treatment. Integral membrane proteins are characterized instead by their strong interaction with the membrane, as indicated by the necessity of using drastic treatment (denaturants, organic solvents) for their removal from the membrane. Moreover, integral membrane proteins are often associated with lipids when isolated and frequently are insoluble in neutral aqueous buffer if free of lipid.

(b) the absolute asymmetry of protein distribution between the two surfaces of the membrane. It is well known that cellular membranes can grow only by the insertion of new components into a preformed membrane vesicle. Because integral membrane proteins have an extremely low rate of transmembrane rotation, and peripheral membrane proteins cannot translocate across the lipid bilayer, membrane protein asymmetry is a consequence of the way in which a newly synthesized protein is assembled in a preformed membrane. To better analyse the problem of protein asymmetry from a biosynthetic point of view, it has been suggested that integral membrane proteins be grouped according to their orientation in the membrane (see Figure 1).

(c) the location of the cellular protein synthetic machinery in the cytosol, where a membrane protein can be translated on free or membrane bound polyribosomes. Thus all membrane proteins displaying at least one domain

101

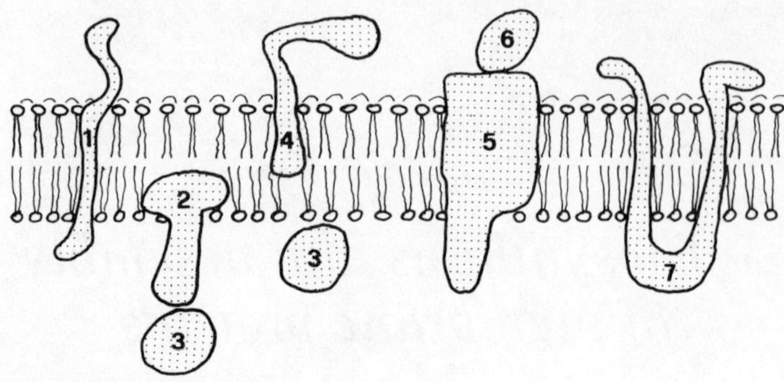

Figure 1. Different types of membrane proteins. According to the fluid mosaic model of the structure of cell membrane (1), proteins n. 3 and 6 are peripheral, whereas proteins n. 1-2-4-5-7 are integral membrane proteins. Two different classification schemes have been proposed for integral membrane proteins. Rothman and Lenard[2] suggested the terms ectoproteins and endoproteins. Ectoproteins have hydrophilic domain(s) beyond the lipid bilayer matrix of the membrane (extracellular domain(s) in the case of plasma membrane proteins, intracisternal domain(s) in the case of endoplasmic reticulum proteins). Endoproteins have their hydrophilic domains associated with the cytoplasmic side of the membrane. Following this classification, if the lower part of the figure represents the cytosol, proteins n. 1-4-5-7 are ectoproteins, whereas protein n. 2 is an endoprotein.

Alternately, Blobel and co-workers[3] proposed that integral membrane proteins be divided into monotopic, bitopic and polytopic groups. This nomenclature indicates whether a protein has a hydrophilic domain on one side (monotopic) or on both sides (bitopic) of the membrane; polytopic proteins have multiple hydrophobic domains embedded in the lipid bilayer and multiple hydrophilic domains on opposite side of the membrane. To apply this classification it is not necessary to specify which side of the membrane is considered. According to this second scheme, proteins n. 2 and 4 are monotopic, n. 1 and 5 are bitopic and protein n. 7 is polytopic. The existence of such polytopic proteins remains to be demonstrated. Throughout this article the terms endo- and ectoproteins will be used. It should be mentioned, however, that this terminology will be difficult to apply for integral membrane proteins of mitochondria and chloroplasts

on the extracellular or intracisternal side of the membrane have to overcome the barrier posed by the lipid bilayer to achieve their final orientation. A further complication is represented by the membrane structure of mitochondria and chloroplasts. These organelles are surrounded by two distinct membranes (outer and inner) which determine a periplasmic and a matrix space. In chloroplasts a third convoluted membrane, the thylakoid, forms another compartment in the matrix space. A few mitochondrial and chloroplast proteins are made by the protein synthetic apparatus present in the matrix space, while the majority are synthesized in the cytoplasm and have, therefore, to cross up to three distinct membranes to reach their specific location.

One consideration emerges immediately from this discussion; that is, the term 'membrane protein' refers to a group of proteins with significantly different properties. Therefore it is difficult to envisage a unified model for their biosynthesis and assembly into a membrane. To illustrate this concept, representative examples of peripheral proteins, endoproteins and of various types of ectoproteins will be presented (see Figure 1). In the last section the possible models, based on the existing evidence will be discussed.

Out of the necessity to keep the subject comprehensible, no attempt has been made to exhaustively review the entire pertinent literature (for recent reviews concerning some aspects of the subject considered in this article see References 2, 3, 4, 11 and 55).

2　BIOSYNTHESIS OF SECRETORY PROTEINS

The biosynthesis of secretory proteins is briefly summarized in this section because the study of the biosynthesis and assembly of membrane proteins is directly derived from the analysis of the early events of the secretory path way.

The evidence accumulated in the study of the biosynthesis of secretory proteins derives mostly from the faithful reconstitution *in vitro* of the translocation of proteins inside a membrane vesicle (Figure 2). This was first accomplished in 1975 with the addition of microsomal membranes to the cell-free protein synthesis system[4] (Figure 2). The basic scheme for the biosynthesis of secretory proteins is summarized in the signal hypothesis[5] (a more recent version is presented in Reference 3). The essential feature is the presence, in the coding region of the mRNA of secretory proteins, of a specific amino terminal signal sequence which triggers the binding of the ribosome to the membrane and initiates the translocation of the nascent chain; at the end of the translocation process, the ribosome reenters the pool of cytosolic ribosomes. The translocation of the nascent chain presumably occurs through the formation of a transient pore in the membrane or by a 'gate' mechanism; in both cases, an active role for a preexistent translocating machinery is postulated. The putative receptors for the ribosome and nascent chain, and the signal peptidase, are considered part of this machinery. The signal sequence is a transient feature of the secretory protein, being removed by a specific enzyme (the signal peptidase) during the translocation. Although the use of the term presecretory protein is quite common, it should be stressed that for the Signal Hypothesis the precursor to the 'authentic' secretory protein is only the nascent chain of the primary translation product. 'Presecretory' proteins entirely translated on free polyribosomes remain in the cytosol and are degraded by cellular enzymes.

An impressive amount of experimental evidence supports the general tenets of the signal hypothesis, both in prokaryotes and in eukaryotes.

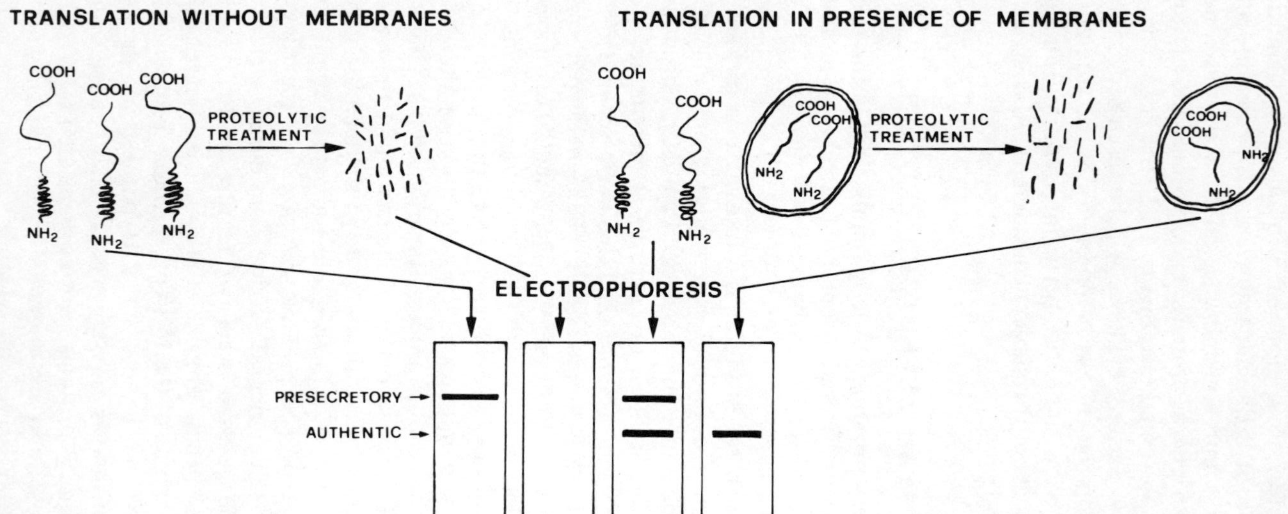

Figure 2. *In vitro* translation of mRNA coding for secretary proteins in the presence or in absence of membranes results in the synthesis of different products. *Presecretory* proteins, containing an animo terminal sequence extension (signal sequence), are the unique products synthesized when the reaction is carried out in the absence of membranes or when membranes are added posttranslationally. Most of the presecretory proteins are converted into 'authentic' secretory proteins if the translation is performed in presence of membranes. Such 'authentic' secretory proteins do not have the amino terminal signal sequence. Only the 'authentic' products are segregated in the membrane vesicles, as shown by selective protection against proteolytic attack. This protection is abolished in presence of detergent. The presecretory proteins usually show a slower electrophoretic mobility of their 'authentic' counterparts, due to the M.W. contribution of the signal sequence, except in the case of glycosylated secretory proteins where the reverse situation is normally found

However, very little is known about the translocating machinery or the source of energy that drives the translocation.

The primary sequence of the signal region of many secretory proteins has been determined. Significant differences in the size (15–29 aa) and primary structure of this region have been found. Common features, however, include a stretch of hydrophobic residues in the middle of the sequence and the presence of polar residues at both ends. It has been proposed that the signals have an helical configuration with a beta turn in proximity of the signal peptidase site.[7] The signal peptidase has been uniquely localized in the luminal region of the rough endoplasmatic reticulum vesicles.[8]

Experimental evidence accumulated after the formulation of the signal hypothesis in 1975 indicates that the mechanism used by the cell to achieve the complete translocation of proteins across the membrane may significantly vary and most importantly, is not unique. In particular: (a) translocation can occur without the removal of the signal sequence. For example, a single amino acid substitution in the aminoterminal signal sequence of *Escherichia coli* lipoprotein prevented the removal of the signal by the signal peptidase, but did not affect the translocation;[9] an internal uncleaved signal sequence has been demonstrated in the case of chicken ovalbumin.[10] (b) In prokaryotes, clear cut evidence of direct binding of ribosomes to the membrane is not available; thus it has been suggested that the ribosomes are attached to the membrane only by their nascent protein chain.[11] (c) Chloroplast and mitochondrial proteins, localized in the matrix or in the intramembrane space and codified by nuclear DNA,[12,13] as well as peroxisomal content proteins,[14] are synthesized in the cytosol on free polyribosomes, and only after synthesis are translocated into the specific organelle by a post-translational translocation mechanism. The chloroplast and mitochondrial proteins examined thus far are synthesized as cytosolic precursors with a sequence extension which is removed during translocation, except mitochondrial cytochrome C which resembles the ovalbumin exception cited above.[15] Despite the existence of two mechanisms by which the cell can achieve the translocation of protein and the several variations on the theme that have been demonstrated, it is nonetheless possible to recognize an important common feature of the biosynthesis of secretory and organelle proteins; that is, at least part of the information necessary for the translocation of a specific protein across a distinct membrane is contained in a discrete portion of the primary sequence of the newly synthesized protein, called the signal sequence. The signal sequence may function only in the nascent chain, thus leading to a cotranslational translocation, or *vice versa* only in the complete, ribosome-free protein, leading to a post-translational translocation. The signal sequence is either a transient or a permanent feature of the protein, respectively.

3 PERIPHERAL MEMBRANE PROTEINS

Two different types of peripheral membrane proteins have to be considered from a biosynthetic perspective: peripheral proteins which interact with the cytosolic side of the cellular membrane and peripheral proteins which are exposed on the extracellular side of a membrane (the cisternal side of an endoplasmic reticulum vesicle being also considered outside the cell). Extracellular peripheral proteins have, in fact, to cross the lipid bilayer barrier to reach their final location, whereas cytosolic peripheral proteins do not. The examples considered in this section are $beta_2$-microglobulin, and E_3 glycoprotein of Semliki Forest Virus, both peripheral extracellular membrane proteins.

3.1 Cytosolic peripheral proteins

Precise studies on the site of synthesis and on the mechanism of assembly of this class of proteins have not been reported. However, because these proteins are linked to the membrane by electrostatic interactions with the hydrophilic portion of integral membrane proteins and/or phospholipids, a particular mechanism of assembly is not '*a priori*' required. By diffusion and steric recognition the peripheral protein could become spontaneously a constituent of a particular membrane, depending on the presence of other specific molecules. For this reason several authors hypothesized that peripheral proteins are synthesized in their mature form on free polyribosomes and become components of distinct membranes post-translationally without the involvement of a specific cellular machinery to drive the process.[2]

3.2 Beta$_2$-microglobulin

Beta$_2$-microglobulin (MW 12,000) is a non-glycosylated, water-soluble, membrane-protein-constituent of the major histocompatibility antigens which are located on the surface of nearly all cells (HL-A in man, H-2 in mice). The histocompatibility antigens consist of an integral membrane protein which completely spans the lipid bilayer (the heavy chain, MW 46,000) and of beta$_2$-microglobulin which is non-covalently linked to the extracellular portion of the heavy chain. Beta$_2$ is also found free in several biological fluids. Homologies have been noted between its structure and that of the immunoglobulin constant region.

Three different laboratories have studied in detail the biosynthesis of this molecule *in vivo* and *in vitro*.[16,17,18] Both beta$_2$ and heavy chain are synthesized on membrane bound polyribosomes; beta$_2$ is cotranslationally segregated in the lumen of the endoplasmic reticulum vesicles whereas the heavy chain is asymmetrically inserted in the membrane. A precursor form of beta$_2$, which contains an amino terminal extension of 19 amino acids has been identified as an *in vitro* translation product when the reaction is carried out in the absence of membrane. This signal sequence is removed during the translocation process in a way directly comparable to that which occurs with secretory proteins. Further, *in vitro* competition experiments suggest a functional similarity between beta$_2$ and secretory protein signal sequences.[18] Interesting results on the membrane assembly of histocompatibility antigens have been obtained by pulse-chase experiments using tissue culture cells. With an antisera which precipitated beta$_2$ only when it was associated with the heavy chain, Dobberstein *et al.*[16] demonstrated that beta$_2$ binds to the heavy chain in the endoplasmic reticulum; moreover, the same experiments suggest the existence in the endoplasmic reticulum of a pool of beta$_2$ molecule competing for the binding with the newly synthesized heavy chain. Since the heavy chains are inserted in the endoplasmic reticulum independently of beta$_2$ synthesis, and cells with a deleted beta$_2$ gene do not express the heavy chain on the cell surface,[19] it was concluded that the early assembly of the histocompatibility antigens is a crucial step for their intracellular migration to the plasma membrane.

3.3 Semliki Forest virus glycoprotein E$_3$

Semliki Forest virus is an enveloped virus of the Alphavirus group. Three different membrane proteins are associated in a monomeric complex with the viral envelope (20): E$_1$ and E$_2$ (MW 50,000) which are integral membrane proteins anchored to the membrane through a short carboxy terminal portion of the molecule, and glycoprotein E$_3$ (MW 10,000), which is non-covalently linked to the extracellular portion of E$_1$–E$_2$. Evidence accumulated by means of several experimental approaches demonstrates that E$_2$ and E$_3$ are initially synthesized as a single integral membrane glycoprotein, P$_{62}$, and that both P$_{62}$ and newly synthesized E$_1$ are translated on bound polyribosomes. E$_3$ corresponds to the amino terminal segment of P$_{62}$, which is located in the lumen of the endoplasmic reticulum vesicle. At the final stage of this intracellular migration, or directly at the plasma membrane level during virus budding, the E$_3$ portion is cleaved from P$_{62}$; E$_3$ remains associated with E$_1$ and E$_2$, and the entire complex probably changes its three dimensional configuration.[21] This complicated assembly of SFV membrane proteins is not yet completely understood at the molecular level; however, the information available indicates that an alternative pathway for the biosynthesis of peripheral 'external' membrane proteins exists. The initial synthesis of an integral membrane protein that eventually results in the additional formation of a peripheral protein can offer some advantages (firm initial anchoring of the future peripheral protein, possibility of regulation of its biological function by modulating the endoproteolytic cleavage, etc.). Another possible example of this alternative pathway is the sucrase–isomaltase enzyme complex of rat small-intestinal microvilli membrane.[22]

4 INTEGRAL MEMBRANE PROTEINS

The biosynthesis and membrane assembly of this class of protein has attracted the interest of many investigators. Integral membrane proteins are intimate structural components of every cell membrane; thus the full elucidation of their assembly will greatly clarify the biogenesis of membranes.

Extensive work on the plasma membrane of eukaryotic cells revealed important roles for integral membrane proteins in receptor function, transmembrane import-export of molecules, the immune response and cell–cell communication; in addition, it is widely recognized that the membrane anchoring of enzymes can offer advantages in the organization and function of complex metabolic routes.

The methodology used in the study of integral membrane protein biosynthesis was largely adapted from earlier work on secretory proteins. Further refinements of the *in vitro* system by membrane biologists, however, led the first demonstration of cotranslational glycosylation of nascent protein *in vitro*.

Integral membrane proteins that will be discussed in detail here include rat liver NADH-cytochrome b_5 reductase, Sindbis virus glycoprotein PE_2 and Semliki Forest virus glycoprotein E_1; as representatives of ectoproteins which span the membrane more than once (see Figure 1 above), *Halobacterium halobium* rhodopsine and erythrocyte band 3 will be briefly mentioned.

4.1 NADH-cytochrome b_5 reductase (b_5 reductase)

b_5 reductase is a constituent enzyme of an electron transport chain involved in fatty acid desaturation and hydroxylation reactions. It consists structurally of a single polypeptide chain of MW 30,000, which is asymmetrically inserted in the membrane; a highly hydrophobic carboxy terminal segment of about 30 aminoacids anchors the enzyme to the membrane,[23] while the bulk of the molecule is exposed in the cytosol. A similar orientation has been shown also for cytochrome b_5.[24] These enzymes can be separated into two components by mild proteolysis. The carboxy terminal 'anchor' segment which is hydrophobic and the rest of the molecule, which retains full biological activity and is characterized as hydrophilic. Moreover these enzymes can be reinserted into various artificial and natural lipid bilayer membranes, but only in their intact form.[24]

Borgese and Meldolesi[25,26,27] carefully studied the intracellular localization and the biosynthesis of b_5 reductase. In rat liver hepatocytes the enzyme appears to be localized in the following cellular compartments: endoplasmic reticulum, Golgi apparatus and mitochondrial outer membrane. Immunological and biochemical evidence strongly suggests the molecular identity of the enzyme isolated from the various subcellular compartments. Studies on the *in vivo* biosynthesis indicate a fast (less than 10 minutes) and probably contemporaneous insertion of the enzyme into the three different target membranes. This finding suggested that the enzyme was synthesized on free polyribosomes. This was confirmed when the products of *in vitro* translation by free and bound rat liver polyribosomes were examined.[28] At least 80 per cent of the enzyme molecules were synthesized on free polyribosomes, under conditions in which clearly different products were translated by the two classes of polyribosomes. The protein synthesized *in vitro* comigrated with the authentic enzyme in the gel system used. Although the authors could not rule out the existence of a precursor form of the enzyme, all the data taken together strongly suggested that b_5 reductase is synthesized in its complete form on free polyribosomes, and it is thereafter quickly inserted into the three different target membranes. This conclusion leads to a very interesting question, that is, why is b_5 reductase inserted into some, but not all, available cellular membranes? As will be pointed out later on, the mechanism by which the cell differentiates its membrane by specifically localizing membrane proteins is a fascinating question to the cell biologist.

It should be mentioned that another very interesting protein seems to belong to this same category, namely the SRC gene product of *Rous sarcoma* virus, a protein kinase of MW about 60,000.[29] Recent evidence indicates that SRC is an authentic integral membrane protein, synthesized on free polyribosomes.[30] Another form of the protein has also been identified. It is 8 kdaltons smaller, is hydrophilic, and has protein kinase activity.[31] The precise cellular location of SRC is still unclear. Immunocytochemical and cell fractionation evidence suggests it is on the cytoplasmic side of the plasma membrane, but a nuclear membrane location has been reported in a different transformed cell line.[31,32,33,34] Based on limited proteolysis data, it has been suggested that SRC protein is asymmetrically bound to the membrane by a hydrophobic amino terminal 8 kdalton segment.[33] (See above.) How the SRC protein mediates neoplastic transformation is unknown.

4.2 Sindbis virus glycoprotein PE$_2$

Sindbis virus (Alphavirus group) is formed by a nucleocapsid surrounded by a lipid bilayer envelope. The nucleocapsid consists of a single stranded genomic RNA molecule, the 42S RNA, complexed with about 300 molecules of a 30 kdaltons protein, C; associated with the lipid bilayer are 300 copies each of two integral membrane glycoproteins, E_1 and E_2 (MW about 50 K). In Semliki Forest virus, 300 molecules of a peripheral glycoprotein, E_3 (see Section 3.3) are also present. PE_2 (MW 60 K) and its equivalent P_{62} of Semliki Forest virus are the intracellular precursor of glycoprotein E_2.

An interesting feature of the biosynthesis of alphavirus structural proteins is that they are all coded for by a single viral mRNA, a 26S RNA.[20] This is particularly surprising because C is a soluble cytosolic protein, whereas E_1, E_2 (and E_3) are membrane proteins. A probable consequence of this situation is the localization of 26S RNA on both free and bound polyribosomes in infected cells.[35] As all eukaryotic mRNA, the 26S RNA is monocistronic; it has only one initiation site for protein synthesis which is located near the 5′ end of the molecule.[36] The structural proteins are generated by proteolysis, and it has been shown that the initial steps of this proteolytic maturation take place cotranslationally.[20] The gene order on the 26S RNA is $C–(E_3)–E_2–E_1$. Recently a further translation product has been identified in infected cells, a 4.2 K peptide (in Semliki Forest virus infected cells, a 6 K peptide).[37] This peptide is localized between E_2 and E_1 on Semliki Forest virus RNA,[38] whereas in Sindbis virus its location is still unknown. Moreover, a transient precursor of glycoprotein E_1, called PE_1, has been demonstrated in Sindbis infected cells.[39] The experiment illustrated in Figure 3 shows a scheme of Sindbis 26S RNA translation. C protein, localized toward the 5′ end of the mRNA, is the first protein to be synthesized, PE_2 appears next, and then a high molecular weight product (about 100 kdaltons) called B_1,[20] together with PE_1. If the translation is performed in the absence of membranes, or if the membranes are added post-translationally, only C and B_1 are synthesized. B_1 contains the primary sequence of both glycoproteins[20] but is itself not a membrane protein; it is localized in the cytosol and is not glycosylated (S. Bonatti unpublished). These findings, taken together, suggest the following model for the translation of 26S RNA (see Figure 4). Translation begins on free polyribosomes, and as soon as the C portion of the nascent chain emerges from the ribosome, C is endoproteolytically removed. The new amino terminus of the nascent chain triggers the binding of the ribosome to the membrane and initiates the translocation of the nascent chain with a mechanism similar to that described for secretory proteins. However, both *in vivo* and *in vitro*, the interaction of the new amino terminus with the membrane may fail; in this case the translation will proceed on free polyribosomes, and C and B_1 will be the only products. If, instead, the translation proceeds on membrane bound ribosomes, a complicated number of events take place cotranslationally in a cascade fashion. The nascent chain is core glycosylated during

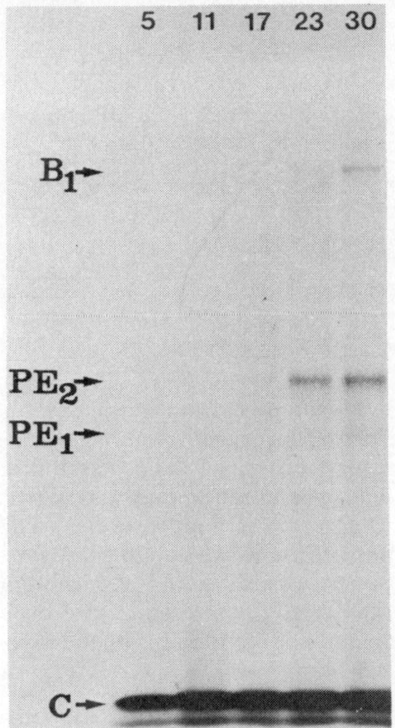

Figure 3. Synchronized translation of Sindbis 26S RNA *in vitro*. Sindbis 26S RNA was added to a Wheat Germ cell-free system incubation mix (40) supplemented with dog pancreas microsomal membranes,[5] and the reaction was started by incubating at 25 °C (time 0). After 30″ Pactamicyn was added (1 micromolar final concentration) to block any further protein synthesis initiation[41] and, at the indicated time points, aliquots of the translation mix were taken and processed for electrophoretic analysis on a 7.5 per cent polyacrylamide slab gel. Translation products smaller than 25 K daltons were run off the gel.

The observed *in vitro* cotranslational endoproteolytic maturation of the nascent chain occurs in all cèll-free systems currently used. Therefore the maturation enzymes are universally distributed in eukaryotic cells or some translation products act as specific proteases. This latter suggestion has been made[42] (S. Bonatti, unpublished result)

translocation, and the PE_2 portion of the growing chain is removed endoproteolytically. In the meantime PE_2 assumes its transmembrane orientation (see below). The translation proceeds further on membrane bound ribosomes with the synthesis and translocation of the third Sindbis 26S RNA product, PE_1. PE_1 is also core glycosylated during translocation. The model presented in Figure 4 favours one of the alternative possibilities for the cleavage at the PE_2/PE_1 site, namely the vesicular location of the cleaving enzyme. However, this aspect of the translocation is still obscure and probably involves the 4.2 K peptide mentioned above. We know much more

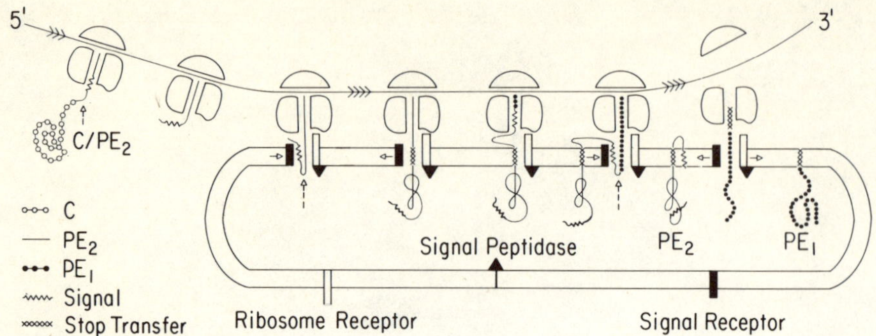

Figure 4. Hypothetical model for the assembly of Sindbis membrane glycoproteins (see text for details). The model incorporates some features originally proposed for the translocation of secretory proteins.[3] The nascent chain containing a signal sequence and the ribosome presumably interact with a signal receptor and a ribosome receptor, both integral membrane proteins. This recognition initiates the formation of a transient pore through which the nascent chain is translocated. The signal peptidase is a component of this complex, although it does not cleave at the PE₂ amino terminus. The stop transfer sequences are proposed to coincide with the membrane embedded domains of the glycoproteins

instead about the assembly of PE_2, because of the possibility of faithfully reconstituting its biosynthesis *in vitro*. It has been shown, in fact, that the PE_2 synthesized *in vitro* is not distinguishable from its *in vivo* labelled counterpart, and it is also similarly core glycosylated.[43] Thus, because of the internal location of PE_2 in the mRNA it was particularly interesting to utilize this *in vitro* system to analyse the initial events involved in the translocation of PE_2. It has been shown that the crucial interaction between the amino terminus of the nascent P_{62} chain and the membrane must occur when the nascent chain is no longer that about 100 amino acids;[44] this had also been shown before with the G glycoprotein of Vesicular Stomatitis virus.[45] A comparison of the amino terminal sequences of PE_2 and B_1 has shown that the putative signal sequence is not removed during the translocation process.[40] The two amino termini have the same sequence; therefore, it is complicate to precisely localize the 'signal acting' internal sequence.[40]

The orientation of newly synthesized PE_2 in the membrane has been also studied in detail. Post-translational proteolytic treatment of PE_2 synthesized *in vitro*[43] or pulse labelled *in vivo*[46] generates a slightly smaller derivative called PE_2'. Therefore a small portion of PE_2 (about 2–3 kdaltons) is exposed on the cytosolic side of the membrane. It is assumed that this 'tail' is located at the carboxy terminus, since the amino terminal sequence of PE_2 is not altered by the proteolytic treatment.[40] Therefore PE_2 spans the membrane having its amino terminus inside the vesicle and a small carboxy terminal portion on the cytosolic side of the membrane. This orientation is maintained by the protein throughout its intracellular migration and also in the released virus by its end product E_2, thus confirming that proteins acquire their orientation in the membrane during the initial assembly.

The study of other transmembrane ectoproteins such as Vesicular Stomatitis virus glycoprotein G, and H-2 and H-LA heavy chains has also indicated that the initial translocation events which occur during the biosynthesis of this type of integral membrane protein are similar to those which have been described for secretory proteins.[16,17,47] Both G and H-2/H-LA are translated on bound polyribosomes, are cotranslationally translocated across the membrane and have an amino terminal signal sequence extension which is removed during translocation. PE_2 however is not the only transmembrane ectoprotein with an uncleaved signal sequence; bovine Opsin is another example.[48]

The fundamental question, unique to this class of membrane proteins, is how the translocation process is specifically interrupted. It has been hypothesized[3,4] that these proteins contain a 'stop transfer sequence', located elsewhere in the sequence, which would halt the translocation by possibly interfering with the ribosome–membrane junction. A possible candidate for this stop transfer sequence could be the hydrophobic segment which spans the membrane (see Figure 4). However, it should be stressed that no experimental evidence exists at present to support the stop transfer sequence model.

The ectoproteins studied in detail thus far show the same transmembrane orientation, i.e. amino terminus on the extracellular side and carboxy terminus on the cytosolic side of the membrane. Some ectoproteins, however, have different orientations; for example, brush border Aminopeptidase is inserted 'upside down'. Various hypotheses to explain its membrane assembly have been suggested,[49] but experimental evidence is not yet available.

In conclusion, all the evidence thus far reported indicates that this class of ectoproteins is synthesized on membrane bound polyribosomes and translocated cotranslationally across the membrane, similarly to secretory proteins, via signal sequences. However, a distinguishing feature of this biosynthetic mechanism is the specific interruption of translocation.

4.3 Semliki Forest virus glycoprotein E_1

Alphavirus glycoprotein E_1 is a possible representative of ectoproteins which are completely restricted to the extracellular side of the membrane, as evidenced (a) by its complete resistance to proteolytic probes added from the cytosolic side of the membrane, and (b) by the nucleotide sequence of the Semliki Forest virus 26S RNA[50] which shows an unique stretch of 24 hydrophobic residues that ends two amino acids from the carboxy terminus. This hydrophobic region probably constitutes the membrane spanning region of E_1. On the basis of the available evidence on the biosynthesis of E_1 (Section 4.2), it could be concluded that this type of membrane protein is synthesized on membrane bound polyribosomes, cotranslationally translocated *via* a signal sequence, and membrane anchored by a stop transfer sequence. As indicated in Figure 4, the only difference would be the extreme carboxy terminal location of the stop transfer sequence. However, alternative biosynthetic models are conceivable. For example, this type of protein could first be completely translocated across the membrane cotranslationally, and subsequently inserted on the extracellular side of the membrane by a mechanism similar to the post-translational insertion of endoproteins (Section 4.1). This mechanism of membrane assembly would be totally independent of 'stop transfer sequence', whereas the initial events of translocation would be similar to those observed with the other ectoproteins (synthesis on membrane bound polyribosomes mediated by signal sequences).

4.4 Halobacterium halobium rhodopsin (bacteriorhodopsin); erythrocyte band 3

The existence of ectoproteins which span the lipid bilayer more than once, leaving discrete hydrophilic domains on both sides, has not been definitively proven (see Figure 1). Possible members of this class of proteins are bacteriorhodopsin and erythrocyte band 3. Bacteriorhodopsin is a transmembrane protein which pumps protons across the membrane by using the light energy absorbed by a retinal group bound to the protein. A detailed model of the membrane orientation of bacteriorhodopsin has been presented, based on pre-existing evidence.[51] The polypeptide chain (247 amino acids) leaves a short amino terminal segment on the extracellular side of the membrane and a carboxyl terminal segment on the cytosolic side, while the bulk of the protein is embedded in the lipid bilayer. In the membrane the protein is organized in 7 parallel helic regions, all perpendicular to the membrane plane with each helical region being oriented in the opposite way from its neighbouring helical region. The 7 helic regions are connected by short segments which are exposed on both sides of the membrane. Unfortunately, nothing is known about the membrane assembly of bacteriorhodopsin, but it is clear that the current model of the biosynthesis of ectoproteins will hardly explain how this unique protein achieve its final orientation.

Erythrocyte band 3 is a glycoprotein (MW 90 kdaltons) which is involved in anion exchange across the membrane. Extensive studies using proteolytic probes and reagents unable to cross from one side of the membrane to the other indicate that the amino terminus of the molecule is located on the cytoplasmic side of the membrane, whereas at least a segment of the carboxy terminal portion is located outside the cell; the same studies suggest that the protein spans the membrane more than once.[52,53] A preliminary report indicates that band 3 is synthesized on membrane bound polyribosomes, thus suggesting a cotranslational insertion in the membrane.[54]

5 PROKARYOTIC MEMBRANE PROTEINS

A considerable amount of experimental evidence supports the view that secretory and membrane proteins of prokaryotic cells are synthesized and translocated across, or inserted into, the membrane by a mechanism similar to that used by eukaryotic cells. Several precursors containing amino terminal signal sequence have been demonstrated, a signal peptidase localized on the periplasmic side of the plasma membrane, and cotranslational translocation of the nascent chain directly proven (for recent reviews see References 11 and 55). This finding is not so surprising considering how highly conserved the mechanism of protein translocation is throughout the animal and plant kingdoms.[4] Prokaryotic systems should be very useful for the study of mechanisms of membrane assembly because of the possibility of genetic approach (see Section 7).

6 MITOCHONDRIA AND CHLOROPLAST MEMBRANE PROTEINS

It has been clearly established that mitochondria and chloroplasts are semiautonomous organelles. They have an endogenous protein synthetic apparatus which, however, synthesizes no more than 10 per cent of the organelle's protein. All other proteins (plus carbohydrates and lipids) are synthesized in the cytoplasm under nuclear control. Thus mitochondria and chloroplasts have a transport system capable of inserting these proteins into their various compartments. In principle, the proteins synthesized in the cytoplasm could be synthesized on (a) polyribosomes bound to the endoplasmic reticulum, (b) polyribosomes bound to mitochondria and chloroplast outer membranes, or (c) free polyribosomes. Extensive studies performed in recent years, however, have excluded both (a) and (b), and pinocytosis was ruled out as a possible mechanism of translocation. Briefly, the following scheme was established (see References 12 and 13 for recent reviews). Citoplasmically made mitochondria and chloroplast proteins are synthesized on free polyribosomes and released into the cytosol as larger precursor forms. These precursors, but not their nascent chains, presumably interact with a specific translocation machinery located on the outer membrane of the organelles; during the translocation, the extra sequence (signal sequence, transit peptide) is endoproteolytically removed, as in the case of secretory and other membrane protein precursors. In the case of the small subunit of ribulose-1,5-biphosphate carboxylase, the extra sequence has been localized at the amino terminus, and its sequence has been determined.[57] The experimental evidence on which the above conclusions are based was obtained with an *in vitro* translation/translocation assay. Post-translational addition of intact mitochondria and chloroplasts to cell-free incubation mixtures led to the incorporation into the organelles of specific newly synthesized protein, with simultaneous removal of the extra sequence. Moreover, this *in vitro* translocation appears to be quite accurate. It has been shown that the small subunit of ribulose-1,5-biphosphate carboxylase not only penetrates but also correctly assembles with the large subunit (synthesized inside the chloroplast) to constitute an active enzyme molecule.[58] Apparently the translocation process requires energy. Transport to the membrane or to the mitochondrial matrix (but not to the intra-membrane space) is ATP dependent.[59]

The biosynthesis and translocation of relatively few cytoplasmically made membrane proteins has been studied up to now; among these are F_1-ATPase and cytochrome c oxidase, peripheral and integral multisubunit enzymes of the mitochondrial inner membrane, respectively. In both cases, precursors larger by 1.5–6 kdaltons than the mature counterparts have been identified and their *in vitro* post-translational translocation demonstrated.[56,60,61] The relevance of this translocation mechanism for the general problem of membrane protein assembly will be discussed in the next section; it should be mentioned here that a translation-independent translocation mechanism that does not involve pinocytosis is used by some plant and microbial proteic toxins (as diphtheria and cholera toxins, colicins etc.) to enter the cells.[62]

7   CONCLUSION

The only conclusion which can be made is that integral and peripheral proteins are synthesized either on free or on membrane bound poly-ribosomes, and assembled in the membrane either post- or cotranslationally. Thus the multiple mechanisms of membrane protein assembly represent a secondary diversification of the two general modes by which a protein can completely or incompletely cross a preformed cellular membrane, the cotranslational and the post-translational mechanisms respectively. Cotranslational assembly is apparently used by ectoproteins (Sections 4.2–4.4); post-translational assembly is used by endoproteins (Section 4.1), probably by peripheral proteins restricted to the cytosolic side of the membrane (Section 3.1), and by all cytoplasmically made peripheral and integral proteins of the mitochondria and chloroplast membrane system (Section 6).

A particular case is represented by the peripheral proteins of the extra-cellular side of the membrane; they cross the membrane cotranslationally but may assemble post-translationally (Section 3.2).

How much is known about these two general mechanisms at the molecu-lar level? To answer this question it is useful to consider separately three aspects of the assembly process: (a) the role of the recipient membrane, (b) the mechanism by which some integral membrane proteins acquire their transmembrane orientation, and (c) the driving force of the assembly pro-cess.

7.1   Role of the recipient membrane

Two alternative hypotheses have been proposed, and available experi-mental evidence does not rule out one or the other. In the signal hypothesis for membrane proteins[3] the assembly step is mediated by a specific trans-location machinery situated in the competent membrane (the signal and the ribosome receptor, and the signal peptidase are part of this structure, see Figure 4); only endoproteins and peripheral proteins on the cytosolic side of the membrane could assemble without such machinery. On the other hand in the membrane trigger hypothesis,[63] the insertion is driven by the inter-action of the lipid bilayer with the protein; no catalytic role for pre-existing membrane machinery is envisaged. Presently, the signal hypothesis has more general support, and it is the author's opinion that the sensitivity of the translocating membrane to high salt and trypsin treatment[58,64,65] and fusion experiments performed on bacterial proteins[55] (see below) fit better with the signal hypothesis.

Quite clearly a crucial role is played by signal sequences in all cases of cotranslational assembly and in mitochondria and chloroplast post-translational assembly. It has been shown in fact that mutations in the signal sequence region abolish the assembly process.[55] Signal sequences are usually located at the amino terminus of the primary translation product (although a carboxy terminal location is certainly conceivable in the case of mito-chondrial and chloroplast proteins), and are normally removed from such precursor forms during assembly. Some membrane proteins are apparently synthesized and inserted without transient precursor forms, but it has been possible to demonstrate the existence of a functional, uncleaved equivalent of the signal sequence in the only case examined.[40,44] Fundamental differ-ences exist however among the signal sequences involved in cotranslational and post-translational assembly. First, the signals are specific for only one of the two possible mechanisms. Second, in cotranslational assembly the infor-mation contained in the signal sequence is expressed only until the nascent chain attains a certain length, at which time it becomes masked,[45] whereas in post-translational assembly the signal sequences either require post-translational modification to be operative or their information is latent until the growing chain is released by the ribosome. The structure of membrane protein signal sequences confirms the basic features first elucidated for secretory protein signals. Although three dimensional models have been presented,[7] the unique and characteristic features of both types of signals are not yet understood.

Endoproteins and peripheral proteins restricted to the cytosolic side of the membrane are apparently synthesized and assembled without transient precursor forms. The interaction of these proteins with the target membrane occurs post-translationally (Sections 3.1, 4.1) and is apparently independent of the translocation machinery. However, definitive proof against the involvement of precursor forms has not been presented. Further, the insertion is specific for the target membrane, and thus a protein–membrane recognition mechanism (other membrane proteins, peculiar lipid composition, etc.) must exist.

7.2   Transmembrane orientation

This part of the assembly process involves ectoproteins and the integral membrane proteins of mitochondria and chloroplast which are synthesized in the cytoplasm, except those restricted to the cytosolic side of the outer membrane. The existence of specific stop transfer sequences which would specifically interrupt the translocation of the chain by interfering with the translocating machinery has been proposed. The presence in the same protein of several signal sequences and a stop transfer sequence could explain how some ectoproteins could span the membrane more than once; such sequences would presumably function by initiating and interrupting, at intervals, the translocation of the polypeptide chain (Reference 43; see Figure 4). The stop transfer sequences could correspond to the membrane embedded portion of the transmembrane proteins, which is an highly hydrophobic segment. The existence of stop transfer sequences has not been proved, and this step of the assembly process is the most obscure.

7.3   Driving force

Not much is known about the energy requirement for the assembly of membrane proteins. This problem is strictly connected with the understanding of the role of the membrane in the process (see Reference 11 for an extensive discussion). Endoproteins and peripheral proteins of the cytosolic side could assemble simply by diffusion and spontaneous binding or insertion of a hydrophobic anchor in the lipid bilayer. For all other membrane proteins however it seems unlikely that the translocation of discrete hydrophilic domain(s) across the lipid bilayer barrier could occur without an energy requirement. In yeast mitochondria the involvement of ATP in the post-translational assembly has been demonstrated, but no other experimental evidence has been reported up to now.[59]

It should be stressed that aspects (a) and (c) discussed above are common to secretory and membrane assembly, whereas the transmembrane orientation is an unique feature of some integral membrane proteins. It is evident that a large part of the assembly mechanism is still unclear but some experimental approaches currently under way in several laboratories seem very promising. Warren and Doberstein[65] and Walter *et al.*[66] reported some success in dissecting the translocating machinery, and purification of the putative factors by conventional biochemical methods is in progress.[66] The genetic approach available in the prokaryotic field seem very useful, in conjunction with genetic engineering technology. The elegant studies of Beckwith[67] and Silhavy group[55] clearly illustrate this point. Mutants defective for translocation of specific proteins have been isolated, and it has been shown that such mutations map in the signal sequence region of the protein; further, hybrid proteins have been constructed having variable amounts of the amino terminal region of an integral membrane protein fused with a large carboxy terminal region of a cytosolic protein, and the cellular localization of these hybrid proteins has been determined. It is therefore most probable that in the near future mutants altered in each step of the translocation process will be isolated and characterized, and that genetic engineering manipulations will be used to test the individual contribution of discrete segments of integral membrane proteins to the assembly process.

The aim of this article was to discuss the problem of membrane protein assembly, focusing on the early events of the biosynthesis of such proteins. It was beyond the scope of the article to discuss subsequent modifications such as glycosylation and acylation, which may indeed be very important for the biological function of these proteins.[68,69,70] Similarly, the general problem of how cells localize specific proteins in different subcellular compartments starting from a common site of synthesis (the sorting problem) was not discussed. However, it is worth mentioning that membrane proteins assembly as well as the biosynthesis of secretory proteins has been proposed to be a particular aspect of the more general problem of protein localization. In a recent hypothesis[71] it has been suggested that all the information necessary to achieve protein localization is contained in the sequence of the primary translation product itself as target-specific 'topogenic' sequences. The same functional sequences would be shared by all different proteins localized in the same cellular compartment. Gene amplification and recombination would explain how a limited number of 'topogenic' sequences could have been acquired by many different proteins.

In conclusion, our knowledge of membrane protein assembly is limited and does not allow theoretical generalizations other than working hypotheses which hopefully will stimulate future experimentation. Whatever the outcome of these experiments, it is likely that significant progress in the understanding of this central problem in cell biology will be made in the near future.

ACKNOWLEDGEMENTS

I wish to thank Drs. R. Cancedda and A. Leone for helpful comments. Supported by 'Progetto Finalizzato Virus' C.N.R., Rome, Italy, grant number 79.0096.84.

REFERENCES

1. Singer, S. J., and Nicholson, G. L. (1972). *Science*, **175**, 720.
2. Rothman, J. E., and Lenard, J. (1977). *Science*, **195**, 743.
3. Blobel, G., Walter, P., Chang, C. N., Goldman, B. M., Erickson, A. H., and Lingappa, V. R. (1979). In *Secretory Mechanism* (ed. Hopkins, C. R., and Duncan, C. J.), Cambridge University Press, Cambridge, p. 9.
4. Sabatini, D. D., and Kreibich, G. (1976). In *The Enzymes of Biological Membranes*, vol. 2 (ed. Martonosi, A.), Plenum Press, New York, p. 531.
5. Blobel, G., and Dobberstein, B. (1975). *J. Cell Biol.*, **67**, 852.
6. Blobel, G., and Dobberstein, B. (1975). *J. Cell Biol.*, **67**, 835.
7. Austen, B. M. (1979). *FEBS* (Fed. Eur. Biochem. Soc.) *Lett.*, **103**, 308.
8. Jackson, R. C., and Blobel, G. (1977). *Proc. Natl. Acad. Sci. U.S.A.*, **74**, 5598.
9. Lin, J. J. C., Kanazawa, H., Ozols, J., and Wu, H. C. (1978). *Proc. Natl. Acad. Sci. U.S.A.*, **75**, 4891.
10. Lingappa, V. R., Lingappa, J. R., and Blobel, G. (1979). *Nature*, **281**, 177.
11. Davis, B. D., and Tai, P. C. (1980). *Nature*, **283**, 433.
12. Chua, N. H., and Schmidt, G. W. (1979). *J. Cell Biol.*, **81**, 461.
13. Schatz, G. (1979). *FEBS* (Fed. Eur. Biochem. Soc.) *Lett.* **103**, 203.
14. Goldman, B. M., and Blobel, G. (1978). *Proc. Natl. Acad. Sci. U.S.A.*, **75**, 5066.
15. Zimmerman, R., Paluch, V., Sprinzl, M., and Neupert, W. (1979). *Eur. J. Biochem.*, **99**, 247.
16. Dobberstein, B., Garoff, H., and Warren, G. (1979). *Cell*, **17**, 759.
17. Ploegh, H., Cannon, L. E., and Strominger, J. L. (1979). *Proc. Natl. Acad. Sci. U.S.A.*, **76**, 2273.
18. Lingappa, V. R., Cunningham, B. A., Jazwinski, S. M., Hopp, T. P., Blobel, G., and Edelmann, G. M. (1979). *Proc. Natl. Acad. Sci. U.S.A.*, **76**, 3651.
19. Bodmer, W. F., Jones, E. A., Barnstable, C. J., and Bodmer, J. G. (1978). *Proc. Roy. Soc. London*, **B202**, 93.
20. Strauss, J. H., and Strauss, E. G. (1977). In *The Molecular Biology of Animal Viruses* (ed. Nayak, D. P.) Marcel Dekker Inc., New York, p. 111.
21. Kalouza, G., and Pauli, G. (1980). *Virology*, **102**, 300.
22. Hauri, H. P., Quaroni, A., and Isselbacher, K. J. (1979). *Proc. Natl. Acad. Sci. U.S.A.*, **76**, 5183.
23. Mihara, K., Sato, S., Sakakihara, R., and Wada, H. (1978). *Biochemistry*, **17**, 2829.

24. Enoch, H. G., Fleming, P. J., and Strittmatter, P. (1979). *J. Biol. Chem.*, **14,** 6483.
25. Borgese, N., and Meldolesi, J. (1980). *J. Cell Biol.*, **85,** 501.
26. Meldolesi, J., Corte, G., Pietrini, G., and Borgese, N. (1980). *J. Cell Biol.*, **85,** 516.
27. Borgese, N., Pietrini, G., and Meldolesi, J. (1980). *J. Cell Biol.*, **86,** 38.
28. Borgese, N., and Gaetani, S. (1980). *FEBS* (Fed. Eur. Biochem. Soc.) *Lett.*, **112,** 216.
29. Brugge, J. S., and Erikson, R. L. (1977). *Nature,* **269,** 346.
30. Lee, J. S., Varmus, H. E., and Bishop, J. M. (1978). *J. Biol. Chem.*, **254,** 8015.
31. Krueger, J. G., Wang, E., and Goldberg, A. R. (1979). *Virology,* **101,** 25.
32. Willingham, M. C., Jay, G., and Pastan, I. (1979). *Cell,* **18,** 125.
33. Krueger, J. G., Wang, E., Garber, E. A., and Goldberg, A. R. (1980). *Proc. Natl. Acad. Sci. U.S.A.*, **77,** 4142.
34. Courtneidge, S. A., Levinson, A. D., and Bishop, J. M. (1980). *Proc. Natl. Acad. Sci. U.S.A.*, **77,** 3783.
35. Martire, G., Bonatti, S., Aliperti, G., De Giuli, C., and Cancedda, R. (1977). *J. Virol.*, **21,** 610.
36. Cancedda, R., Villa-Komaroff, L., Lodish, H. F., and Schlesinger, M. J. (1975). *Cell,* **6,** 215.
37. Welch, W. J., and Sefton, B. M. (1979). *J. Virol.*, **29,** 1186.
38. Welch, W. J., and Sefton, B. M. (1980). *J. Virol.*, **33,** 230.
39. Bonatti, S. (1980). *Eur. J. Cell Biol.*, **22,** 159.
40. Bonatti, S., and Blobel, G. (1979). *J. Biol. Chem.*, **254,** 12261.
41. Cohen, L. B., Goldberg, I. H., and Herner, A. E. (1969). *Biochemistry,* **8,** 1312.
42. Aliperti, G., and Schlesinger, M. J. (1979). *Virology,* **90,** 366.
43. Bonatti, S., Cancedda, R., and Blobel, G. (1979). *J. Cell Biol.*, **80,** 219.
44. Garoff, H., Simons, K., and Dobberstein, B. (1978). *J. Mol. Biol.*, **124,** 587.
45. Rothman, J. E., and Lodish, H. F. (1977). *Nature,* **269,** 775.
46. Wirth, D. F., Katz, F., Small, B., and Lodish, H. F. (1977). *Cell,* **10,** 253.
47. Katz, F., Rothman, J. E., Lingappa, V. R., Blobel, G., and Lodish, H. F. (1977). *Proc. Natl. Acad. Sci. U.S.A.*, **74,** 3278.
48. Schechter, I., Burstein, Y., Zernell, R., Zin, E., Kantor, F., and Papermaster, D. S. (1979). *Proc. Natl. Acad. Sci. U.S.A.*, **76,** 2654.
49. Desnuelle, P. (1979). *Eur. J. Biochem.*, **101,** 1.
50. Garoff, H., Frischouf, A. M., Lerhac, H., and Simons, K. (1980). *Eur. J. Cell Biol.*, **22,** 159.
51. Engelman, D. M., Henderson, R., McLachlan, A. D., and Wallace, B. A. (1980). *Proc. Natl. Acad. Sci. U.S.A.*, **77,** 2023.
52. Drickamer, L. K. (1978). *J. Biol. Chem.*, **253,** 1242.
53. Steck, T. L., Koziarz, J. J., Singh, M. K., Reddy, G., and Kohler, H. (1978). *Biochemistry,* **17,** 1216.
54. Sabban, E., Sabatini, D. D., Adesnik, M., and Marchesi, V. (1979). *J. Cell Biol.*, **83,** SP2524.
55. Emr, S. D., Hall, M. N., and Silhavy, T. J. (1980). *J. Cell Biol.*, **86,** 701.
56. Mihara, K., and Blobel, G. (1980). *Proc. Natl. Acad. Sci. U.S.A.*, **77,** 4160.
57. Schmidt, G. W., Devillers-Thiery, A., Desruisseaux, H., Blobel, G., and Chua, N. H. (1979). *J. Cell Biol.*, **83,** 615.
58. Chua, N. H., and Schmidt, G. W. (1978). *Proc. Natl. Acad. Sci. U.S.A.*, **75,** 6110.
59. Nelson, N., and Schatz, G. (1979). *Proc. Natl. Acad. Sci. U.S.A.*, **76,** 4365.

60. Maccecchini, M. L., Rudin, Y., Blobel, G., and Schatz, G. (1979). *Proc. Natl. Acad. Sci. U.S.A.*, **76,** 343.
61. Lewein, A. S., Gregor, I., Mason, T. L., Nelson, N., and Schatz, G. (1980). *Proc. Natl. Acad. Sci. U.S.A.*, **77,** 3998.
62. Neville, Jr., D. M., and Chang, T. M. (1978). *Curr. Top. Membranes Transp.*, **10,** 65.
63. Wickner, W. (1979). *A. Rev. Biochem.*, **48,** 23.
64. Walter, P., Jackson, R. C., Marcus, M. M., Lingappa, V. R., and Blobel, G. (1979). *Proc. Natl. Acad. Sci. U.S.A.*, **76,** 1795.
65. Warren, G., and Dobberstein, B. (1978). *Nature*, **273,** 569.
66. Meyer, D. I., and Dobberstein, B. (1980). *Eur. J. Cell Biol.*, **22,** 153.
67. Silhavy, T. J., Bassford, Jr., P. J., and Beckwith, J. R. (1979). In *Bacterial Outer Membranes: Biogenesis and Function* (ed. Inouye, M.) John Wiley & Sons. Inc., New York.
68. Tabas, I., Schlesinger, S., and Kornfeld, S. (1978). *J. Biol. Chem.*, **253,** 716.
69. Robbins, P. W., Hubbard, S. C., Turco, S. J., and Wirth, D. F. (1977). *Cell*, **12,** 893.
70. Schmidt, M. F. G., and Schlesinger, M. J. (1980). *J. Biol. Chem.*, **255,** 3334.
71. Blobel, G. (1980). *Proc. Natl. Acad. Sci. U.S.A.*, **77,** 1496.

Hormone Receptors
Edited by L. D. Kohn
© 1982, John Wiley & Sons, Ltd

6 *The purification and characterization of interferons*

K. C. Zoon and M. E. Smith

1 INTRODUCTION

Since the discovery of interferon in 1957,[1] there have been continued efforts to isolate and chemically characterize the molecules responsible for inhibition of viral replication. The requirement of functional protein and mRNA synthesis for interferon production, together with inactivation of antiviral activity by proteolytic enzymes, and stability of this activity to the action of other types of hydrolytic enzymes (nucleases, glycosidases) supported the assumption that interferons were indeed proteins. Further characterization of interferon was not achieved for a number of years due to the lack of sufficient quantities of crude starting material for purification. At one point, the purification of chick interferon to homogeneity was reported;[2,3] however, further studies revealed that this preparation was actually purified ovalbumin which was binding interferon. The tendency for interferons to associate with various carrier proteins such as ovalbumin has suggested homogeneity of other preparations in which interferon was only a very small fraction of the total protein. Eventually, purification methods improved and the observed range of specific activities became well established as 10^8–10^9 units/mg protein which was very near the calculated theoretical maximum value of 2.5×10^9.[4]

Classification of interferons has been dependent upon both the producer cell and the nature of the induction process. Two general types of interferons have been defined. The first, type I, is the predominant product of cells induced by viruses or synthetic polynucleotides. These interferons are stable at pH 2 and consist of two antigenically distinct classes, Le (leukocyte type), and F (fibroblast type). The second class, type II or immune interferon, is a product of lymphocytes induced with T cell mitogens or sensitized lymphocytes exposed to antigen. These interferons are labile at pH 2 and are antigenically different from the first class.

Successful purification of all interferons requires production methods capable of generating 10^8–10^9 units with comparative ease and reproducibility. This quantity has been shown to be the minimum starting material necessary for obtaining a significant amount of purified product. Two cell culture systems are utilized as sources of interferon—non-transformed cells in primary cultures and transformed cells propagated in long term cultures.

Normal human blood leukocytes, obtained from whole blood, are put in primary culture and induced by exposure to Sendai virus. Obtaining a sufficient quantity of interferon requires a source of large amounts of whole blood. For production of 1–2×10^8 units of crude interferon, 40 litres of whole blood must be processed.[5] A second source of non-transformed interferon-producing cells is primary cultures of human or mouse neonatal fibroblasts in primary culture. These primary cultures have the ability to multiply *in vitro* and therefore the amounts of interferon produced can be significantly increased. Culture systems producing 10^8 interferon units usually contain on the order of 5 to 20 litres of monolayer culture supernatant.[6,7]

Interferon is produced in significant quantities by various transformed human and mouse cell lines, of both lymphoid and fibroblast origin. These production systems are limited only by the quantity of culture that can be conveniently handled. Up to 800 litres of human lymphoblastoid cell culture have been prepared and induced with Newcastle disease virus producing 10^9 interferon units. Other transformed cells have been grown in large quantities and are able to generate significant quantities of interferon.[8] Finally, lymphocytes from patients with chronic myelogenous leukaemia have produced large quantities of interferon when induced with Newcastle disease virus.[9]

Purification of the various types of interferon is essential for the understanding of its biological properties and its potential value as a therapeutic tool. Pure preparations of interferon are necessary for the dependable interpretation of biological effects and the mechanism of action in generating the 'antiviral state' in a protected cell. Other biological activities have been ascribed to interferon (cytotoxic activity,[10] immune modulation[11,12] and growth regulation[13]) and only by investigations with purified material can these properties be confirmed. Following the determination of the sequence, interferon can be made in several synthetic or biosynthetic ways. Considerable effort is now underway to isolate the interferon gene(s), incorporate them into a plasmid or visus *in vitro* and transfer them into bacterial cells where they can be perpetuated and an active interferon protein can be synthesized.[14,15] Alternatively, the chemical synthesis of interferon is under consideration. Both routes may prove to be a source of sufficient amounts for clinical or experimental applications.

Many procedures have been used for the purification of interferons. Initially the conventional techniques of differential precipitation, gel filtration, and ion exchange chromatography were used by almost all investigators.[3,16] Results with these methods yielded a reasonable degree of purification; however, the overall scheme resulted in unacceptably low yields because of the large number of steps involved. One or another affinity chromatography procedure is now generally employed during purification. Immunoadsorbant chromatography using antibodies of high specificity, and various novel types of affinity adsorbants such as bovine serum albumin.[17] Concanavalin A,[18,19] hydrophobic ligands,[20,21,22] Cibacron blue F3GA,[23] and polynucleotides[24] have provided significant advances in many purifications. High performance liquid chromatography (HPLC) has also been used for the purification of interferon with great success.[25]

The primary consideration for purification should be to generate a scheme with the minimum number of steps that can yield the maximum amount of homogenous material. A list of successful purification schemes is shown in Table 1. These procedures represent the most successful attempts to purify various types of interferon to date and several laboratories have produced minute quantities of apparently homogeneous material.

Variations in the specific activities of the purified interferons reflect both error in the bioassay for antiviral activity and inadequacy of methods for determination of minute quantities of protein. The most accurate method for the quantitation of protein in purified interferon preparations is amino acid analysis.

Excellent recoveries of homogeneous human fibroblast[26] and mouse C-243[24] cell interferons have been achieved using a two-step purification procedure. High performance liquid partition chromatography was used successfully for the purification of one species of interferon from both normal leukocytes and leukocytes from a patient with chronic myelogenous leukaemia.[9,25] Immunoabsorbant-affinity chromatography is essential to the successful purification of human lymphoblastoid,[27] mouse C-243 cell[24] and mouse L cell[28] interferons. This method results in large increases of the specific activities with recoveries ranging from 70–100 per cent. Berg et al.[29] have reported the purification of five species of human leukocyte inteferon using precipitation, gel filtration, copper-chelate-, blue-dextran- and immunoadsorbant-affinity chromatography (specific activity of 2×10^9 interferon units/mg protein). Human fibroblast has been purified by zinc-chelate-affinity chromatography to a specific activity of 3×10^8 interferon units/mg protein.[30]

Human fibroblast interferon has also been prepared as a translation product of its mRNA in recticulocyte lysates.[31] A 23,000 dalton polypeptide radiolabelled with ^{35}S-methionine was immunoprecipitated by antiserum of rabbits immunized with partially purified interferon, and characterized by SDS polyacrylamide gel electrophoresis.

At present neither mouse or human immune interferons have been purified to homogeneity.[32,33]

Improvements in the detection and analysis of picomole quantities of amino acids,[34,35] peptides and proteins[25,36,37,38] have permitted studies of

Table 1. Purification of interferons[a]

Type	Purification	Molecular weight	Specific activity (interferon units/ mg protein)	Total recovery (%)	References
Normal cell lines					
Human leukocyte	pH 4 supernatant trichloroacetic acid precipitation (1.5%), Triton X-100/acetic acid supernatant, trichloro-acetic acid precipitation (4%), Sephadex G100, high performance liquid chromatography using Lichrosorb RP-8 and Lichrosorb diol.	17,500	$2\text{–}4\times10^8$	3	25
Human fibroblast	Blue-Sepharose (twice) SDS PAGE	20,000	5×10^8	35–50%	26
Human immune (type II)	Matrex blue unbound, controlled pore glass, Ultrogel ACA 54.	40,000–46,000 65,000–70,000	$2\times10^{6\,a}$	NR	33
Mouse immune (Type II)	Ammonium sulphate pre-cipitation, BSA-Affi-gel 10.	70,000–90,000 40,000	$2.4\times10^{5\,a}$	NR	32
Transformed cell lines					
Mouse Ehrlich Ascites Tumor Cell	Controlled Pore glass, CM-Sephadex, Phospho-cellulose, Sephacryl S-200, Phosphocellu-lose (concentration), isoelectric focusing (octyl-Sepharose)	33,000 26,000 20,000	$2\text{–}3\times10^{9\,b}$	11–20[b]	53 60
Mouse L Cell	Zinc acetate precipi-tation, ammonium sul-phate precipitation, immunoabsorbant-affin-ity chromatography, DEAE Sephadex, CM-Sephadex, Biogel P-60, Biogel P-100.	40,000 24,000	$4.0\times10^{8\,c}$ 7.3×10^8	0.09 0.78	28
Mouse C-243	Poly U-Sepharose, immunoabsorbant-affinity chroma-tography.	35,000 22,000	$2.4\times10^{9\,d}$	52[d]	24
Human Fibro-blastoid	Centrifugation 100,000×g (super-natant) ammonium sul-phate precipitation, controlled-pore glass, concanavalin A-Sepharose, two-dimensional gel elec-trophoresis (a) acid–urea (b) SDS PAGE.	20,000	$2\text{–}5\times10^8$	10	8

Table 1. (*Continued*)

Type	Purification	Molecular weight	Specific activity (interferon units/ mg protein)	Total recovery (%)	References
Leukocyte (chronic myelogenous leukaemia patients)	Purification same as human leukocyte interferon.	16–21,000	2.4×10^8	10	9
Human lymphoblastoid	Trichloroacetic acid precipitation, Sephadex G-25, immunoadsorbant-affinity chromatography, Sephadex G-150, SP-Sephadex, L-tryptophyl-L-tryptophan-affi-gel 10, SDS PAGE.	18,500 21,000	2–2.5×10^8 7.7×10^7–2.1×10^8	3–7[e]	27

[a] partially purified.
[b] Mixture of 3 species.
[c] estimate of specific activity after SDS PAGE, 2.6×10^9 units/mg protein.
[d] Mixture of the 35,000 dalton and 22,000 dalton polypeptides.
[e] Mixture of 2 species.
N.R., not reported.

the chemical properties of several interferons. The amino acid compositions of human fibroblast,[26] fibroblastoid,[8] leukocyte[25] and lymphoblastoid interferons[39] and band C mouse Ehrlich ascites tumour cell interferon[40] are presented in Figure 1. A comparison of the various interferon compositions

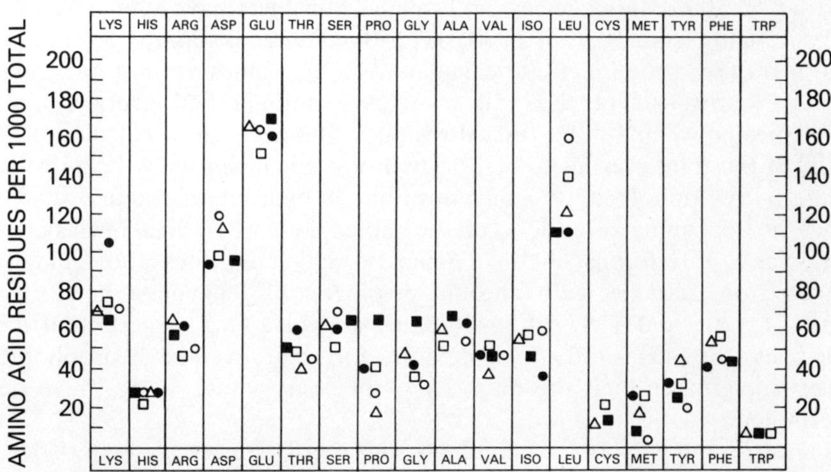

Figure 1. Quantitation of protein in purified interferon preparations ●, mouse (b and c); □, human leukocyte (MW 17,500); ■, human lymphoblastoid (MW 18,500); △, human fibroblast; ○, human fibroblastoid

Table 2. Average hydrophobicities ($H\Phi_{ave}$) of several mouse and human interferon

Interferon	$H\Phi_{ave}$ (cal/residue)	Reference
Human lymphoblastoid (18,500 dalton)	1050	39
Human fibroblast	1080	26
Mouse Ehrlich ascites (band C)	1050	53
Leukocyte (normal)	1210	25
Leukocyte (CML)[a]	1190	9

shows that they are strikingly similar. Many of these interferons have shown similar characteristics in their ability to bind hydrophobic ligands. This binding may be a result of the overall hydrophobicity of the molecule (intrinsic hydrophobicity) or due to a hydrophobic site (apparent hydrophobicity). The intrinsic hydrophobicity of a protein is directly related to its average hydrophobicity, $H\Phi_{ave}$.[41,43] The $H\Phi_{ave}$ is related to the change in free energy during transfer of each amino acid side chain from ethanol to water[43] and the amino acid composition of the protein. The values of this parameter for each of the five purified species of interferon are presented in Table 2. Three of the 5 interferons, mouse Ehrlich ascites band C, human lymphoblastoid 18,500 dalton species, and human fibroblast have similar values of $H\Phi_{ave}$ ranging from 1050 to 1080 cal/residue. When compared to data from over 150 other proteins, these calculated $H\Phi_{ave}$ values are not particularly high.[41] Nearly half of the 150 proteins examined had average hydrophobicities between 1000–1100 cal/residues. Based on these calculations the ability of these interferons to bind to hydrophobic ligands may be a result of apparent hydrophobicity, a concentration of hydrophobic amino acid residues in one or more regions of the molecule, rather than intrinsic hydrophobicity. Interferons obtained from normal leukocytes and leukocytes derived from patients with chronic myelogenous leukaemia have $H\Phi_{ave}$ values of 1205 and 1187 cal/residues respectively. This suggests that these molecules are intrinsically hydrophobic, for their average hydrophobicity values are greater than the majority of proteins whose $H\Phi_{ave}$ have been determined.

As more information accumulates on the chemical characteristics of interferons from various sources, there appear to be several varieties responsible for antiviral activity. Many cell lines secrete more than one type of interferon. These interferons differ in their antigenic cross-reactivities, apparent molecular weights (usually estimated by SDS polyacrylamide gel electrophoresis), the ability to bind to a variety of affinity columns, or, most convincingly, in amino-terminal sequences. This heterogenity probably results from post-translational modification (proteolytic degradation or glycosylation) and multiple interferon genes.

The amino terminal sequences of several human and mouse interferons are presented in Figure 2. Some homology between human fibroblast interferon[26] and mouse Ehrlich ascites tumour cell interferon band A (33,000 daltons)[40] exists. Three of the first 13 amino acid residues are identical and all but one (position 7) of the other amino acid residues have codon(s) which differ by a single base. More striking is the homology between the major component of human lymphoblastoid interferon (18,500 daltons) and band C of mouse Ehrlich ascites tumour cell interferon (20,000 daltons). Thirteen of the first 20 amino acid residues are identical and all the other residues have codons differing by only single base substitutions. No homology exists between band C of the mouse interferon[26,40] and human fibroblast interferon nor between mouse interferon band A and the 18,500 dalton component of human lymphoblastoid interferon.[39,40] Preliminary amino acid sequence comparison between the native human fibroblast (20,000 dalton) and the human lymphoblastoid interferon(s) (18,500 dalton) show no significant homology. No antigenic cross-reactivity has been observed between the species. This strongly indicates two distinct primary structures for human fibroblast interferon and the 18,500 dalton form of lymphoblastoid interferon.

Various methods have been utilized to prepare deglycosylated interferons. Chemical cleavage with periodate[44,45] or biosynthesis in the presence of metabolic inhibitors of carbohydrate synthesis, 2-deoxy-D-glucose, D-glucosamine[46,47] or tunicamycin[48,49] (which inhibits the glycosylation of protein[50]), yield interferons with less carbohydrate than the native forms. These

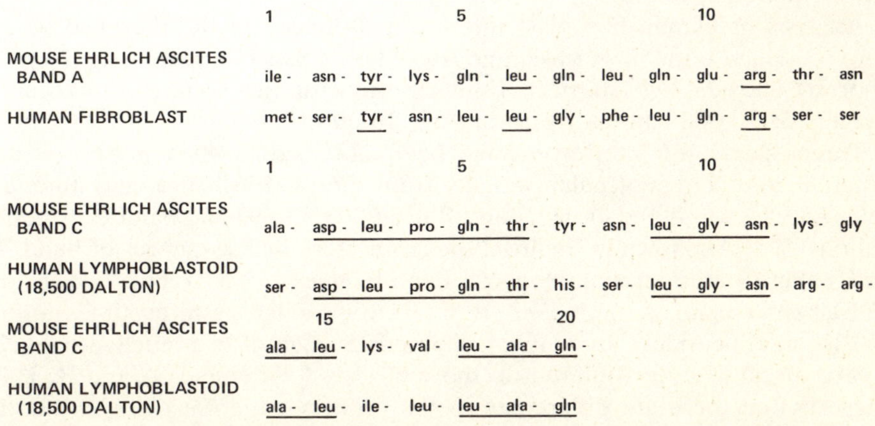

Figure 2. Amino terminal sequences of some human and mouse interferons

preparations will allow analysis of the contribution of carbohydrate to antigenicity and species specificity of antiviral activity.

Kawade and coworkers[28] have found that the two species of interferon from mouse L cells induced with Newcastle disease virus, S (40,000 daltons) and F (24,000 daltons) are antigenically distinct. Treatment of the S component with glycosidases lowers the apparent molecular weight by 5–10,000 daltons but does not change the antigenicity. Mouse L cell interferon produced in the presence of tunicamycin, has an apparent molecular weight of 18,000 and is antigenically similar to the F component.[48] These investigators also observed that the deglycosylated mouse L cell interferon is neutralized by anti-human leukocyte interferon.[51] A comparison of the tryptic digest fragments of the S and F species showed that 3 out of 4 ^{35}S-methionine-labelled peptides were chromatographically identical and 3 out of 6–7 ^{125}I-labelled peptides were also chromatographically identical.[52] These data suggest that the two interferon species possess some common polypeptide structure.

Stewart *et al.*[44] have observed two species of interferon produced by Newcastle disease induced mouse L_{929} cells by SDS polyacrylamide gel electrophoresis, one with an apparent molecular weight of 38,000 (80 per cent) and the other 22,000 (20 per cent). Treatment of a mixture of both species with acidic periodate eliminated the 38,000 dalton species and resulted in an increased amount of the 22,000 dalton species. Further incubation with periodate resulted in the emergence of a 15,000 dalton species. Mouse L cell interferon produced in the presence of 2-deoxy-D-glucose or D-glucosamine had normal levels of antiviral activity; however, less than 10 per cent of the activity was found in either the 38,000 dalton or the 22,000 dalton species and approximately 90 per cent of the antiviral activity was observed in the 15,000 dalton species. The 15,000 dalton component was neutralized by antiserum prepared against the 38,000 and 22,000 dalton species.

Analysis of human fibroblast interferon produced in the presence of 2-deoxy-D-glucose or D-glucosamine by SDS polyacrylamide gel electrophoresis showed two species of interferon with the apparent molecular weights of 20,000 (the native form) and 16,000.[47]

Three species of interferon have been obtained purified in form with different apparent molecular weights from mouse Ehrilich ascites tumour cells, band A (33,000 daltons), band B (26,000 daltons) and band C (20,000 daltons).[40,53] As shown in Figure 2 the amino terminal sequences of band A and band C are not homologous. Peptide maps of a tryptic digests of ^{125}I-labelled band A and band B had very similar patterns (6–7 major ^{125}I-labelled peptides). In contrast, the peptide map of ^{125}I-labelled band C tryptic digest was very different to those observed with bands A and B. This suggests that there are differences in the primary amino acid sequences of the interferon molecules.

Human lymphoblastoid interferon is composed of two antigenically distinct interferons; 13 per cent is the F (fibroblast) type and the remainder is the Le (leukocyte) type.[54] Human leukocyte[55] and lymphoblastoid (Namalwa)[56] interferons fractionate into two components with the apparent molecular weights of 21,000–21,500 and 15,000–18,000 after electrophoresis on polyacrylamide gels in the presence of SDS. Treatment of the 21,500 dalton species of these interferons with mixture of glycosidases results in their conversion to the lower molecular weight form and the loss of a majority of the charge heterogeneity as determined by isoelectric focusing.[56,57,58] Treatment of human leukocyte interferon with acidic periodate for 4 hr results in the detection of only one species of interferon with an apparent molecular weight of 15,000.[45] This component focuses as a single band with a pI of 5.7.

The production of active mouse L cell interferon and human immune interferon in the presence of tunicamycin indicates that the carbohydrate is not essential for secretion or antiviral activity.[48,49] Experiments performed after removal of a portion of the carbohydrate from several interferons with enzymes or chemicals also suggests the carbohydrate is not critical for antiviral activity.[44,45,57] Human immune type interferon (type II) produced in the presence of tunicamycin is not significantly different in either molecular weight or antiviral activity indicating no contribution by glycosylation to either of these parameters.[49] However, the apparent hydrophobicity of the immune interferon, i.e. the ability to bind hydrophobic ligands is altered.[49] Type II human immune interferon is also antigenically distinct from either of type I human interferon types.[33]

Most purified human leukocyte, lymphoblastoid or mouse interferon preparations have two or more active components with different apparent molecular weights as determined by SDS polyacrylamide gel electrophoresis. Differences in the carbohydrate content has been demonstrated by various methods and can account for a significant portion of this molecular weight heterogenity. These components also exhibit different levels of antiviral activity with heterologous target cells indicating some changes in the degree of species specificity. This observation is consistent with the assumption that some of the species specificity found in these interferons is related to the presence of carbohydrate. Also, treatment of human leukocyte interferon with α chymotrypsin, yielded a 16,500 dalton interferon species which is more active on bovine cells than human cells.[59] This study, along with the surprising amino terminal sequence homology observed between human lymphoblastoid interferon (18,500 daltons) and mouse Ehrlich ascites band C (20,000 daltons) and a lesser though significant homology between human fibroblast (20,000 daltons) and mouse Ehrlich ascites band A (33,000 daltons),[60] implies evolutionary conservation of the amino acid sequences which express antiviral activity as well as substantial contributions by carbohydrate to modulate species specificity.

2 SUMMARY

Significant advances have been made in the last four years on the purification of interferons. Components of several human and mouse interferons have been purified to homogeneity. Partial characterization of these molecules has been achieved (i.e. amino acid composition, analytical maps of radiolabelled tryptic peptides, and amino terminal amino acid sequence). In addition, preliminary evidence has been accumulated which implies that the carbohydrate portion of the molecules may play a significant role in species specificity of interferons. At present, with minute quantities of homogenous interferons available, limited structural information has been obtained. Using current production and purification schemes, sufficient material can be obtained for further amino acid sequence studies and examination of the carbohydrate portion of the molecules. Additional structural characterization of the interferons will facilitate both chemical synthesis and isolation of the interferon gene(s). Furthermore, as sequence homologies have been observed between human and mouse interferons, all new structural information will help to establish the interrelationships in the family of interferon proteins.

REFERENCES

1. Isaacs, A., and Lindenmann, J. (1957). Virus Interference I. The interferon. *Proc. Roy. Soc., Ser. B.*, **147,** 258–267.
2. Fantes, K. H. (1965). Further purification of chick interferon. *Nature*, **207,** 1298. (1966).
3. Fantes, K. H. (1966). Purification, concentration, and physico-chemical properties of interferons. In *Interferons* (Finter, N., ed.) pp. 119–180, North Holland, Amsterdam.
4. Ng, M. H., and Vilcek, J. (1972). Interferons: physiochemical properties and control of cellular synthesis. In *Advances in Protein Chemistry* (Anfinsen, C. B., Edsall, J. T., and Richards, F. M., eds.), **26,** 173–241.
5. Cantell, K., and Hirvonen, S. (1977). Preparation of human leukocyte interferon for clinical use. In *The Interferon System: A Current Review to 1978.* (Baron, S., and Dianzani, F., eds.) 35, pp. 138–144, University of Texas Medical Branch, Galveston, Texas.
6. Knight, E., Jr. (1976). Interferon: purification and initial characterization from human diploid cells. *Proc. Nat. Acad. Sci. U.S.*, **73,** 520–523.
7. Billau, A., Van Damme, J., Van Leuven, F., Edy, V. G., De Ley, M., Cassiman, J., van der Berghe, H., and De Somer, P. (1979). Human fibroblast interferon for clinical trials: production, partial purification, and characterization, *Antimicrob. Agents and Chemother.*, **16,** 49–55.
8. Tan, Y. H., Barakat, F., Berthold, W., Smith-Johannsen, H., and Tan, C. (1979). The isolation and amino acid/sugar composition of human fibroblastoid interferon. *J. Biol. Chem.*, **254,** 8067–8073.
9. Rubenstein, M., Rubenstein, S., Familletti, P. C., Brink, L. D., Herschberg, R. D., Gutterman, J., Nester, J., and Pestka, S. (1979). *Peptides: Structure and Biological* (Gross, E., and Meinhoffer, J., eds.) Rockford Ill, 99–103.

10. Lindhal, P., Leary, P., and Gresser, I. (1973). Enhancement by interferon of the specific cyctotoxicity of sensitized lymphocytes. *Proc. Nat. Acad. Sci. U.S.*, **70**, 2785–2788.

11. Johnson, H. M., and Baron, S. (1976). The nature of the suppressive effect of interferon and interferon inducers on the *in vitro* immune response, *Cell. Immunol.*, **25**, 106–109.

12. Johnson, H. M., Smith, B. G., and Baron, S. (1975). Inhibition of the primary *in vitro* antibody response by interferon preparations. *J. Immunol.* 114, 403–409.

13. Gresser, I., Bandu, M. T., Tovey, M., Bodo, G., Paucker, K., and Stewart II, W. E. (1973). Interferon and cell division VII. Inhibitory effect of highly purified interferon preparations on the multiplication of leukemia L1210 cells. *Proc. Soc. Exp. Biol. Med.*, **142**, 7–10.

14. Taniguchi, T., Sakai, M., Fujii-Kuriyama, Y., Muramatsu, M., Kobayashi, S., and Sudo, T. (1979). Construction and identification of a bacterial plasmid containing the human fibroblast interferon gene sequencer. *Proc. Jap. Acad.*, **55**, 464–469.

15. Revel, Michel (1980). Personal communication.

16. Stewart II, W. E. (1979). *The Interferon System*, Springer-Verlag, New York, 134–183.

17. Huang, J. W., Davey, M. W., Hejna, C. J., von Muenchhausen, W, Sulkowski, E., and Carter, W. A. (1974). Selective binding of human interferon to albumin immobilized on agarose. *J. Biol. Chem.*, **249**, 4665–4667.

18. Davey, M. W., Huang, J. W., Sulkowski, E., and Carter, W. A. (1974). Hydrophobic interaction of human interferon with concanavalin A-agarose. *J. Biol. Chem.*, **1974**, 6354–6355.

19. Davey, M. W., Sulkowski, E., and Carter, W. A. (1976). Binding of human fibroblast interferon to concanavalin A-agarose. Involvement of carbohydrate recognition and hydrophobic interaction. *Biochemistry*, **15**, 704–713.

20. Sulkowski, E., Davey, M. W., and Carter, W. A. (1976). Interaction of human interferons with immobilized hydrophobic amino acids and dipeptides. *J. Biol. Chem.*, **251**, 5381–5385.

21. Davey, M. W., Sulkowski, E., and Carter, W. A., Jr. (1976). Purification and characterization of mouse interferon with novel affinity sorbents. *Virology*, **17**, 439–445.

22. Davey, M. W., Huang, J. W., Sulkowski, E., and Carter, W. A. (1975). Hydrophobic binding sites on human interferon. *J. Biol. Chem.*, **250**, 348–349.

23. Jankowski, W. J., von Muenchhausen, W., Sulkowski, E., and Carter, W. A. (1976). Binding of human interferons to immobilized cibracron blue F3GA: The nature of molecular interaction. *Biochemistry*, **15**, 5182–5187.

24. De Mayer-Guignard, J., Tovey, M. G., Gresser, I., and De Mayer, E. (1978). Purification of mouse interferon by sequential affinity chromatography on poly(U)- and antibody-agarose columns. *Nature*, **271**, 622–625.

25. Rubinstein, M., Rubinstein, S., Familletti, P. C., Miller, R. S., Waldman, A. A., and Pestka, S. (1979). Human leukocyte interferon: production, purification to homogeneity, and initial characterization. *Proc. Nat. Acad. Sci. U.S.*, **76**, 640–644.

26. Knight, Jr., E., Hunkapiller, M., Korant, B. D., Hardy, R. W. F., and Hood, L. E. (1980). Human fibroblast interferon: amino acid analysis and N-terminal amino acid sequence. *Science*, **207**, 525–526.

27. Zoon, K. C., Smith, M. E., Bridgen, P. J., zur Nedden, D., and Anfinsen, C. B. (1979). Purification and partial characterization of human lymphoblastoid Interferon. *Proc. Nat. Acad. Sci. U.S.*, **76**, 5601–5605.

28. Iwakura, Y., Yonehara, S., and Kawade, Y. (1978). Purification of mouse L cell interferon, *J. Biol. Chem.*, **253**, 5074–5079.

29. Berg, K., Osther, K., and Heron, I. (1980). The complete properties of five species of human leukocyte interferon. In *Regulatory Functions of Interferons: Annals of the New York Academy of Sciences* (Vilcek, J., ed.), New York Academy of Sciences, **350**, 594–595.

30. Edy, V. G., Billiau, A., and De Somer, P. (1977). Purification of human fibroblast interferon by zinc chelate affinity chromatography. *J. Biol. Chem.*, **252**, 5934–5935.

31. Weissenbach, J., Zeevi, M., Landau, T., and Revel, M. (1979). Identification of the translation products of human fibroblast interferon mRNA in recticulocyte lysates. *Eur. J. Biochem.*, **98**, 1–8.

32. Georgiades, J. A., Langford, M. P., Stanton, G. J., and Johnson, H. M. (1979). Separation of immune interferon and MIF. *Proc. Soc. Exp. Biol. Med.*, **161**, 167–170.

33. Langford, M. P., Georgiades, J. A., Dianzani, F., Stanton, G. J., and Johnson, H. M. (1979). Large scale production and physicochemical characterization of human immune interferon. *Infection and Immunity*, **26**, 36–41.

34. Hare, P. E. (1977). Subnanomole-range amino acid analysis. In *Methods in Enzymology* (Hirs, C. H. W., and Timasheff, S. N., eds.) **47**, pp. 3–18, Academic Press, New York.

35. Benson, J. R., and Hare, P. E. (1975). o-Phthalaldehyde: fluorogenic detection of primary amines in the picomole range. Comparison with fluorescamine and ninhydrin. *Proc. Nat. Acad. Sci. U.S.*, **72**, 619–622.

36. Johnson, N. D., Hunkapiller, M. W., and Hood, L. E. (1980). Analysis of phenylthiohydantoin amino acids by high performance liquid chromatography on DuPont Zorbax CN columns, *Anal. Biochem.*, **100**, 335–338.

37. Hunkapiller, M. W., and Hood, L. E. (1978). Direct microsequence analysis of polypeptides using an improved sequenator, a nonprotein carrier (polybrene) and high presure liquid chromatography. *Biochemistry*, **17**, 2124–2133.

38. Hunkapiller, M. W., and Hood, L. E. (1980). A new protein sequenator with increased sensitivity. *Science*, **207**, 523–525.

39. Zoon, K. C., Smith, M. E., Bridgen, P. J., Anfinsen, C. B., Hunkapiller, M., and Hood, L. E. (1980). *Science*, **207**, 527–528.

40. Taira, H., Broeze, R. J., Jayaram, B. M., Lengyel, Hunkapiller, M., and Hood, L. E. (1980). Mouse interferons: NH_2-terminal amino acid sequences of various species. *Science*, **207**, 528–530.

41. Bigelow, C. C. (1967). On the average hydrophobicity of proteins and the relation between it and protein structure. *J. Theoret. Biol.*, **16**, 187–211.

42. Sulkowski, E., Vastola, K., and Lee, H. V. (1980). Hydrophobic properties of interferons. In *Regulatory Functions of Interferons: Annals of the New York Academy of Sciences* (Vilcek, J., ed.), New York Academy of Sciences, **350**, 339–346.

43. Tanford, C. (1962). Contribution of hydrophobic interactions to the stability of the globular conformation of proteins. *J. Amer. Chem. Soc.*, **84**, 4240–4247.

44. Stewart II, W. E., Chudzio, T., Lin, L. S., and Wiranowska-Stewqrt, M. (1978). Interferoids: *In vitro* and *in vivo* conversion of native interferons to lower molecular weight forms. *Proc. Nat. Acad. Sci. U.S.*, **75**, 4814–4818.

45. Stewart, W. E., Linn, L. S., Wiranowska-Stewart, M., and Cantell, K. (1977). Elimination of size and charge heterogeneities of human leukocyte interferons by chemical cleavage. *Proc. Nat. Acad. Sci. U.S.*, **74**, 4200–4204.

46. Cundliffe, B., and Morser, J. (1979). The effect of inhibitors of glycosylation on

interferon production in human lymphoblastoid cells. *J. Gen. Virol.*, **43**, 457–461.
47. Havell, E., Yamazaki, S., and Vilcek, J. (1977). Altered molecular species of human interferon produced in the presence of inhibitors of glycosylation. *J. Biol. Chem.*, **252**, 4425–4427.
48. Fujisawa, J., Iawkura, Y., and Kawade, Y. (1978). Nonglycosylated mouse L cell interferon produced by the action of tunicamycin. *J. Biol. Chem.*, **253**, 8677–8679.
49. Mizrahi, A., O'Malley, J. A., Carter, W. A., Takatsuki, A., Tamura, G., and Sulkowski, E. (1978). Glycosylation of interferons: effects of tunicamycin on human immune interferon. *J. Biol. Chem.*, **253**, 7612–7615.
50. Struck, D. K., and Lennarz, W. J. (1977). Evidence for the participation of saccharide-lipids in the synthesis of the oligosaccharide chain of ovalbumin. *J. Biol. Chem.*, **252**, 1007–1013.
51. Kawade, Y., Yammoto, Y., Fujisawa, J., and Watanabe, Y. (1980). Antigenic relationships among various interferon species. In *Regulatory Functions of Interferon: Annals of the New York Academy of Sciences* (Vilcek, J., ed.), New York Academy of Sciences, **350**, 422–427.
52. Iwakura, Y., Yonehara, S., and Kawade, Y. (1978). Presence of a common structure in the two molecular species of mouse L cell interferon. *Biochem. Biophys. Res. Commun.*, **84**, 557–563.
53. Cabrer, B., Taira, H., Broeze, R. J., Kempe, T. D., Williams, K., Slattery, E., Konigsberg, W. H., and Lengyel, P. (1979). Structural characteristics of interferons from mouse Ehrlich ascite tumor cells. *J. Biol. Chem.*, **254**, 3681–3684.
54. Havell, E. A., Yip, Y. K., and Vilcek, J. (1977). Characteristics of human lymphoblastoid (Namalva) interferon. *J. Gen. Virol.*, **38**, 51–59.
55. Stewart II, W. E., and Desmyter, J. (1975). Molecular heterogeneity of human leukocyte interferons: two populations differing in molecular weights, requirements for renaturation and cross-species antiviral activity. *Virology*, **67**, 68–78.
56. Bridgen, P. J., Anfinsen, C. B., Corley, L., Bose, S., Zoon, K. C., and Ruegg, U. T. (1977). Human lymphoblastoid interferon, large scale production and partial purification. *J. Biol. Chem.*, **252**, 6585–6587.
57. Bose, S., Gurari-Rotman, D., Ruegg, U. T., Corley, L., and Anfinsen, C. B. (1976). Apparent dispensability of the carbohydrate moiety of human interferon for antiviral activity. *J. Biol. Chem.*, **251**, 1659–1662.
58. Zoon, K. C. (1980). Unpublished results.
59. Braude, I. A., Lin, L. S., and Stewart II, W. E. (1979). Differential inactivation and separation of homologous and heterologous antiviral activity of human leukocyte interferon by a proteolytic enzyme. *Biochem. Biophys. Res. Commun.*, **89**, 612–619.
60. Kawakita, M., Cabrer, B., Taira, H., Rebello, M., Slattery, E., Weideli, H., and Lengyel, P. (1978). Purification of interferon from mouse Ehrlich ascites tumor cells, *J. Biol. Chem.*, **253**, 598–602.

7 Carbohydrates and receptor recognition

O. Gabriel

1 INTRODUCTION

1.1 General features of receptor models

Message transformation between cells and appropriate response to various 'signals' is at present a central issue of modern biology. How cell membranes perform this function is believed to be understood at least in general terms: A membrane receptor is composed of several components; the one exposed on the cell surface initiates the attachment of a 'signal' to the cell surface and is termed the 'discriminator'. The major function of the discriminator is its ability to be 'stereoselective' in the discrimination of various messages. The outside asymmetric location of the receptor binding site is required for the proper function of the discriminator to permit ready access of the message to the binding site. For thermodynamic reasons polar and hydrophobic regions of the discriminator will assure an orientation away from the lipid bilayer of the membrane. The 'transducer' a second component of cellular receptors has primarily a 'signal-response' coupling function that indicates that binding of a proper signal has occurred which will elicit a response in the third element of cellular receptors referred to as 'effector'. The effector will amplify the 'signal' binding of the first messenger to the discriminator binding site by activation of an enzyme resulting in formation of a second messenger thereby producing an intracellular signal. Physiological response requires fast, sensitive, selective and reversible message transmission. In order to fulfil these requirements the interaction between binding site and ligand will have to be non-covalent. Models for this specific type of interaction are enzyme–substrate or antigen–antibody binding which have some of the required features such as stereospecificity and high-affinity binding. The analogy between cellular receptors and the model of a strict complementary nature in stereochemical terms between receptor and ligand was recognized to be only partially true.

The thinking evolved parallel to that for enzyme–substrate binding where the simple 'lock–key' concept of Fischer[1] was later refined by the 'induced fit' model of Koshland.[2] The principle that specific binding of a substrate–ligand can induce conformational changes in the enzyme protein is fully applicable to receptor–ligand interaction. In other words, the specific binding of a ligand will induce conformational changes in the receptor. Conversely, the analogous principle is also applicable to the ligand[3] where binding of the ligand to the receptor can result in a conformational ligand change. Actually, 'receptor recognition' may result in conformational changes of both components, the receptor and the ligand which could account for some of the observed binding phenomena that are not consistent with a simple 'lock–key' model. The location of the receptor as a component in the lipid bilayer may provide an additional element of complexity that is at present not fully appreciated. The fluidity of the membrane, mobility and rearrangement of receptors within the membrane, induced conformational changes of receptor or neighbouring components all may alter significantly surface topography with profound effects upon binding and signal transmission. These complexities are at present in the realm of speculation and precise answers are not as yet available. Present simplistic models of receptor binding will have to be extensively refined to account for the complexities of message transmission observed in nature.

1.2 Carbohydrates as specificity determining components

As was just described, the selectivity of cellular recognition requires the specific interaction between two components such as a receptor and a ligand.

As illustrated schematically in Figure 1, covalently attached carbohydrate oligomers can be a component of either receptor or ligand. Validity of this model implicitly assumed that either one of the two interacting components (receptor or ligand) is a protein since only proteins are endowed with the required degree of binding specificity to recognize a specific carbohydrate oligomer sequence. For example, a ligand bound to a receptor protein can be either another protein or a structure covalently linked to an oligomeric heterosaccharide structure. Conversely, the receptor can be a carbohydrate containing membrane component such as a glycoprotein or glycolipid and the interacting ligand is a protein with carbohydrate binding specificity. Understanding of the specificity exhibited by carbohydrate binding proteins requires detailed structural studies of the oligosaccharide moieties on glycoproteins and/or glycolipids. Considerable research activity in this area also indicates the need for more detailed structural knowledge of cell surface carbohydrate oligomers.

Concomitant with rapidly expanding understanding of structural features of glycoproteins and glycolipids, increased interest centres on the study of

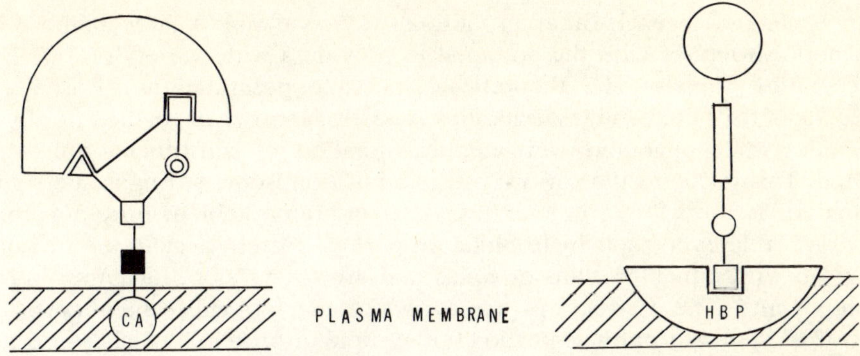

Figure 1. Schematic representation of examples for receptor-ligand interaction involving lectins. Glycolipid G_{M1} (CA = ceramide) acting as part of the plasma membrane receptor and is recognized by the β-subunit of cholera toxin as described in the text (left side of diagram). Hepatic binding protein (HBP) is the receptor for asialoglycoproteins (right side of the diagram). Multiple binding sites and conformational changes are not considered

various types of carbohydrate binding proteins. Several major groups of carbohydrate binding proteins have been observed. A clear difference in the mechanism of binding among all those groups may not remain when we start to appreciate the molecular events involved in the binding process. At present there are at least five individual groups of carbohydrate binding proteins:

1. glycoprotein hormones such as chorionic gonadotropin and thyrotropin[4]
2. bacterial toxins (cholera and botulinum toxins)[5]
3. glycohydrolases[6] and glucosyl-transferases[7]
4. anticarbohydrate antibodies[8,9]
5. lectins[10]

Research on the properties of carbohydrate binding proteins and their interaction with carbohydrate containing cell-surface components provide additional important information about the structure and architecture of cell surface components containing carbohydrate oligomers. More importantly, various biological phenomena such as cell aggregation, message transmission for growth, differentiation and development in normal and malignant transformed cells appear to be triggered by specific binding of proteins to cell surface components.

2 SURFACE HETEROSACCHARIDES ON CELLULAR AND SUBCELLULAR SURFACES

Carbohydrates have two unique properties that qualify these structures to be especially useful in processes of cellular recognition. First, carbohydrates can be linked to each other in various ways, yielding a large number of isomeric structures with the potential to provide a wide variety of different recognition signals. The theoretically possible permutations of isomeric structures for hetero-oligosaccharides is vastly larger than for hetero-oligo-peptides when oligomers with identical numbers of constituents are compared. This is due to the various possible position isomers, ring isomers and anomeric isomers for carbohydrates. A second remarkable feature of carbo-hydrates relates to their hydrophilic properties. Heterosaccharides are con-jugated with either proteins or lipids and are found as cellular membrane components. These structures are a part of the biological membranes by virtue of their hydrophobic portion (either lipid or protein), which is inserted into the membrane lipid bilayer. By contrast, the heterosaccharide portion due to its hydrophilic properties will assume an orientation away from the membrane lipid environment to be the most distal molecular structure encountered by any reagent approaching a cellular surface. Both, glyco-proteins and glycolipids are asymmetrically oriented in cellular membranes. All this evidence suggests that both groups of glycoconjugates glycolipids and glycoproteins, are logical candidates to be involved in cell-surface related molecular events.

For a discussion of various glycoconjugates it is useful to identify three regions in the hetero-oligosaccharides attached to either lipids or proteins: (1) A linkage region involved in establishing a covalent linkage to form the glycoconjugate. (2) A core region contiguous to the linkage region charac-teristic for N-glycosidic type glycoproteins and glycolipids. (3) A distal region containing the reducing end(s) of the hetero-oligosaccharide. Most structural variation is in this terminal sequence of the non-reducing end(s). As discussed above, this region is the most distal part of structure associated with cellular membranes and is most likely to be the first one to be contacted when a reagent, such as a carbohydrate binding protein approach-ing the cell. Most of our interest will centre on this region of glyco-conjugates.

It appears important to describe some general structural features of glycoproteins and glycolipids to point out some of their similarities and differences. The attachment of heteropolysaccharides to the protein back-bone is either N-glycosidic or O-glycosidic and some glycoproteins (e.g. fetuin) contain both types of linkages. The core portion of carbohydrate chains in N-glycosidic heteropolymers consists mainly of N-acetyl-glucosamine and mannose moieties and has similar structure in various glycoproteins. By contrast, O-glycosidic heteropolymers do not possess a distinct core region and the 'variable' distal heteropolysaccharide is linked directly to either serine or threonine. Glycolipids have generally a core that consists of the disaccharide lactose which is linked to the lipid moiety.

Terminal sequences of glycoproteins and glycolipids show considerable variation and unique structures and may therefore qualify as 'recognition signal'. For example, there are glycolipids and glycoproteins with the sequence Gal β 1 → 4 Glc NAc β 1- linked to the core region. This structure can occur with various substitutions such as NANA or L-fucose on the galactose terminus and/or L-fucose on the N-acetylglucosamine moiety. Some of the N-glycosidic heterosaccharides of proteins contain structures terminating with NANA and are referred to as 'acidic'. Many of these structures occur also as structural analogues linked to glycolipids. Substitution of Gal β 1 → 4 Glc NAc β 1- with galactose and/or hexosamines leads to various 'neutral' heterosaccharides, many of them with blood group antigenic properties (type 1 chains). Structures devoid of sialic acid and characteristic to occur in many O-glycosidic heterosaccharides contain the β 1 → 3 isomeric structure Gal β 1 → 3 Glc NAc β 1-. Substitution with L-fucose, galactose of N-acetylgalactosamine leads to neutral heterosaccharides with ABH and Lewis antigenic properties (type 2 chains).

A third group of heterosaccharides with various substitutions on Gal β 1 → 3 Gal NAc- can be distinguished and is found in gangliosides especially as NANA substituted derivatives. These structures can also occur in glycoproteins and direct linkage to the protein without core region was observed.

Many types of identical heterosaccharide structures can be found which are linked to either proteins or lipids. The degree of similarity between glycoproteins and glycolipids appears to be a function of the sugar chain length. Short chains with only a few carbohydrate units are structurally entirely different (e.g. the core regions of glycoconjugates). By contrast, longer heterosaccharides have identical terminal structures when glycoproteins and glycolipids are compared. This may indicate that the specificity of some glycosyltransferases is limited to recognize only a short terminal sequence such as Gal β 1 → 4 Glc NAc β 1- common to glycoproteins and glycolipids; e.g. NANA will be transferred to all glycoconjugates with the above sequence by the same sialotransferase into α 2 → 3 linkage to galactose resulting in identical monosialo heterosaccharide termini. Thus, differences in either the core region or the lipid, protein portion of the acceptor are not recognized by the transferase, a fact that appears to be established for ABH blood group antigens.

The structural similarities of terminal carbohydrate sequences in glycoconjugates may not only be confined to glycosyltransferases but may also reflect selective recognition of terminal sequences in catabolic reactions. Glycosidases capable to remove sequentially specific carbohydrates from the non-reducing end(s) of heterosaccharides may act on glycolipids and glycoproteins as well. In certain glycosidase deficiency resulting in accumulation of gangliosides, glycoprotein derived heterosaccharides with identical structures to the gangliosides accumulated.

The objective of this general discussion of surface membrane-structures occurring as glycoconjugates is to point out the great similarities that were found when terminal sugar sequences in glycoproteins and glycolipids are compared. This structural similarity may be a consequence of identical specificity requirements of glycosyltransferases and glycohydrolases limited only to a short recognition sequence on the most distal part of the molecular structure and enzymatic reactions will proceed irrespective of the rest of the molecule being linked to a protein or a lipid moiety. Both types of glycoconjugates carbohydrates linked to either lipids or proteins with identical terminal heterosaccharides can serve as ligands for carbohydrate binding proteins of identical specificity. Consequently, from a structural viewpoint cell surface associated binding is shared by both lipid- and protein-linked carbohydrates.

The informational role of carbohydrates in cell surface constituents is by now well established but many aspects of specific cell recognition processes such as they occur during cellular differentiation and other cell surface related phenomena are not yet understood. Various examples of interaction between cell surface components that involve binding of carbohydrate containing structures will be presented. The choice of subjects discussed is by no means comprehensive and was primarily based on the author's desire to illustrate general principles. It should be noted that the criteria for the selection of examples undoubtedly reflect interests and limitations of the author and numerous significant contributions will not be discussed due to the large scope of the subject and space limitations.

A general survey of cellular recognition was published recently.[11]

3   EXAMPLES FOR CARBOHYDRATE RECOGNITION SYSTEMS

Select examples of carbohydrate binding proteins and their role in carbohydrate mediated cell surface recognition processes will be presented. Special emphasis will be placed upon binding specificity of the glycoconjugate ligand. Discussion of metabolic sequelae following binding will be minimal.

3.1   Recognition involving *N*-acetyl-neuraminic acid

A large number of 'acidic' heteropolymeric carbohydrates contain NANA as a constituent and are part of glycoproteins and glycolipids. Many of these 'acidic' glycoconjugates were identified as cell surface components. In all instances reported NANA occupies a terminal position of the heteropolysaccharide chains. Consequently, if indeed non-reducing carbohydrate termini serve as recognition signals, NANA qualifies very well for this function.

3.1.1   *Bacterial toxins and glycoprotein hormones*

A more detailed account of hormonal receptors and the mechanism of information transfer into the cell will be presented elsewhere in this volume

$$\text{Ceramide} \rightarrow 1\,\text{Glc}\,4 \rightarrow \beta\,1\,\text{Gal}\,4 \rightarrow \beta\,1\,\text{Gal NAc}\,3 \rightarrow \beta\,1\,\text{Gal}$$
$$3$$
$$\downarrow$$
$$\alpha\,2\,\text{NANA}$$

Figure 2. Structure of G_{M1}

(Chapter 9). A short discussion will focus on the role of NANA in the binding of bacterial toxins and glycoprotein hormones to cell receptors. Cholera toxin has been used as a model to study cellular recognition and its interaction with cellular receptors was examined extensively. The toxin is known to be composed of A and B protein subunits. The A protein contains non-identical subunits while the B protein was shown to be a pentamer of identical subunits.[12,13] A NANA containing ganglioside G_{M1} (Figure 2) was identified as the most likely cell surface glycoconjugate to be part of the membrane receptor and responsible for its specific interaction with cholera toxin. The evidence to substantiate specific binding was based upon measurement of biological response resulting in cyclic AMP formation and included various experimental approaches: Competition with exogenous gangliosides blocked the biological effect[14] and transformed cells without G_{M1} did not respond to cholera toxin.[15] When various gangliosides with structures other than G_{M1} were taken up into the cell membrane to substitute for G_{M1} no cyclic AMP formation was observed when cells were exposed to cholera toxin.[16] The specific interaction between membrane G_{M1} and cholera toxin was inferred from protection of G_{M1} against chemical modification. G_{M1} is susceptible to galactose oxidase treatment resulting in conversion of its terminal galactose to the corresponding 6-aldehydo-derivative. Similarly, NANA can be converted to a mixture of 7- and 8-aldehydoderivatives by mild periodate oxidation. Reduction with [^{3}H]-NaBH$_4$ introduces [^{3}H] covalently linked to G_{M1}, which provides a sensitive assay for accessible cell surface glycoconjugates with galactose or NANA termini. When cells were subjected to oxidation by galactose oxidase or periodate followed by [^{3}H]-NaBH$_4$, the presence of cholera toxin bound to the cell protected specifically G_{M1} while other membrane glycolipids were highly tritium labelled.[17] Cholera toxin has multiple binding sites as suggested by aggregation of G_{M1} containing erythrocytes. Direct binding studies indicate one binding site for each B subunit, thus cholera toxin has five binding sites. The binding of G_{M1} results in a conformational change of the B subunits as evidenced by fluorescence studies[18] and the oligosaccharide alone is as effective as intact G_{M1} to cause the fluorescence change.[19] The specific binding of cholera toxin to receptor G_{M1} is accompanied by a conformational change of the toxin resulting in a perturbation of the cellular membrane. This in turn is thought to facilitate dissociation and penetration of A subunits which triggers activation of adenylcyclase.

The mechanism of the just described action of cholera toxin upon cell receptors appears to be similar to the interaction between several glycoprotein hormones and their appropriate target organs. The glycoprotein hormones thyrotropin, lutropin and human chorionic gonadotropin consist of two types of subunits, α and β, and analogous to the action of cholera toxin, it was though that the β subunit confers target tissue specificity. Solubilization of thyrotropin receptor from bovine thyroid plasma membranes permitted identification of a glycoprotein with thyrotropin binding activity.[20] This observation prompted investigators to propose a two component receptor model containing a glycoprotein as well as a ganglioside component.[21] The functional receptor may be even more complex consisting of a glycoprotein, glycolipid and membrane phospholipids.[22,23] It is thought that the receptor glycoprotein may be the primary binding site for the hormone resulting in perturbation of receptor lipid components which is followed by message transmission. A clear correlation between target specificity of hormones and carbohydrate structures of various target receptors have yet to be established. The difficulties encountered to establish a correlation between unique heterosaccharides of membrane glycoconjugates and their role in specific binding to the discriminator component of membrane receptors is apparent from the examples just described.

3.1.2 *Influenza virus receptor on erythrocytes*

An early example for the role of carbohydrates as recognition signal are the by now classical studies by Burnet, Gottschalk and Hirst.[24] These investigators demonstrated the absolute requirement for the presence of terminal sialic acid residues on the erythrocyte surface for the attachment of influenza virus. The attachment of virus particles to the erythrocyte surface was found to be accompanied by agglutination of erythrocytes (virus hemagglutination). The major sialoglycoprotein of the erythrocyte membrane, glycophorin, was identified as the receptor site for viral attachment. Virus attachment to the target cell is a prerequisite for the infection of host cells by virus particles.

The detailed analysis of glycophorin[25] reveals a complex structure and the presence of two types of heterosaccharides. The majority of heterosaccharide chains (fifteen), are in O-glycosidic linkage while one chain per mole of glycophorin is linked in an N-glycosidic linkage. The O-glycosidic chains are directly attached to the protein (Figure 3). By contrast, the

$$\alpha\,2\,\text{NANA}$$
$$\uparrow$$
$$6$$
$$\text{Thr/Ser}\ O \rightarrow 1\ \text{Gal NAc}\ 3 \rightarrow \beta\ 1\ \text{Gal}\ 3 \rightarrow 2\ \alpha\ \text{NANA}$$

Figure 3. Structure of O-linked oligosaccharides in glycophorin

$$\text{Asn} \rightarrow \text{Core} \rightarrow \text{Glc NAc } 4 \rightarrow \beta\,1\,\text{Gal (3)6} \rightarrow \alpha\,2\,\text{NANA}$$

Figure 4. Structure of *N*-linked oligosaccharide in glycophorin

N-glycosidic heteropolymer contains in addition to the terminal sequence a core region (Figure 4). As can be seen from these structures, sialic acid can be linked $\alpha\,2 \rightarrow 3$ or $\alpha\,2 \rightarrow 6$ to galactose while the linkage to *N*-acetylgalactosamine was found to be only $\alpha\,2 \rightarrow 6$. Recently, the purification and isolation of three specific sialotransferases was described[26,27] each one capable to link sialic acid to one of the just mentioned terminal positions. Using asialoglycophorin as starting material in combination with these specific sialotransferases, the three possible monosialoposition isomers of glycophorin were prepared and tested for their ability to serve as a receptor for various viruses. It was found that Sendii, Newcastle disease and influenza/equine I virus have absolute specificity for heterosaccharides terminated with NANA $\alpha\,2 \rightarrow 3$ Gal $\beta\,1 \rightarrow 3$ Gal NAc.... Myxovirus hemagglutinins were shown to discriminate NANA position isomers as well as *N*- and *O*-acyl substituents such as *N*-glycolylneuraminic acid or 4-*O*-acetyl NANA.[28] From these studies it is apparent that virus surface receptors display structural discrimination of sialic acid position isomers and various *N*- and *O*-substituents.

3.1.3 *MN-blood group antigens*

The example just described is an illustration that presence, linkage and substitution of sialic acid determines specific virus attachment to red blood cells and glycophorin, the major glycoprotein of erythrocytes is the cellular receptor responsible for it. The same glycoprotein is also known to carry MN blood antigens. As was mentioned earlier, antigen–antibody interaction is used in many instances as a model to explain ligand–discriminator interaction. The MN blood group antigens can serve as an example to illustrate the dangers encountered in the interpretation of binding data. For a long time the importance of terminal sialic acid for the expression of MN blood group antigens was known and conversion of N to M specificity was considered to be related to sialic acid content.[29] Recent experiments demonstrated that conversion of N- and M-antigen is unrelated to sialic acid termini, since the serotype of erythrocytes treated with neuraminidase and resilylated with specific sialotransferases corresponded always to the phenotype from which the cells originated, irrespective of the type of sialotransferase used.[30] This indicates that neuraminidase treated erythrocytes retain their M or N specificity and indeed it was found the difference between these blood group antigens is located in a difference of two amino acids in the polypeptide chain. The role of sialic acid was therefore suggested to be involved in maintenance of proper antigen conformation and absence of the charged sialic acid results in a conformational change that prevents antigen–antibody binding. This example is applicable to ligand–discriminator interaction and illustrates the difficulties one may encounter to establish the role of sialic acid as a signal for cellular recognition.

3.2 Recognition involving mannose-6-phosphate

3.2.1 *Uptake of lysosomal enzymes by fibroblasts*

Extensive discussion of receptor mediated uptake of lysosomal enzymes with mannose-6-phosphate as recognition signal will be presented elsewhere in this volume (Chapter 8). A few remarks will focus on the role of carbohydrates in this process and will provide some of the experimental evidence that supports the notion that covalently linked mannose-6-phosphate mediates uptake of hydrolytic enzymes into lysosomal particles. Studies on the catabolism of proteoglycans revealed the stepwise and ordered degradation of these macromolecules as a prerequisite for normal steady state levels of proteoglycans in tissues. Several genetic disorders were found to have a specific catabolic enzyme missing (e.g. α-L-iduronidase) resulting in accumulation of mucopolysaccharides.[31] Fibroblasts cultured from patients with such genetic defects were shown capable to restore normal levels of mucopolysaccharides by the addition of 'corrective factors' to the medium. These 'corrective factors' were purified and isolated from the urine of normal individuals.[32] The proteins taken up by the cells were identified as the catabolic enzyme causing the genetic defect.

Lysosomal enzymes like α-L-iduronidase and β-D-glucuronidase appear to have mannose-6-phosphate covalently attached to these enzymes. The evidence that this function as a recognition signal is still indirect but fairly strong: uptake of corrective enzymes is strongly inhibited by mannose-6-phosphate, fructose-1-phosphate which has structural resemblance to mannose-6-phosphate, mannose-6-sulphate,[33,34] and phosphorylated mannans.[35] Also, neoglycoproteins such as albumin covalently linked to mannose-6-phosphate was found to be inhibitory.[36] Naturally occurring glycoproteins isolated from urine were found to be potent inhibitors of enzyme uptake, and as expected, mild periodate oxidation or preincubation with alkaline phosphatase abolished competition of enzyme uptake.

Evidence for the presence of a mannose-6-phosphate receptor in the fibroblast membrane is mostly indirect and attempts for its purification and isolation are in progress.

A suggested function for the presence of mannose-6-phosphate as a covalently attached component of these catabolic enzymes suggests that during the synthesis of many secretory proteins specific glycosylation occurs in the Golgi apparatus which provides an effective way for selective uptake by certain cells (e.g. fibroblasts) and subcellular organelles such as lysosomes.[37] Recent work on the uptake and fate of hydrolytic enzymes indicates that the properties of the mannose-6-phosphate system just described appears to be more complex. As expected, mannose-6-phosphate was found to be a competitive inhibitor of both enzyme uptake and enzyme binding. Surprisingly, mannose-6-phosphate also facilitates dissociation of bound enzyme in an as yet unexplained way.

Experiments that were carried out under conditions that exclude *de novo* biosynthesis of proteins showed a continued uptake of enzyme, by far exceeding the amount of receptors present on the fibroblast surface as measured by direct binding. Excluding resynthesis of receptors during this uptake process, it is postulated that during internalization the enzyme–receptor complex dissociates and free receptor is recycled to the plasma membrane.[38] Another important feature that could explain various metabolic fates of hydrolases relates to the fact that the recognition signal of mannose-6-phosphate, can result in quite different affinity to the receptor as a function of the number of and/or steric position of sugar-phosphates residues. In model experiments using phosphomannosyl fragments of various size a large potentiation of the inhibition of enzyme uptake by high molecular fragments was reported as compared to a pentamer[35] implicating multivalent attachment to the receptor as an important factor for binding to the receptor.

3.3   Recognition involving galactose

3.3.1   *Uptake of asialoglycoproteins by the hepatic binding protein*

One of the best documented studies of carbohydrate recognition was described by Ashwell and Morell.[39] It is the only example where detailed molecular information is available for both the receptor protein (hepatic binding protein) as well as for the carbohydrate ligand (asialoglycoprotein) and interestingly, both components turned out to be glycoproteins. The fundamental observation was that only glycoproteins with all their sialic acid termini intact remain in the circulatory system with a half life time of almost 60 hours, while asialoglycoproteins are removed from circulation within minutes. The initial experiments were carried out with ceruloplasmin, a copper containing glycoprotein with 10 NANA termini per mole of protein.[40] The rapid uptake was demonstrated to be generally applicable to a wide variety of asialoglycoproteins. The binding of circulating asialoglycoproteins to the hepatic receptor was attributed to galactose, the penultimate carbohydrate of glycoproteins. This interpretation was experimentally verified by competition with oligomeric structures containing galactose termini and by chemical modification of the heterosaccharide structure of asialo-glycoproteins. For example, galactose oxidase treatment of removal of galactose by β-galactosidase abolished the binding phenomenon.[41] A quantitative assessment of the number of galactose termini required to cause binding of asialoceruloplasmin revealed that only 2 out of the theoretically possible 10 galactose residues have to be unmasked to be removed from circulation.[42] The locus of binding was identified as the liver plasma membranes.[43] This was followed by complete degradation to constituent amino acids and carbohydrates in lysosomal particles.[44] The hepatic binding protein was identified to be a sialoglycoprotein and preparation of asialo-binding protein resulted in complete loss of binding activity. When the galactose residues of the hepatic binding protein were either removed by enzymatic hydrolysis or were reconverted with CMP-NANA and sialo-transferase to the native protein, the original binding activity was completely restored.[45,46]

The hepatic receptor responsible for the selective binding of asialoglyco-proteins was isolated by affinity chromatography[47] and was purified to homogeneity. The proteins had the ability to form aggregates with itself. Identification and structural analysis of the carbohydrate moiety showed it to contain the usual terminal sequence sialic acid–galactose–hexosamine.[48] This structure is in agreement with the finding mentioned above that asialo-binding protein is inactive by recognizing and binding its own galactose residues.

The purified hepatic binding protein caused aggregation of erythrocytes[49] and was able to elicit mitogenic response of lymphocytes.[50] Recognition and binding of specific carbohydrate residues, haemagglutination, mitogenic response of lymphocytes all these properties are usually associated with lectins which were considered to be unique proteins found in plants only. These findings prompted Ashwell to propose that the use of the term 'lectin' should be extended to all biological cellular systems[51] and he suggested a more general function of lectins during differentiation and growth of cellular systems.

This concept of general significance of lectin like receptors on cell surfaces was further strengthened when subcellular organelles such as membranes of the Golgi apparatus and lysosomes were found to contain the identical binding protein identified in liver plasma membranes. Immunological methods and binding specificity were used to establish identity between the binding proteins located in different cellular compartments.[52]

A sensitive competition assay system using [^{125}I]-asialoorosomucoid as the ligand and various unlabelled glycoproteins as competitors was employed to measure binding quantitatively[43,53] using this assay system, binding was found to be calcium dependent with highest affinity to α-methyl-N-acetyl-D-galactosamine and in decreasing affinity followed by β-methyl-N-acetyl-D-galactosamine, β-methyl-D-galactose. All glycosides tested had the pyranose configuration. Surprisingly, the hepatic binding protein can not discriminate between D-galacto and D-gluco- configurations.[54]

The occurrence of apparently the same binding protein in various membranes of cellular organelles raised the possibility of binding protein transfers from the plasma membrane to lysosomal particles. This provides for the first time a rational molecular basis to explain receptor mediated pynocytosis. The internalization of asialoglycoprotein facilitated by specific binding to the plasma membrane binding protein is much more effective and selective than simple trapping within pinocytotic vesicles. To test the orientation of the binding protein in organelles advantage was taken of antibodies directed against the hepatic receptor protein to block binding of [^{125}I]-asialoorosomucoid. All the data obtained[55] were consistent with the orientation of a hepatic receptor oriented at the external cytosolic surface of lysosomal membranes, while membranes from the Golgi apparatus had the binding protein oriented on the inner luminal surface. At present the exact fate of the hepatic binding protein during pinocytosis is subject to speculation and one attractive possibility consistent with other data is that the internalized receptor ligand complex undergoes a reorientation after insertion into the lysosomal particle whereby the binding protein can function as a transport vehicle for asialoglycoprotein while the ligand is degraded inside the lysosomal particle. From all the evidence, it appears that the role of the hepatic binding protein mediates the physiological turnover of circulatory glycoproteins. It should be noted that the concepts illustrated by the elegant studies of Ashwell and his coworkers has far-reaching consequences into many aspects of cellular physiology.

3.3.2 *Specific bacterial adherence to host tissues mediated by lectin interaction*

A system analogous to the one just described but with quite different physiological objectives provides a rational basis to explain specific adherence of certain bacteria to mucosal surfaces. The involvement of lectin like binding proteins located on bacterial surfaces were shown to have the ability to recognize specific sugar residues on the host tissue cell surface. This mechanism of adherence is of special interest to explain the specific attachment of bacteria in areas like the oral cavity where adhesion to host tissue is a prerequisite to initiate bacterial colonization and propagation. Interest to study the molecular events that account for specific adherence of *Actinomyces viscosus* to oral cell surfaces was prompted by several observations. These microorganisms are considered to be important factors in the development of dental plaque, dental caries and periodontal disease[56] produce extracellular neuraminidase[57] and were shown to agglutinate human A, B and O erythrocytes.[58]

It is generally accepted that interaction between receptor and ligand that involves binding of a lectin to a carbohydrate ligand located in a surface structure is most readily observed and detected by studying cell agglutination. A few words of caution concerning the study of cell–cell interaction using aggregation techniques are in order. The apparently simple phenomenon of visible cell aggregation is complex and is affected by various factors many of them poorly understood. Therefore the study of binding and specificity characteristics of a bacterial lectin is just one of the many factors that influence cell–cell aggregation. For example, cell surface properties such as number, distribution and accessibility of receptor sites, mobility of interacting sites and membrane fluidity all play an important role in cell–cell aggregation. As one may expect, the number of binding sites on the cell surface will be of importance and several investigators reported that a critical minimal number of cell surface receptors and carbohydrate ligands must interact before cell aggregation does occur.[59,60] This threshold response is enhanced by a cooperative effect that takes place when multiple binding occurs.[61]

Using A, B or O erythrocytes as a model system to study lectin mediated interaction of *A. viscosus* with mammalian cell surfaces the action of bacterial neuraminidase and the progress of haemagglutination were found to be related and interdependent. For both, the progress of bacteria mediated haemagglutination and extracellular bacterial neuraminidase activity 2-deoxy-2,3-dehydro-*N*-acetyl-neuraminic acid was demonstrated to be a competitive inhibitor.[58] A further observation relates to the fact that haemagglutination was reversed by 0.01 M lactose. This was considered to be suggestive evidence for binding between a bacterial lectin and galactose termini of RBC receptors such as asialoglycoproteins or asialoglycolipids. Electron microscopic studies demonstrated the location of the binding activity on fimbriae of *A. viscosus*.[62]

Quantitative studies on neuraminidase treated erythrocytes indicated that only a small amount of the total cell surface neuraminic acid has to be released to result in cell aggregation. Attempts to modify the exposed galactose residues by treatment with galactose oxidase or removal with β-galactosidase did not abolish agglutination. This suggested that either galactose termini were not participating in the cell aggregation or alternatively that neither of the reactions employed to modify or remove galactose were quantitative and sufficient galactose residues remained to cause aggregation.

This difficulty was overcome by using instead of RBC Latex beads that were coated with glycoproteins of well defined structure such as α_1-acid glycoprotein or fetuin. In this model system, neuraminidase dependent aggregation between *A. viscosus* cells and glycoprotein coated latex beads was demonstrated. Unmasked galactose residues converted by galactose oxidase to the 6-aldelydo-derivative were no longer capable to cause cell

Table 1. Examples for carbohydrate mediated recognition phenomena

Receptor recognition signal	Organism cell type	Recognition phenomenon or physiological response	Reference
N-Acetylneuraminic acid	Virus	Attachment to erythrocyte surface	64
N-Acetylneuraminic acid	Various mammalian target cells	Hormone binding results in intracellular cyclic AMP production	4
D-Mannose-6-phosphate	Human fibroblasts	Uptake of lysosomal glyco-hydrolases—pinocytosis	32, 33
D-Mannose	Bacteria	Adherence to mammalian cells	65
D-Mannose or N-acetyl-D-glucosamine	Mammalian retitculo-endothelial cells	Uptake of glycoproteins?	66, 67
D-Galactose or N-acetyl-D-galactosamine	Bacteria	Adherence to mammalian cells	57
D-Galactose or N-acetyl-D-galactosamine	Eucaryotic cells slime molds	Intracellular recognition and differentiation	68
D-Galactose	Electric organ of eel	Intracellular recognition and differentiation	69
D-Galactose	Avian brain	Development and cell differentiation	70
D-Galactose	Avian embryo muscle	Development and cell differentiation	71
D-Galactose	Mammalian heart and lung	Development and cell differentiation	72
D-Galactose	Mammalian hepatocyte	Uptake of asialoglycoproteins pinocytosis?	50
N-Acetyl-D-glucosamine	Avian hepatocyte	Uptake of agalactohepatocytes pinocytosis?	73
L-Fucose	Mammalian hepatocyte	Uptake of transferrin?	74

aggregation, while addition of $NaBH_4$ to this suspension of cells caused immediate aggregation. This clearly implicated galactose termini as ligand in the lectin binding. The same system of latex beads coated with various glycoproteins was used to study some properties of the *A. viscosus* lectin. In addition to the recognition of β-D-galacopyranoside residues, α-*N*-acetyl-D-galactosamine was found to bind with even greater affinity and binding was dependent upon the presence of calcium.[63] The evidence presented so far is consistent with the view that *A. viscosus* interaction with mammalian cells requires two sequential and distinct steps: first, removal of terminal sialic acid to unmask carbohydrate termini with galactose configuration and second binding of the bacterial lectin located on surface fimbriae of the bacteria to the galactose containing receptor. Thus, the production of extracellular neuraminidase is a prerequisite for *A. viscosus* to attach to mammalian cells. When different *Actinomycetes* strains were examined for their neuraminidase activity, all the strains isolated from animal sources were found to be neuraminidase producing while isolates from other sources were non-producers. Neuraminidase production was found to be inducible.[58] Summarizing the information presented on biological recognition involving D-galactose, it is remarkable to note that bacterial and mammalian systems employ identical molecular events to carry out different physiological functions. Removal of terminal sialic acid and exposure of penultimate galactose residues followed by binding leads to internalization and catabolism of glycoproteins while the same sequence of events results in recognition mediated adherence of bacteria to mammalian tissues. These findings support the notion of the ubiquitous occurrence and physiological significance of lectins in all biological cellular systems. It also poses the interesting question as to whether these quite different biological recognition phenomena are just superfically similar or are a reflection of a common evolutionary origin.

4 BIOLOGICAL SYSTEMS WITH ESTABLISHED CARBOHYDRATE RECEPTOR RECOGNITION

There are many well documented examples of cell surface related recognition phenomena where carbohydrates are known to play a significant role in the specificity of this process that were not described in the text. Some additional carbohydrate mediated recognition systems are listed in Table 1.

REFERENCES

1. Fischer, E. (1894). Einfluss der konfiguration and die wirkung der enzyme, *Ber. Deutsch. Chem. Ges.*, **27**, 2985–2993.
2. Koshland, D. E., and Neet, K. E. (1968). The catalytic and regulatory properties of enzymes. *Ann. Rev. Biochem.*, **37**, 359–410.

3. Burgen, A. S. V., Roberts, G. C. K., and Feeney, J. (1975). Binding of flexible ligands to macromolecules. *Nature*, **253**, 753–755.

4. Kohn, L. D. (1978). Relationships in the structure and function of receptors for glycoprotein hormones, bacterial toxin and interferon. In *Receptors and Recognition* (Cuatrecasas, P., and Greaves, M. F., eds.), Chapman and Hall, London, Series A, Vol. 5, pp. 134–212.

5. Brady, R. O., and Fisherman, P. H. (1979). Biotransducers of membrane mediated information. *Adv. Enzymol.*, **50**, 303–323.

6. Flowers, H. M., and Sharon, N. (1979). Glycosidases—Properties and application to the study of complex carbohydrates and cell surfaces. *Adv. Enzymol.*, **48**, 29–95.

7. Shur, B. D., and Roth, S. (1975). Cell surface glycosyltransferases. *B.B.A.*, **415**, 473–512.

8. Glaudemans, C. P. T. (1975). The interaction of homogeneous murine myleoma immunoglobulins with polysaccharide antigens. *Adv. Carbohydrate Chem. Biochem.*, **31**, 313–346.

9. Smith, D. F., and Ginsburg, V. (1979). Antibodies against sialyloligosaccharides coupled to protein. *J. Biol. Chem.*, **255**, 55–59.

10. Goldstein, I. J., and Hayes, C. E. (1978). The lectins: carbohydrate-binding proteins of plants and animals. *Adv. Carbohydr. Chem. and Biochem.*, **35**, 127–340.

11. Frazier, W., and Glaser, L. (1979). Surface components and cell recognition. *Am. Rev. Biochem.*, **48**, 491–523.

12. Lai, C. Y. (1977). Determination of the primary structure of cholera toxin B subunit. *J. Biol. Chem.*, **252**, 7249–7256.

13. Kurosky, A., Markel, D. E., and Peterson, J. W. (1977). Covalent structure of the β chain of cholera enterotoxin. *J. Biol. Chem.*, **252**, 7257–7264.

14. Vanheyningen, W. E., Carpenter, C. C. L., Pierce, N. F., and Greendugh, W. B. III (1971). Deactivation of cholera toxin by ganglioside. *J. Infect. Dis.*, **124**, 415–418.

15. Moss, J., Fishman, P. H., Manganiello, V. C., Vaughan, M., and Brady, R. O. (1976). Functional incorporation of ganglioside into intact cells: Induction of choleragen responsiveness. *Proc. Natl. Acad. Sci., U.S.*, **73**, 1034–1037.

16. Fishman, P. H., Moss, J., and Vaughan, M. (1976). Uptake and metabolism of gangliosides in transformed mouse fibroblasts. Relationship of ganglioside structure to choleragen response. *J. Biol. Chem.*, **251**, 4490–4494.

17. Moss, J., Manganiello, V. C., and Fishman, P. H. (1977). Enzymatic and chemical oxidation of gangliosides in cultured cells: effects of choleragen. *Biochemistry*, **16**, 1876–1881.

18. Moss, J., Osborne, J. C., Jr., Fishman, P. H., Brewer, H. B., Jr., Vaughan, M., and Brady, R. O. (1977). Effect of gangliosides and substrate analogous on the hydrolysis of nicotinamide adenine dinucleotide by choleragen. *Proc. Natl. Acad. Sci. U.S.*, **74**, 74–78.

19. Fishman, P. H., Moss, J., and Osborne, J. C., Jr. (1978). Interaction of choleragen with the oligosaccharide of ganglioside G_{M1}: Evidence for multiple oligosaccharide binding sites. *Biochemistry*, **17**, 711–716.

20. Tate, R. L., Holmes, J. M., Kohn, L. D., and Winand, R. J. (1975). Characteristics of a solubilized thyrotropin receptor from bovine thyroid plasma membranes. *J. Biol. Chem.*, **250**, 6527–6533.

21. Kohn, L. D. (1977). Characterization of the thyrotropin receptor and its involvement in Graves disease. In *Horizons in Biochemistry and Biophysics*, Vol.

3 (Quagliariello, E., ed.), pp. 123–163, Addison-Wesley Publishing Co., Reading, Massachusetts.

22. Omodeo-Sale, F., Brady, R. O., and Fishman, P. H. (1978). Effect of thyroid phospholipids on the interaction of thyrotropin with thyroid membranes. *Proc. Natl. Acad. Sci. U.S.*, **75,** 5301–5305.

23. Sharom, F. J., and Grant, C. W. M. (1978). A model for ganglioside behaviour in cell membranes. *Biochem. Biophys. Acta,* **507,** 280–293.

24. Burnet, F. M. (1951). Mucoproteins in relation to virus action. *Physiol. Rev.,* **31,** 131–150.

25. Tomita, M., and Marchesi, V. T. (1975). Amino acid sequence and oligosaccharide attachment sites of human erythrocyte glycophorin. *Proc. Natl. Acad. Sci. U.S.,* **72,** 2964–2968.

26. Sadler, J. E., Rearick, J. I., Paulson, J. C., and Hill, R. L. (1979a). Purification to homogeneity of a β-galactoside $\alpha 2 \rightarrow 3$ sialytransferase and partial purification of an α-N-acetyl-galactosaminide $\alpha 2 \rightarrow 6$ sialytransferases from porcine submaxillary glands. *J. Biol. Chem.,* **254,** 4434–4443.

27. Rearick, J. R., Sadler, J. E., Paulson, J. C., and Hill, R. L. (1979). Enzymatic characterization of β-D-galactoside $\alpha 2 \rightarrow 3$ sialyltransferases from porcine submaxillary gland. *J. Biol. Chem.,* **254,** 4444–4451.

28. Paulson, J. C., Sadler, J. E., and Hill, R. L. (1979). Restoration of specific myxovirus receptors to asialoerythrocytes by incorporation of sialic acid with pure sialyltransferases. *J. Biol. Chem.,* **254,** 2120–2124.

29. Springer, G. F., and Desai, P. R. (1975). Human blood-group MN and precursor specificities: structural and biological aspects. *Carbohydrate Res.,* **40,** 183–192.

30. Sadler, J. E., Paulson, J. C., and Hill, R. L. (1979). The role of sialic acid in the expression of human MN blood group antigens. *J. Biol. Chem.,* **254,** 2212–2119.

31. McKusick, V. A., Neufeld, E. F., and Kelly, T. E. (1978). The mucopolysaccharide storage diseases. In *The Metabolic Basis of Inherited Diseases* (Stanbury, J. B., Wyngaarden, J. B., and Fredrickson, D. S., eds.), pp. 1282–1307, McGraw-Hill Book Company, New York.

32. Neufeld, E. F., and Cantz, M. J. (1971). Corrective factors for inborn errors of mucopolysaccharide metabolism. *Ann. N.Y. Acad. Sci.,* **179,** 580–587.

33. Kaplan, A., Achord, D. T., and Sly, W. S. (1977). Phosphohexosyl components of a lysosomal enzyme are recognized by pinocytosis receptors on human fibroblasts. *Proc. Natl. Acad. Sci. U.S.,* **74,** 2026–2030.

34. Sando, G. N., and Neufeld, E. F. (1977). Recognition and receptor mediated uptake of a lysosomal enzyme, α-L-iduronidase, by cultured fibroblasts. *Cell,* **12,** 619–627.

35. Fischer, H. D., Natowicz, M., Sly, W. S., and Bertthaver (1980). Fibroblast receptor for lysosomal enzymes mediates pinocytosis of multivalent phosphomannan fragment. *J. Cell Biol.,* **84,** 77–86.

36. Karson, E. M., Neufeld, E. F., and Sando, G. N. (1980). p-Isothiocyanatophenyl 6-phospho α-D-mannopyranoside coupled to albumin. A model compound recognized by the fibroblast lysosomal enzyme uptake system. 2-Biological properties. *Biochem.,* **19,** 3856–3860.

37. Hickman, S., and Neufeld, E. F. (1972). A hypothesis for I-cell disease: defective hydrolases that do not enter lysosomes. *Biochem. Biophys. Res. Comm.,* **49,** 992–999.

38. Rome, L. H., Weissmann, B., and Neufeld, E. F. (1979). Direct demonstration of binding of a lysosomal enzyme α-L-iduronase, to receptors on cultures fibroblasts. *Proc. Natl. Acad. Sci. U.S.,* **76,** 2331–2334.

39. Ashwell, G., and Morell, H. G. (1974). The role of surface carbohydrates in the hepatic recognition and transport of glycoproteins. *Adv. Enzymol.*, **41,** 99–128.

40. Jamieson, G. A. (1965). Studies on glycoproteins. I. The carbohydrate portion of human ceruloplasmin. *J. Biol. Chem.*, **240,** 2019–2027.

41. Morell, A. G., Van den Hamer, C. J. A., Scheinberg, I. H., and Ashwell, G. (1966). Physical and chemical studies on ceruloplasmin IV. Preparation of radioactive, sialic acid-free ceruloplasmin labeled with tritium on terminal D-galactose residues. *J. Biol. Chem.*, **241,** 3748–3749.

42. Van den Hamer, C. J. A., Morell, A. G., Scheinberg, I. H., Hickman, J., and Ashwell, G. (1970). Physical and chemical studies on ceruloplasmin IV. The role of galactosyl residues in the clearance of ceruloplasmin from the circulation. *J. Biol. Chem.*, **245,** 4397–4402.

43. Pricer, W. E., Jr., and Ashwell, G. (1971). The binding of desialylated glycoproteins by plasma membranes of rat liver. *J. Biol. Chem.*, **246,** 4825–4833.

44. Gregoriadis, G., Morell, A. G., Sternlieb, I., and Scheinberg, I. H. (1970). Catabolism of desialylatede ceruloplasmin in the liver. *J. Biol. Chem.*, **245,** 5833–5837.

45. Paulson, J. C., Hill, R. L., Tanabe, T., and Ashwell, G. (1977). Reactivation of asialo-rabbit binding protein by resilylation with β-D-galactoside $\alpha 2 \rightarrow 6$ sialyltransferase. *J. Biol. Chem.*, **252,** 8624–8628.

46. Stockert, R. J., Morell, A. G., and Scheinberg, I. H. (1977). Hepatic binding protein: The protective role of its sialic acid residues. *Science*, **197,** 667–668.

47. Hudgin, R. L., Pricer, W. E., Jr., Ashwell, G., Stockert, R. J., and Morell, A. G. (1974). The isolation and properties of a rabbit liver binding protein specific for asialoglycoproteins. *J. Biol. Chem.*, **249,** 5536–5543.

48. Kawasaki, T., and Ashwell, G. (1976). Chemical and physical properties of an hepatic membrane protein that specifically binds asialoglycoproteins. *J. Biol. Chem.*, **251,** 1296–1302.

49. Stockert, R. J., Morell, A. G., and Scheinberg, I. H. (1974). Mammalian hepatic lectin. *Science*, **186,** 365–366.

50. Novogrodsky, A., and Ashwell, G. (1977). Lymphocyte mitogenesis induced by a mammalian liver protein that specifically binds desialylated glycoproteins. *Proc. Natl. Acad. Sci. U.S.*, **74,** 676–678.

51. Ashwell, G. (1977). A functional role for lectins. *Trends Biol. Sci.*, **2,** N186.

52. Pricer, W. E., Jr., and Ashwell, G. (1976). Subcellular distribution of a mammalian hepatic binding protein specific for asialoglycoproteins. *J. Biol. Chem.*, **251,** 7539–7544.

53. Van Lenten, L., and Ashwell, G. (1972). The binding of desialylated glycoproteins by plasma membranes of rat liver: development of a quantitative inhibition assay. *J. Biol. Chem.*, **247,** 4633–4640.

54. Stowell, C. P., and Lee, Y. C. (1978). The binding of D-glucosyl-neoglycoproteins to the hepatic asialoglycoproteins receptor. *J. Biol. Chem.*, **253,** 6107–6110.

55. Tanabe, T., Pricer, W. E., Jr., and Ashwell, G. (1979). Subcellular membrane topology and turnover of a rat hepatic binding protein specific for asialoglycoproteins. *J. Biol. Chem.*, **254,** 1038–1043.

56. Van der Hoeven, J. S., Miks, F. H., Konig, K. G., and Plassghaert, A. J. M. (1974). Plaque formation and caries in gnotobiotic and SPF rats with *Actinomyces viscosus*. *Caries Res.*, **8,** 211–223.

57. Gabriel, O. (1978). Cell wall associated neuraminidase in strains of *Actinomyces viscosus* T14. *J. Dent. Res.*, **57,** 202.

58. Costello, A. H., Cisar, J. O., Kolenbrander, P. E., and Gabriel, O. (1979). Neuraminidase dependent hemagglutination of human erythrocytes by human strains of *Actinomyces viscosus* and *Actinomyces naeslundii*. *Infection and Immunity*, **26,** 563–572.

59. Rando, R. R., Orr, G. A., and Bangerter, F. W. (1979). Threshold effects on the concavalin A mediated agglutination of modified erythrocytes. *J. Biol. Chem.*, **254,** 8318–8323.

60. Weigel, P. H., Schnaar, R. L., Kuhlen Schmidt, M. S., Schmell, E., Lee, R. T., Lee, Y. C., and Roseman, S. (1979). Adhesion of hepatocytes to immobilized sugars. A threshold phenomenon. *J. Biol. Chem.*, **254,** 10830–10838.

61. Crothers, D. M., and Metzger (1972). The influence of polyvalency on the binding properties of antibodies. *Immunochemistry*, **9,** 341–351.

62. Cisar, J. O., Vatter, A. E., and McIntire, F. C. (1978). Identification of the virulence associated antigen on the surface fibrils of *Actinomyces viscosus* T14. *Infection and Immunity*, **19,** 312–319.

63. Costello, A. H. (1980). Neuraminidase dependent cell recognition. Dissertation, Department of Biochemistry, Georgetown University, Washington, D.C.

64. Marchesi, V. T., and Andrews, E. P. (1971). Glycoproteins: isolation from cell membranes with lithium diiodosalicylate. *Science*, **174,** 1247–1248.

65. Eshdat, Y., Ofek, I., Yashauv-Gan, Y., Sharon, N., and Mirelman, D. (1978). Isolation of a mannose-specific lectin from *Escherichia coli* and its role in the adherence of the bacteria to epithelial cells. *Biochem. Biophys. Res. Comm.*, **85,** 1551–1559.

66. Achord, D. T., Brot, F. E., Bell, C. E., and Sly, W. S. (1978). Human S-glucuromidase: *in vivo* clearance and *in vitro* uptake by a glycoprotein recognition system on reticuloendothelial cells. *Cell*, **15,** 269–278.

67. Stahl, P. D., Rodman, J. S., Miller, M. J., and Schlesinger, P. H. (1978). Evidence for receptor-mediated binding of glycoproteins, glycoconjugates and lysosomalglycosidases by alveolar macrophages. *Proc. Natl. Acad. Sci. U.S.*, **75,** 1399–1403.

68. Frazier, W. A., Rosen, S. D., Reitherman, R. W., and Barondes, S. H. (1975). Purification and comparison of two developmentally regulated lectins from *dicotyostelium discoideum*. *J. Biol. Chem.*, **250,** 7714–7721.

69. Teichberg, V. I., Silman, I., Beitsch, D. D., and Resheff, C. (1975). A β-D-galactoside binding protein from electric organ tissue of *Electrophorus electricus*. *Proc. Natl. Acad. Sci. U.S.*, **72,** 1383–1387.

70. Gremo, F., Kobilerd, and Barondes, S. H. (1978). Distribution of an endogenous lectin in the developing chick optic tectum. *J. Cell. Biol.*, **79,** 491–499.

71. Den, H., and Malinzak, D. A. (1977). The isolation and properties of a β-D-galactoside specific lectin from chick embryo thigh muscle. *J. Biol. Chem.*, **252,** 5444–5448.

72. DeWaard, A., Hickman, S., and Kornfeld, S. (1976). Isolation and properties of β-galactoside binding lectins of calf heart and lung. *J. Biol. Chem.*, **251,** 7581–7587.

73. Lunney, J., and Ashwell, G. (1976). A hepatic receptor of avian origin capable of binding specifically modified glycoproteins. *Proc. Natl. Acad. Sci. U.S.*, **73,** 341–343.

74. Prieels, J. P., Pizzo, S. V., Glascow, L. R., Paulson, J. C., and Hill, R. L. (1978). Hepatic receptor that specifically binds oligosaccharides containing fucosyl $\alpha 1 \rightarrow$ 3 N-acetyl glycosamine linkages. *Proc. Natl. Acad. Sci. U.S.*, **75,** 2215–2219.

8 Receptor-mediated alterations in membrane potential and internal pH as intracellular signals

E. F. Grollman

1 INTRODUCTION

This chapter will stress the possibility that electrochemical ion gradients play a role in receptor-mediated ligand signalling and the functional response of target cells. Only a few of the many examples will be given where an alteration in resting membrane potential and an effect on ion fluxes has been an integral part of the interaction of a ligand with its receptor. These examples stress the point that many of the properties of excitable cells are in part shared by peripheral cells which includes: stimulus induced changes in membrane potential, electrical coupling between cells and membrane potential dependent secretory release. Non-neuronal cells even have 'sodium channels' that can interact with neurotoxins like veratridine and tetrodotoxin to affect sodium entry.[1,2]

An alteration of membrane potential induced by a ligand-receptor interaction can affect many diverse transport functions which in turn can alter the internal solute concentration and the uptake of nutrients. Since the ionic environment optimizes the enzymatic and metabolic operations within the cell, effects on protein and DNA synthesis may be consequences of changes in the intracellular solute composition initiated at the time of ligand interactions with its receptor.

2 TRANSPORT MECHANISMS RELATED TO ALTERATIONS IN MEMBRANE POTENTIAL

The ionic content of most mammalian cells is characterized by being relatively rich in potassium and poor in sodium and chloride as compared to

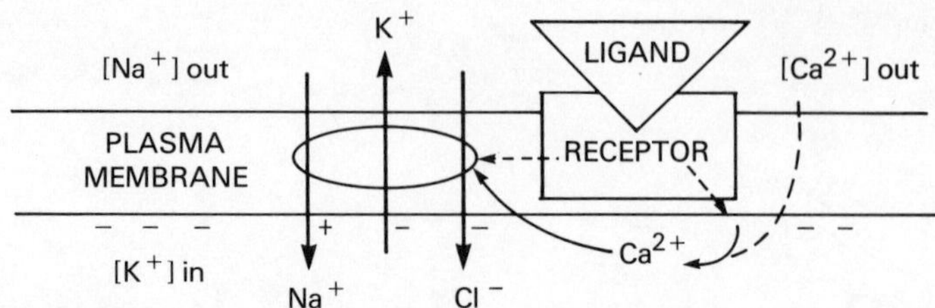

Figure 1. Schematic representation of the plasma membrane showing a ligand–receptor interaction coupled to changes in ion permeability. These changes can be induced by release of calcium from the inner surface of the membrane, or may be linked directly to the ligand–receptor interaction

the extracellular concentration of these ions. Ion selective channels provide specific dissipative routes for ion gradients which ATP fuelled pumps maintain across the plasma membrane. Figure 1 shows how a change in plasma membrane permeability to one or more ions, with the movement of this ion(s) down its chemical gradient can alter the resting membrane potential. Thus an increase in the plasma membrane permeability to potassium, or an increase in the permeability to chloride, will result in an increase in the transmembrane potential (inside negative); the influx of sodium will have an opposing effect, that is, to depolarize. The number of ions involved need not be large to alter the membrane potential, but the effect will be retarded by any opposing effect of the newly established electrochemical gradient, or by the neutralizing effect of the freely permeable ions. In the stimulus induced hyperpolarization of L cells and macrophages cited as examples below, the triggering event for the permeability changes leading to hyperpolarization is the release of calcium from the inside of the membrane; the stimulus induced changes being mimicked by intracellular injections of calcium. Ligand–receptor interactions could as well directly affect the membrane permeability to specific ions without a calcium transient; examples discussed below, are the conductance changes induced by cholera toxin and thyrotropin in artificial membranes with incorporated receptor components.

Figure 2 illustrates the central role of sodium in mammalian cell transport events. These concentration differences, a high intracellular K$^+$ relative to Na$^+$ and high extracellular concentration of Na$^+$ relative to K$^+$, are maintained by transport events at the plasma membrane such as the operation of the Na$^+$–K$^+$ ATPase wherein energy (ATP) is used to eject 3 Na$^+$ ions from the cell simultaneously with the reentry of 2 K$^+$ ions. The unequal translocation of these ions by the ATPase contributes to the measured (inside negative) membrane potential of mammalian cells. There is therefore maintained an electrochemical gradient for the reentry of sodium, and it is this inward movement of sodium that is coupled to many specialized transport functions

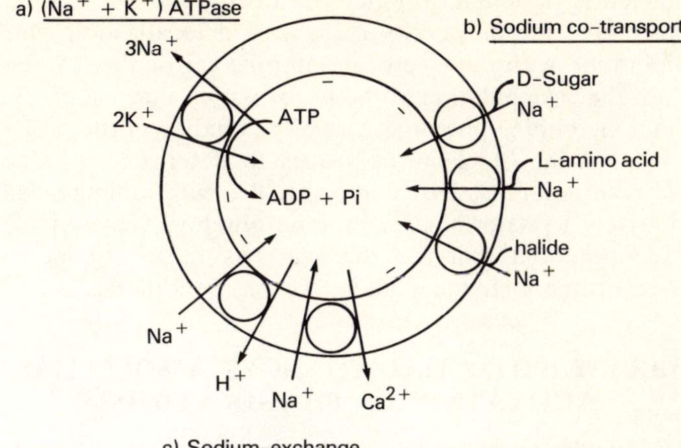

Figure 2. Diagrammatic representation of the central role of sodium in many of the transport processes operative in the plasma membrane of mammalian cells

of cells. This is exemplified in (b), where sodium entry is coupled to the uptake of sugars, amino acids and iodide uptake by thyroid cells in (c); sodium entry is coupled to calcium extrusion in neuronal cells and cardiac sarcolemma; another example of sodium counter-transport is the more universal sodium–proton exchange. Sodium counter or co-transport is rheo- or electrogenic depending on the number and charge of the transported species. It is clear that an affect on membrane potential will effect sodium entry and thereby sodium dependent transport events such as nutrient uptake and ion exchange reactions. It should be emphasized that these sodium transport mechanisms can in turn alter membrane potential.

3 METHODS USED TO MEASURE MEMBRANE POTENTIAL

The measurement of membrane potential has traditionally relied on standard electrophysiological techniques that require impaling of selected cells with microelectrodes. This remains the most sensitive and direct method available. Although this technique is limited by the size of most peripheral mammalian cells, it can be utilized to record continuous changes such as oscillations of membrane potential.[3]

An alternative biochemical method for determining membrane potential involves the use of permeant lipophilic cations such as tetraphenylphosphonium$^+$ (TPP$^+$) or triphenylmethylphosphonium$^+$ (TPMP$^+$). This class of compounds was first introduced ten years ago to measure electrical potentials across biological membranes and have now been applied successfully to a variety of systems. The ions are constructed in such a fashion that they are sufficiently lipophilic to enter the hydrophobic core of the membrane but also have the propensity for charge delocalization which allows passive equilibration with the electrical potential across the membrane. The contribution of the accumulation of the probe within internal organelles that also maintain a membrane potential, as well as binding of the probe must be considered when using biochemical probes to determine a value for the plasma membrane potential. Distribution studies with radiolabelled cations, used quantitatively to determine membrane potential, have yielded values for membrane potential that are the same as those obtained by using intracellular recording techniques under various conditions.[4]

4 MEMBRANE POTENTIAL CHANGES ASSOCIATED WITH ACTIVATION OF PHAGOCYTOSIS

Changes in membrane potential are the earliest event in ligand–receptor induced activation of phagocytic cells. This membrane potential response can be separated from, and precedes in time, the metabolic and morphological events of phagocytosis; chemotaxis, degranulation, enzyme secretion and superoxide (O_2^-) radical generation.[5,6] The response is induced by the addition of a number of chemoattractants including the synthetic peptide *N*-formyl-methionyl-leucyl-phenylalanine, phorbol myristate acetate and concanavalin A.[5,6,7,8]

Using intracellular microelectrode recording techniques, human and mouse macrophages have a resting membrane potential of 14 to 26 mV (inside negative) and show spontaneous and stimulus-induced hyperpolarization responses of 10 to 50 mV associated with an increased permeability to potassium.[9,10,11] Studies of membrane potential have been extended to other phagocytic cells such as polymorphonuclear leukocytes. Since thse cells are too small to be impaled by microelectrodes, membrane potential changes have been measured using the equilibrium distribution of TPMP$^+$ and the fluorescent cyanine dyes.[5,7,12] Using different methods for membrane potential determinations, the direction of stimulus induced polarization changes have not been entirely consistent, although some differences may relate to the different concentrations and types of fluorescent cyanine dyes used. What is clear, however, in these studies, is that changes in membrane potential precede in time, and can be separated from the subsequent metabolic events that occur with phagocytic activation.

Changes in membrane potential in response to chemoattractants appears to be a necessary prerequisite for phagocytic activation, since both responses are absent in polymorphonuclear leukocytes from patients with chronic granulomatous disease.[6,7]

Changes in membrane potential of mouse spleen lymphocytes also occurs during mitogenic stimulation as monitored using tetraphenylphosphonium$^+$ (TPP$^+$).[13] Addition of lipopolysaccharide, concanavalin A, or foetal calf serum results in a systematic alteration in membrane potential according to the following pattern: depolarization 1 to 4 hours after mitogen addition; repolarization within 6–8 hours; and differentiation into blast cells over the last 24 hours accompanied by hyperpolarization. There is a direct correlation between the increase in membrane potential and the rate of DNA synthesis as judged by thymidine incorporation.

5 STIMULUS INDUCED HYPERPOLARIZATION IN L CELLS

L cells, an established line of mouse fibroblastic origin, have a resting membrane potential of 16–20 mV (inside negative) as measured using standard intracellular recording techniques of non-dividing or x-irradiated giant L cells. These cells generate an active electrical membrane response to mechanical, electrical or chemical (acetylcholine) stimuli.[14,15] The response consists of a hyperpolarization of about 30 mV due primarily to a large increase in membrane permeability to potassium. Intracellular injection of calcium elicits the same response suggesting that the increased potassium permeability is mediated by an increase in cytosolic calcium.[16] This active electrical response seen in L cells is transmitted between contiguous cells by an unknown mechanism. Although atropine blocks the effect of acetylcholine in generating the electrical response, it does not block the response in the neighbouring cells.[17]

6 THE EFFECT OF INTERFERON ON THE MEMBRANE POTENTIAL OF L CELLS

L cells are a target tissue for the antiviral protective effect of mouse interferon and have therefore been used for the study of the early biochemical events associated with interferon addition. L cells that are particularly sensitive to the antiviral effects of interferon, are notable in having large amounts of the monosialoganglioside GM_2 not normally present in L cells.[18] Gangliosides have been implicated as a component of the cell surface receptor for mouse interferon[19] (see also Chapter 1 of this volume).

Membrane potential in L cells has been monitored using the distribution measurement of the cation TPP^+. L cells accumulate TPP^+ when suspended in physiological media, and the uptake is reduced if the sodium in the media is replaced with potassium, or when the mitochondria are inhibited with agents such as rotenone or azide. Addition of 100–500 reference units of mouse interferon per ml leads to a dose dependent increase in TPP^+ uptake within 5 minutes, and increase which is maintained for at least 30 minutes; but 4000 units decreases the uptake.[20] The effect of interferon on TPP^+ uptake by mouse L cells is apparent in the presence of ouabain and oligomycin, and is blocked by preincubation of the interferon with the antibody to interferon. This effect of interferon on TPP^+ uptake is specific since it is not seen with cholera toxin, an agent which inhibits interferon bioactivity in L cells but which has no measurable biological effect.[21] Measured under the same conditions, interferon has no effect on TPP[+] uptake by KB cells, a human cell line that binds interferon, but does not convey antiviral protection.[22]

A comparable effect of interferon is seen on TPMP[+] uptake by plasma membrane vesicles prepared from L cells with an artificially imposed sodium chloride gradient (out > in) which is consistent with an effect of interferon on the plasma membrane permeability to chloride and/or sodium.[18] As noted above, increased permeability and hyperpolarization in L cells is associated with an increase in intracellular ionized calcium. Interferon has no effect on the uptake of $^{45}Ca^{2+}$ by either L cells in suspension or plasma membrane vesicles. There is however, a release of ionized calcium from L cell membrane preparations treated with interferon, detected using the spectral change in the calcium sensitive arsenazo III dye.[23] The calcium-arsenazo III difference spectra is characterized by absorbance peaks at about 585 and 650 nm. Although the method is not quantitative as other ions affect the magnitude of the response, the spectral changes are characteristic only for calcium. As seen in Figure 3, the release of calcium from L cell membranes

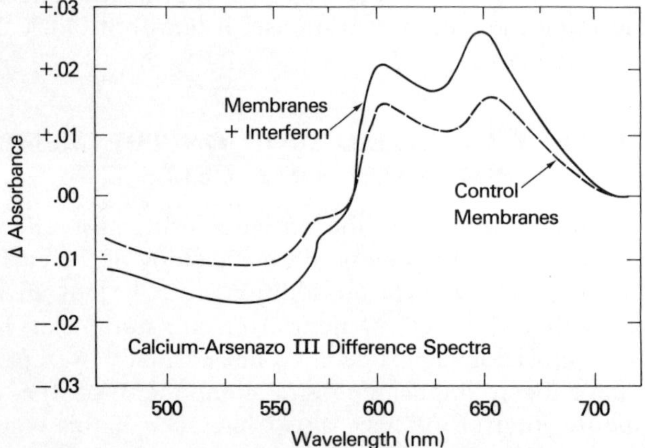

Figure 3. Difference spectrum of membranes plus arsenazo III dye (dashed curve). Tanden double cuvettes were used. One compartment of the sample cuvette contained 150 μg of membrane protein, 10 μg of purified arsenazo III dye in 1.0 ml of 0.020 M Tris-Hepes buffer, pH 7.4. The other compartment contained 1.0 ml of 0.020 M Tris-Hepes Buffer, pH 7.4. One compartment of the blank cuvette contained 150 μg of membrane protein in 1.00 ml of 0.020 M Tris-Hepes Buffer, pH 7.4, and the other compartment contained 10 μg of arsenazo III dye in 1.0 ml of the same buffer. Difference spectrum of Interferon-treated membranes plus arsenazo dye (solid curve). The arrangement of solutions was as described above except that membranes were pretreated with 200 reference units of interferon before adding to the cuvettes

is enhanced in the presence of interferon as indicated by the calcium-arsenazo III difference spectra. The amount of calcium present in the membranes prior to interferon addition is the same in the control and +interferon membranes. The addition of a related ligand TSH or an excess of interferon alone has no affect on the spectra.

Besides its antiviral activity, there is very convincing evidence that interferon also has other effects on cells, such as inhibition of cell and tumor growth, and regulation of the immune system (reviewed by Friedman[24]). As discussed below, the activation of cell growth has been associated with alterations in ion fluxes, in particular the influx of sodium as well as changes in intracellular pH. One speculation is that the changes in membrane potential induced by interferon reflect changes in ion fluxes, that could culminate in an altered intracellular composition responsible for some of the biological effects of interferon.

7   STIMULANT-EVOKED MEMBRANE POTENTIAL CHANGES AND SECRETION IN PANCREATIC CELLS

Mouse pancreatic cells have a resting membrane potential of $-20\,mV$ for islet β-cells and -41 for acinar cells as measured using intracellular recording techniques. The membrane potential of islet β-cells depends on the extracellular concentration of D-glucose and in the absence of glucose, the membrane is hyperpolarized; the membrane potential of acinar cells is not affected by external glucose concentrations.[25] In addition D-glucose triggers the release of insulin from islet β-cells which may depend on the metabolism of glucose or alternatively sodium influx and depolarization may initiate the glucose-induced release of insulin.[26,27]

The secretory response of various peripheral tissues is very reminiscent of transmitter release at nerve endings. The common triggering event appears to be depolarization of the plasma membrane. As in the nervous system, a rise in intracellular ionized calcium concentrations seems critical for secretory release, perhaps involved in membrane fusion at the time of exocytosis.

The secregogues bombesin, acetylcholine and cholecystokinin evoke quantitatively identical membrane responses with respect to maximum depolarization, resistance reduction and equilibrium potential, as well as the same maximal and dose-responses for amylase secretion in pancreatic acinar cells.[28] There is evidence that these three agents do not share the same receptor sites, nor do they all stimulate the adenylate cyclase in pancreatic plasma membrane preparations. It is generally accepted that calcium ions play an essential role in 'stimulus-secretion' coupling and pancreatic amylase secretion can be stimulated without receptor activation by raising the cytosol-ionized calcium concentration with the divalent cation ionophore A23187 or direct intracellular calcium injection. Secregogues are thought to induce a rise in the intracellular calcium concentration by increasing the uptake of calcium from the extracellular medium or alternatively by increasing sodium conductance, evoking release of membrane bound calcium.

8   THE EFFECT OF TSH ON MEMBRANE POTENTIAL

One effect of thyrotropin (TSH) stimulation of thyroid tissue is an alteration in membrane potential as measured directly using microelectrodes. By this method the membrane potential response to TSH stimulation varies and can be diphasic, depolarized or hyperpolarized.[29,30,31,32] The effect of TSH on membrane potential is also evident in studies using the biochemical probes for membrane potential, the lipophilic cation triphenyl-methylphosphonium$^+$ (TPMP$^+$). Thyroid cells grown in tissue culture accumulate TPMP$^+$ and TSH stimulates the uptake of the cation three-fold, but this stimulatory effect of TSH on TPMP$^+$ accumulation is abolished if the TSH receptor activity of the cells is destroyed by treatment with trypsin. Analogous effects are observed with thyroid plasma membrane vesicles. TPMP$^+$ uptake and stimulation by TSH occurs when NaCl, KCl or Tris-HCl concentration gradients are artifically imposed across the vesicle membrane which is consistent with TSH either increasing the permeability of the membrane to anions or decreasing its permeability to cations.[33]

9   THE EFFECT OF MEMBRANE POTENTIAL ON IODIDE UPTAKE BY THYROID CELLS

It is well established from studies of dispersed thyroid cells and thyroid slices that iodide is actively transported across the basal membrane of thyroid epithelial cells by a sodium dependent process; and therefore depends on a functioning Na$^+$–K ATPase.[34] Agents which inhibit its function (ouabain) or indirectly by limiting the available ATP (mitochondrial inhibitors such as rotenone and oligomycin) decrease iodide accumulation. It is apparent that TSH could affect iodide uptake by altering the electrochemical gradient for sodium and/or the sodium dependent carrier for iodide. The role of membrane potential in the uptake of iodide can be evaluated using thyroid cell vesicles which have the advantage that the sodium gradient and membrane potential can be imposed independently. These studies support the view that iodide accumulation is dependent on the electrical as well as the chemical gradient for sodium. Thyroid cell vesicles equilibrated in KP$_i$, take up TPP$^+$ when suspended in KCl as seen in Figure 4. The accumulation of TPP$^+$ is the same when NaCl replaces KCl, although not shown. When the vesicles are diluted into NaCl in the presence of valinomycin, TPP$^+$ uptake increases indicating that K$^+$ efflux (in $\rightarrow$ out) hyperpolarizes the vesicle (inside more negative). Valinomycin has no affect in the absence of a

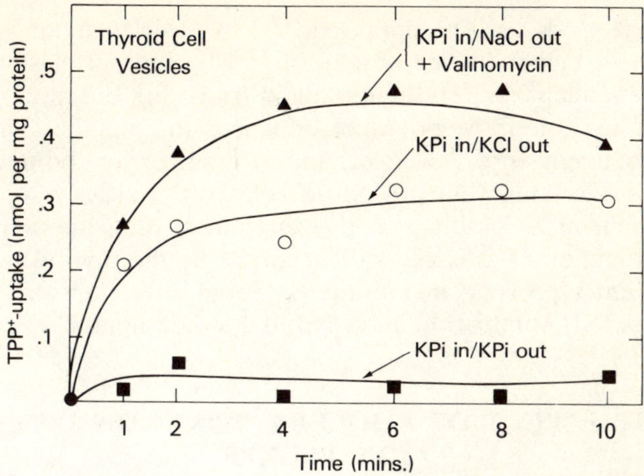

Figure 4. TPP$^+$ uptake by thyroid cell vesicles. Vesicles were prepared from functional rat thyroid cells grown in tissue culture (Ambesi-Impiombato[52]) TPP$^+$ uptake was determined by rapid filtration as described in Ramos[53] The vesicles (15 mg/ml) prepared as described in Grollman[20] were equilibrated in 0.1 M KP$_i$, pH 7.4, 1 mM Mg Cl$_2$ and diluted 20 fold in the same buffer (squares); diluted in 0.1 M NaCl or KCl, 1 mM MgCl$_2$, 5 mM Tris-HCl, pH 7.4 (circles); or diluted in 0.1 M NaCl, 1 mM MgCl$_2$, 5 mM Tris-HCl, pH 7.4, and 2.5 μM valinomycin (triangles)

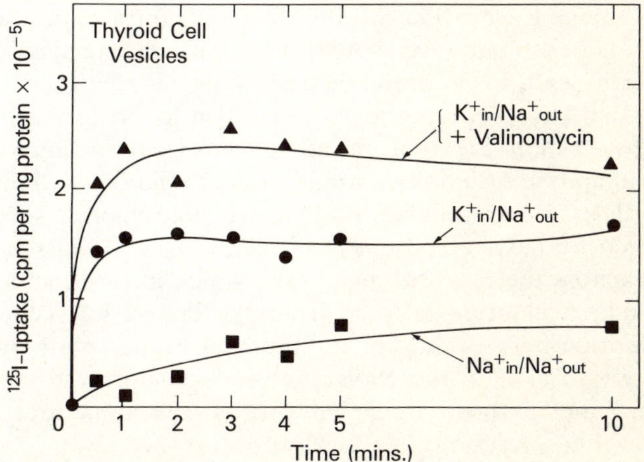

Figure 5. ^{125}I uptake by thyroid cell vesicles. Vesicles and assay conditions were the same as described in Fig. 4 with 1 μCi carrier-free ^{125}I replacing TPP$^+$. Vesicles were equilibrated in 0.1 M NaP$_i$, pH 7.4, 1 mM MgCl$_2$ and suspended in 0.1 M NaCl, 1 mM MgCl$_2$, 5 mM Tris-HCl, pH 7.4 (squares); equilibrated in 0.1 M KP$_i$, pH 7.4, 1 mM MgCl$_2$, and suspended in 0.1 M NaCl, 1 mM MgCl$_2$, 5 mM Tris-HCl, pH 7.4 (circles); or equilibrated in 0.1 M KP$_i$, pH 7.4, 1 mM MgCl$_2$ and suspended in 0.1 M NaCl, 1 mM MgCl$_2$, 5 mM Tris-HCl, pH 7.4 with 2.5 μM valinomycin (triangles)

K^+ gradient ($K^+_{in} = K^+_{out}$). The uptake of ^{125}I by vesicles under similar conditions is seen in Figure 5. The uptake of ^{125}I by thyroid plasma membrane vesicles is dependent on a sodium gradient (out > in), but more importantly, the increase in membrane potential enhances this uptake. The uptake of iodide is dependent on the electrochemical gradient for sodium rather than iodide, whose accumulation in thyroid cells and vesicles is opposed to its own electrochemical gradient and exemplifies the situation wherein a specialized function of a tissue (iodide uptake by the thyroid) is existent on the sodium entry process; membrane potential alterations can provide the link between TSH stimulation and thyroidal iodide uptake.

10 THE EFFECT OF CHOLERA TOXIN ON INTESTINAL ION FLUXES

Although the exact mechanism of cholera toxin action on ion fluxes remains uncertain, it exemplifies the situation wherein the transport features that subserve the specialized functions of a tissue, determine the type of response to external stimuli. The transport characteristics of the junctional complex/lateral intercellular space in 'loose epithelium', contributes to the unique response of the ilium to cholera toxin.

Cholera toxin added to membranes prepared from most types of cells induces a sustained elevation in the level of intracellular cyclic 3'-5'-adenosine monophosphate (cyclic-AMP); but only in intestinal preparations does the toxin lead to a dramatic secretion of electrolytes and fluid. Mechanisms that have been proposed to explain the action of cholera toxin on intestinal secretion are that raised levels of intracellular cyclic-AMP result in (1) inhibition of influx or stimulation of efflux of sodium chloride at the brushborder;[35,36] (2) stimulation of electrogenic chloride secretion which is accompanied by passive sodium movements;[37] or (3) increased chloride conductance across the mucosal membrane which in turn increases the rate of passive sodium chloride leakage from the lateral intercellular space.[38] Increased chloride permeability of the mucosal border is the only requirement necessary to explain the choleragen-induced increase in electrogenic Cl^- secretion, and in short-circuit current, as well as neutral secretion of NaCl and net fluid secretion.[38]

11 LIGAND-RECEPTOR INTERACTIONS FORM ION-CONDUCTING CHANNELS IN ARTIFICIAL MEMBRANES

The role of gangliosides as components of the receptors for cholera toxin and thyrotropin (TSH) is detailed in a prior chapter in this volume. Cholera toxin and TSH can interact with lipid bilayers containing specific gangliosides to form channels which allow ions present in the aqueous solution to traverse the membrane. Planar membranes for these studies are composed of a mixture of glycerol monooleate and gangliosides. The addition of cholera toxin to the aqueous solution on one side of a bilayer containing the monosialoganglioside G_{M1} but not G_{M2} results in an abrupt and substantial increase in membrane conductance.[39] The increase in conductance is dependent on the amount of cholera toxin added and the concentration of G_{M1} incorporated into the bilayer. Planar membranes that do not contain gangliosides, do not respond to the addition of cholera toxin with a change in conductance.

TSH similarly interacts with glycerol monooleate planar membranes containing the trisialoganglioside G_{T1} leading to a sixteen fold increase in conductance.[40] TSH does not effect the conductance in bilayers made

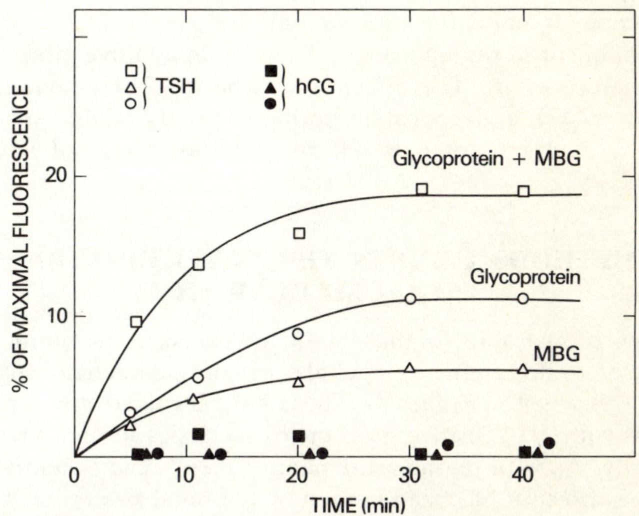

Figure 6. Effect of TSH (open symbols) or hCG (closed symbols) on the release of 6-carboxyfluorescein (6-CF) from liposomes containing 200 mM of the dye entrapped in the interlamellar space and containing (i) the glycoprotein component of the TSH receptor embedded in the lipid bilayer (circles); or (ii) mixed brain gangliosides (MBG) (triangles); or (iii) a combination of the two—glycoprotein receptor and mixed brain gangliosides (squares). The measurements were started with the addition of either TSH or hCG at a final concentration of 2.8×10^{-6} M. Fluorescence was measured at 22 °C with the excitation and emission monochromators set at 480 and 519 nm, respectively and followed by repeated fluorescence measurements over time. Results are expressed as per cent of maximal, that is the total release of trapped 6-CF achieved by adding Triton X-100 to the sample. (Figure reproduced from Beguinot,[41] 1982, with permission of the authors)

without gangliosides or those containing the ganglioside G_{M1}. The trisialo-ganglioside G_{T1} is a much more potent inhibitor than G_{M1} of TSH binding to thyroid membranes.[22]

Increase permeability incident to TSH-receptor interactions are also suggested by studies using large multilamellar liposomes with receptor components incorporated in which the water-soluble fluorescent dye, 6-carboxyfluorescein is entrapped.[41] The binding of these liposomes to iodinated TSH is unaffected by the dye. The dye is concentrated within the liposomes (200 mM) and is self-quenched, but will fluoresce when released into the suspending medium. TSH stimulated 6-carboxyfluorescein release is seen in Figure 6. TSH at a concentration of 2.8×10^{-6} M causes a time dependent increase of the fluorescence signal that increases for 30 minutes, and stabilizes thereafter. TSH does not cause significant fluorescent changes in a suspension of liposomes lacking receptor components. The magnitude of the TSH effect on 6-carboxy fluorescein release is dependent on the concentration of the hormone. As seen in Figure 6, TSH mediated release of 6-carboxy-fluorescein depends upon the presence of the ganglioside or glycoprotein receptor component in the liposomes. There is an additive effect when both components are present. The effect is specific for TSH; human chorionic gonadotropin (hCG), a glycoprotein hormone closely related structurally to TSH but whose target organ is not the thyroid, does not stimulate the release of the dye.

12   METHODS USED IN THE DETERMINATION OF INTRACELLULAR pH

Techniques comparable to those used to measure membrane potential have been used to determine intracellular pH of mammalian cells (reviewed in Boron[42]) (reviewed in Gillies[43]). These have included the impalement of selected cells with pH-sensitive glass microelectrodes to measure intracellular pH directly. As with the measurement of membrane potential, although this method is direct, it has been limited by the small size of most peripheral cells. The value for intracellular pH obtained by the biochemical techniques, uses the distribution measurement of a weak acid or base, and rests on the assumption that only the uncharged form of a weak acid or weak base probe is freely permeable to the membrane and is present in equal concentations on both sides of the membrane. The equilibrium distribution of the weak acid/base is therefore pH dependent and independent of the membrane potential. Some disadvantages of the biochemical method is the need to determine a value for the internal space in order to calculate the intracellular pH and more importantly the probes will also distribute within intracellular organelles. The most widely used probes are the weak acid 5,5-di-methyl-2,4-oxazolidinedione (DMO), and the weak base methylamine since both are available radiolabelled. The values obtained for intracellular pH using the measurement of the distribution of a weak acid or base have been directly comparable in the same systems where direct measurements with pH-sensitive microelectrodes were also used. The values obtained using a weak acid or base may not represent the true cytoplasmic pH since the probes distribute as well within internal compartments which maintain proton gradients.

13 IONIC MECHANISMS INVOLVED IN THE MAINTAINANCE OF TRANSMEMBRANE PROTON GRADIENTS

It is clear that mammalian cells maintain a proton gradient across the plasma membrane as well as across internal organelles. Cytoplasmic pH when measured in peripheral cells is in general more alkaline and differs from a purely passive distribution of protons in response to the measured plasma membrane potential. The mechanism for maintaining and altering proton gradients incident to ligand-receptor interactions are comparable in type to the mechanisms relating to membrane potential described above. These include proton countertransport across the plasma membrane; proton-ATPases; and altered permeability of the membrane to protons. To date, the best characterized proton extruding mechanisms are the Na^+–H^+ exchange found in excretory and secretory tissues, and driven by the transmembrane sodium gradient; and Cl^-–HCO_3^- exchange found in erythrocytes and gastric tissue that requires metabolic energy.[44]

14 THE USE OF LYSOMOTROPHIC AGENTS TO ALTER INTRACELLULAR pH

Intracellular pH is relatively insensitive to altered external pH using non-permeant buffers. However, exposure of cells to permeant weak acids or bases produces large and rapid changes in internal pH. The rationale for using weak bases or acids for altering intracellular pH is the same as their use as radiolabelled probes in the measurement of intracellular pH. Since weak bases have a propensity for the low pH of the lysosomal compartment, these agents are referred to as 'lysomotrophic weak bases', and include chloroquine, amantidine, methylamine, tributylamine and the most widely used ammonium chloride.

The exposure of cells for example to NH_4Cl at constant pH, produces a concentration dependent rise in internal pH. This alkalinization is due to the influx of NH_3 which combines with internal protons to produce NH_4^+, and continues until ammonia concentrations are equal inside and out. Since NH_4^+ is also permeant at a very low level, at high concentrations it will acidify.

Just as exposure to NH_4^+ salts lead to a rapid intracellular alkalinization, exposure to CO_2 causes a rapid, reversible acidification. CO_2 enters the cell readily where it becomes hydrated and dissociates to form H^+ and HCO_3^-. As in the case with exposure to NH_4Cl, the internal pH change produced by CO_2 depends on the external concentration of the uncharged member of the conjugate pair. The higher levels of CO_2 produces greater degrees of acidification. Intracellular acidification can also be observed with the weak acid DMO and lactic acid.

The use of 'lysomotrophic weak bases' have emphasized the very important role of pH in ligand internalization.[45] Ammonium salts, by an unknown mechanism, also block the antiviral state of interferon.[46]

15 THE EFFECT OF INTRACELLULAR pH ON INTERCELLULAR COMMUNICATION

Specialized regions of contact known as gap junctions exist between opposed plasma membranes of communicating cells; these junctions are believed to mediate and regulate the passage of ions and small molecules between the cell interiors. The chemical nature of these molecules is unknown, but candidates that have been suggested are cyclic-AMP, protons and calcium. Another form of communication between cells is through an electrical response generated in neighbouring cells. Variations in intracellular pH have marked effects on the electrical coupling between neighbouring cells. Increasing the intracellular pH by exposing pancreatic and lacrimal acinar cells to CO_2, markedly increases the junctional resistance and electrical coupling in these cells.[47] The effect of 100 per cent CO_2 is not mimicked by anoxia or by changes in external pH. The electrical uncoupling effect of acetylcholine in these tissues is immediately and fully reversed with 10 mM NH_4Cl, a procedure expected to increase the intracellular pH. Calcium also has an important role in intercellular communication, and it is not clear whether changes in internal pH are coupled to localized changes in intracellular calcium.

Another example of intercellular communication is the demonstration that interferon-induced viral resistance in L cells can be transmitted by an unknown mechansim to co-cultivated heterologous cells that do not have a cell surface receptor for interferon.[48]

16 THE EFFECT OF INTRACELLULAR pH ON PROTEIN SYNTHESIS IN SEA URCHIN EGG ACTIVATION

From the very eloquent studies of the biochemical events at the time of fertilization of the sea urchin egg, it is clear that in this system, the activation of protein synthesis is dependent on the elevation of intracellular pH.[49]

During normal fertilization of the sea urchin egg, the sequence of events are: a transient elevation in cytosolic calcium concentration, followed at one minute by protons being released from the egg in exchange with sodium, resulting in an increase in intracellular pH of 0.45 pH units, and in an increase in protein synthesis at 5 minutes. Treatment of the egg with 10 mM ammonium chloride can also increase the intracellular pH, thus bypassing the calcium transient, and lead to an increase in protein synthesis as measured by radiolabelled amino acid incorporation into the egg. The calcium transient has no affect on the rate of protein synthesis alone, but in conjunction with the intracellular pH change, protein synthesis will be maximally stimulated.[50] In the few mammalian systems that have been studied, stimulation of protein synthesis has been correlated with sodium entry rather than internal pH. Serum and growth factors for example activate a sodium channel in the plasma membrane of fibroblasts grown in culture.[51] Serum and purified growth factors more than double the rate of $^{22}Na^+$ entry and the accumulation of sodium in the cell when the exit of this ion is prevented by ouabain. The sodium channel in the plasma membrane that is activated is sensitive to amilioride, an agent that blocks Na^+–H^+ exchange.

As stated above, membrane potential changes correlate with DNA synthesis during mitogenic stimulation of lymphocytes. It has yet to be determined whether these changes in membrane potential are also associated with changes in sodium/proton fluxes.

In summary, one means a cell can respond to external stimuli in its environment is by changes in its internal solute composition with profound effects on function. One manifestation of ion and proton fluxes is an alteration in membrane potential. Membrane potential can in turn influence transport events as well as be a rapidly transmitted signal.

17 ACKNOWLEDGEMENTS

The author gratefully acknowledges the inspiration and support of Drs. Ronald H. Kaback and Leonard D. Kohn in the development of the concepts expressed in this chapter.

REFERENCES

1. Romey, G., Jacques, Y., Schweitz, H., Fosset, M., and Lazdunski, M. (1979). The sodium channel in non-impulsive cells interaction with specific neurotoxins. *Biochim. Biophys. Acta*, **556**, 344–353.
2. Munson, R., Westermark, B., and Glaser, L. (1979). Tetrodotoxin-sensitive sodium channels in normal human fibroblasts and normal human glia-like cells. *Proc. Nat. Acad. Sci. U.S.A.*, **76**, 6425–6429.
3. Okada, Y., Doida, Y., Roy, G., Tsuchiya, W., Inouye, K., and Inouye, A. (1977). Oscillations of membrane potential in L cells. *J. Membrane Biol.*, **35**, 319–335.

4. Lichtschtein, D., Kaback, H. R., and Blume, A. J. (1979). Use of a lipophilic cation for determination of membrane potential in neuroblastoma-glioma hybrid cell suspensions. *Proc. Nat. Acad. Sci. U.S.A.*, **76,** 650–654.

5. Korchak, H. M., and Weissmann, G. (1978). Changes in membrane potential of human granulocytes antecede the metabolic responses to surface stimulation. *Proc. Nat. Acad. Sci. U.S.A.*, **75,** 3818–3822.

6. Whitin, J. C., Chapman, C. E., Simons, E. R., Chovaniec, M. E., and Cohen, H. J. (1980). Correlation between membrane potential changes and superoxide production in human granulocytes stimulated by phorbol myristate acetate. *J. Biol. Chem.*, **225,** 1874–1878.

7. Seligmann, B. E., and Gallin, J. I. (1980). Use of lipophilic probes of membrane potential to assess human neutrophil activation. *J. Clin. Invert.*, **66,** 493–503.

8. Utsumi, K., Sugiyama, K., Mijahara, M., Naito, M., Awai, M., and Inoue, M. (1977). Effect of Concanavalin A on membrane potential of polymorphonuclear leukocyte monitored by fluorescent dye. *Cell. Struct. Funct.*, **2,** 203–209.

9. Dos Reis, G. A., and Oliveira-Castro, G. M. (1977). Electrophysiology of phagocytic membranes. *Biochim. Biphys. Acta*, **469,** 257–263.

10. Dos Reis, G. A., Persechini, P. M., Ribeiro, M. C., and Oliveira-Castro, G. M. (1979). Electrophysiology of phagocytic membranes. *Biochim. Biophys. Acta*, **552,** 331–340.

11. Gallin, E. K., and Gallin, J. I. (1977). Interaction of chemotactic factors with human macrophages. *J. Cell. Biol.*, **75,** 277–289.

12. Seligmann, B. E., Gallin, E. K., Martin, D. L., Shain, W., and Gallin, J. I. (1980). Interaction of chemotactic factors with human polymorphonuclear leukocytes: studies using a membrane potential-sensitive cyanine dye. *J. Membrane Biol.*, **52,** 257–272.

13. Kiefer, H., Blume, A. J., and Kaback, H. R. (1980). Membrane potential changes during mitogenic stimulation of mouse spleen lymphocytes. *Proc. Nat. Acad. Sci. U.S.A.*, **77,** 2200–2204.

14. Nelson, P. G., and Peacock, J. H. (1972). Acetylcholine responses in L cells. *Science*, **177,** 1005–1007.

15. Nelson, P. G., Peacock, J., and Minna, J. (1972). An active electrical response in fibroblasts. *J. Gen. Physiol.*, **60,** 58–71.

16. Henkart, M. P., and Nelson, P. G. (1979). Evidence for an intracellular calcium store releasable by surface stimuli in fibroblasts (L cells). *J. Gen. Physiol.*, **73,** 655–673.

17. Nelson, P. G., and Peacock, J. H. (1973). Transmission on an active electrical response between fibroblasts (L cells) in cell culture. *J. Gen. Physiol.*, **62,** 25–36.

18. Grollman, E. F., Lee, G., Ramos, S., Lazo, P. S., Kaback, R. H., Friedman, R. M., and Kohn, L. D. (1978). Relationships of the structure and function of the interferon receptor to hormone receptors and establishment of the antiviral state. *Cancer Res.*, **38,** 4172–4185.

19. Besancon, F. and Ankel, H. (1976). Specificity and reversibility of interferon ganglioside interaction. *Nature*, **259,** 576–578.

20. Grollman, E. F., Friedman, R. M., and Kohn, L. D. (1982). Interferon induced changes in tetraphenylphosphonium$^+$ ion uptake by L cell. Manuscript in preparation.

21. Friedman, R. M. and Kohn, L. D., (1976). Cholera toxin inhibits interferon action. *Biochem. Biophys. Res. Commun.*, **70,** 1078–1084.

22. Kohn, L. D., Friedman, R. M., Holmes, J. M., and Lee, G. (1976). Use of thyrotropin and cholera toxin to probe the mechansim by which interferon initiates its antiviral activity. *Proc. Nat. Acad. Sci. U.S.A.*, **73,** 3695–3699.

23. Kendrick, N. C., Ratzlaff, R. W., and Blaustein, M. P. (1977). Arsenazo III as an indicator for ionized calcium in physiological salt solutions: its use for determination of the CaATP dissociation constant. *Analytical Biochem.*, **83**, 433–450.
24. Friedman, R. M. (1977). Antiviral activity of interferons. *Bact. Reviews*, **41**, 543–567.
25. Dean, P. M., and Matthews, E. K. (1970). Glucose-induced electrical activity in pancreatic islet cells. *J. Physiol.*, **210**, 255–264.
26. Lowe, D. A., Richardson, B. P., Taylor, P., and Donatsch, P. (1976). Increasing intracellular sodium triggers calcium release from bound pools. *Nature*, **260**, 337–338.
27. Donatsch, P., Lowe, D. A., Richardson, B. P., and Taylor, P. (1977). The functional significance of sodium channels in pancreatic beta-cell membranes. *J. Physiol.*, **267**, 357–376.
28. Iwatsuki, N., and Petersen, O. H. (1978). In vitro action of bombesin on amylase secretion, membrane potential, and membrane resistance in rat and mouse pancreatic acinar cells. *J. Clin. Invest.*, **61**, 41–46.
29. Gorban, E. N. (1979). Age and the effect of thyrotropic hormone on the membrane potential of thyroid gland cells. *Fiziol. Zh.*, **25**, 395–401.
30. Williams, J. A. (1970). Effect of TSH on thyroid membrane properties. *Endocrinology*, **86**, 1154–1158.
31. Batt, R., and McKenzie, J. M. (1976). Hyperpolarization of thyroid cells *in vitro* by thyrotropin and cyclic AMP. *Am. J. Physiol.*, **231**, 65–767.
32. Pace, C. S., and Gaitan, E. (1979). The differential effect of thyrotropin on the electrical responses of thyroid cells in monolayer cultures of varying duration. *Biochimica et Biophysica Acta*, **555**, 111–118.
33. Grollman, E. F., Lee, G., Ambesi-Impiombata, F. S., Meldolesi, M. F., Aloj, S. M., Coon, H. G., Kaback, H. R., and Kohn, L. D. (1977). Effects of thyrotropin on the thyroid cell membrane; hyperpolarization induced by hormone–receptor interaction. *Proc. Nat. Acad. Sci. U.S.A.*, **74**, 2352–2356.
34. Bagchi, N., and Fawcett, D. M. (1973). Role of sodium ion in active transport of iodide by cultured thyroid cells. *Biochimica et Biophysica Acta*, **318**, 235–251.
35. Nellans, H. N., Frizzell, R. A., and Schultz, S. G. (1973). Coupled sodium-chloride influxes across the brush border of rabbit ileum. *Am. J. Physiol.*, **225**, 467–475.
36. Nellans, H. N., Frizzell, R. A., and Schultz, S. G., (1974). Brushborder processes and transepithelial Na and Cl transport of rabbit ileum. *Am. J. Physiol.*, **226**, 1131–1141.
37. Field, M. (1979). Mechanisms of action of cholera and *Escherichia coli* entertoxins. *Amer. J. Clin. Nutrition*, **32**, 189–196.
38. Naftalin, R. J., and Simmons, N. L. (1979). The effects of theophylline and choleragen on sodium and chloride ion movements within isolated rabbit ileum. *J. Physiol.*, **290**, 331–350.
39. Tosteson, M. T., and Tosteson, D. C., (1978). Bilayers containing gangliosides develop channels when exposed to cholera toxin. *Nature*, **275**, 142–144.
40. Poss, A., Deleers, M., and Ruysschaert, J. M., (1978). Evidence for a specific interaction between GT_1 ganglioside incorporated into bilayer membranes and thyrotropin. *FEBS Lett.*, **86**, 160–162.
41. Beguinot, F., Formisano, S., Consiglio, E., Kohn, L. D., and Aloj, S. M. (1982). Thyrotropin induced perturbation of liposomes containing components of the thyrotropin receptor. *J. Biol. Chem.*, in press.
42. Boron, W. F., (1979). Intracellular pH Regulation. *Curr. Top. Membranes and Transport.* **13**, 3–22.

43. Gillies, R. J., and Deamer, D. W. (1979). Intracellular pH: Methods and applications. *Curr. Top. Bioenerg.*, **9,** 63–87.
44. Aickin, C. C., and Thomas, R. C. (1977). An investigation of the ionic mechanism of intracellular pH regulation in mouse soleus muscle fibers. *J. Physiol.*, **273,** 295–316.
45. White, J., and Helenius, A. (1980). pH-dependent fusion between the Semliki Forest virus membrane and liposomes. *Proc. Nat. Acad. Sci. U.S.A.*, **77,** 3273–3277.
46. Commoy-Chevalier, M. J., Robert-Galliot, B., and Chany, C. (1978). Effect of Ammonium salts on the interferon-induced antiviral state in mouse L cells. *J. Gen. Virol.*, **41,** 541–547.
47. Iwatsuki, N., and Peterson, O. H. (1979). Pancreatic acinar cells: the effect of carbon dioxide, ammonium chloride, and acetylcholine on intercellular communication. *J. Physiol.*, **291,** 317–326.
48. Blalock, J. E., and Baron, S. (1977). Interferon-induced transfer of viral resistance between animal cells. *Nature*, **269,** 422–425.
49. Shen, S. S., and Steinhardt, R. A. (1978). Direct measurement of intracellular pH during metabolic derepression of the sea urchin egg. *Nature*, **272,** 353–354.
50. Winkler, M. M., Steinhardt, R. A., Grainger, J. L., and Minning, L. (1980). Dual ionic controls for the activation of protein synthesis at fertilization. *Nature*, **287,** 558–561.
51. Smith, J. B., and Rozengurt E. (1978). Serum stimulates the Na$^+$, K$^+$ pump quiescent fibroblasts by increasing Na$^+$ entry. *Proc. Nat. Acad. Sci. U.S.A.*, **75,** 5560–5564.
52. Ambesi-Impiombato, F. S., Parks, L. A. M., and Coon, H. G. (1980). Culture of hormone-dependent functional epithelial cells from rat thyroids. *Proc. Nat. Acad. Sci. U.S.A.*, **77,** 3455–3459.
53. Ramos, S., Grollman, E. F., Lazo, P. S., Dyer, S. A., Habig, W. H., Hardegree, M. C., Kaback, H. R., and Kohn, L. D. (1979). Effect of tetanus toxin on the accumulation of the permeant lipophilic cation tetraphenylphosphonium by guinea pig brain synaptosomes. *Proc. Nat. Acad. Sci. U.S.A.*, **76,** 4783–4787.

9 *Interactions between the cAMP and calcium messenger systems*

H. Rasmussen

1 INTRODUCTION

In writing this article for *Horizons in Biochemistry and Biophysics*, I have taken the point of view that this article should look out from our present plateau of knowledge to the dimly perceived horizons of the shape of things to come. In my presentation I shall depart from the standard or usual type of article to consider as the central question the following: when differentiated, highly specialized animal cells are called upon to perform their specialized function for the organism is there a universal system which couples stimulus to response, or a large number of different systems. I will propose that on the basis of our present knowledge one must conclude there is a single universal system that involves the interrelated second messenger functions of calcium ion and cAMP. I propose that this universal control system be designated *synarchy* to indicate that two (or more) messengers act together to rule the intracellular domain and thereby determine specialized cellular responses.

There are four aspects of this system which will be considered: the first is that of how it is organized; the second the definition of stimulus–response coupling; the third the validity of considering excitable as distinct from non-excitable cell; and the fourth the problem of whether one can consider as universal a system which exhibits such a host of particular variations.

From an historical perspecive, the two major fields of physiological enquiry of relevance to our discussion are the mechanisms of synaptic transmission and excitation–response coupling in muscle, i.e. neuromuscular physiology; and the mechanisms of action of peptide and amine hormones, molecular endocrinology.

From work in these areas over the period of 1950–70 two general concepts developed: the first that nerve and muscle are distinctive physiologically because they display action potentials and are, therefore,

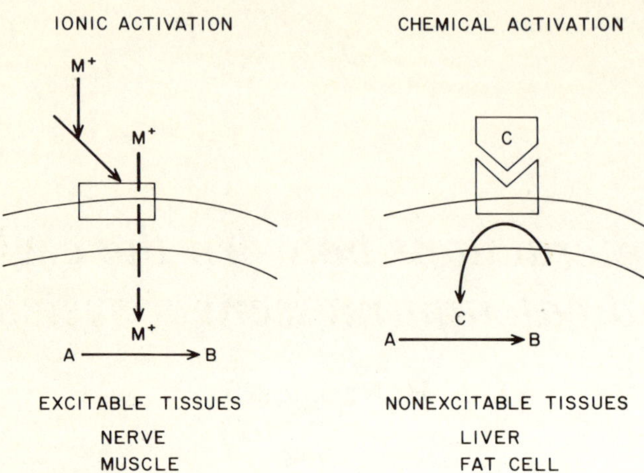

Figure 1. Schematic representations of stimulus-response coupling in excitable and non-excitable tissues

excitable and employ calcium ion to couple stimulus to response; and the second that the majority of peptide and amine hormones act upon non-excitable cells to elict metabolic or transport responses and employ cAMP as a second messenger (Figure 1). Hence, on the one hand were excitable cells displaying ionic currents on their cell surface that gave rise in turn to an ionic (calcium) messenger that elicited response; and on the other, cells without such currents being activated by chemical extracellular messengers which when acting on the cell surface gave rise to a chemical intracellular messenger. A certain symmetry seemed to exist in the two systems and therefore they appeared distinct modalities by which specialized cells became activated. Furthermore, the concept of stimulus–response coupling was generally restricted to events in the so-called excitable tissues.

From the time of the crystallization of these ideas in the period from 1955–1970 there were facts that did not quite fit with this apparently rather sharp separation and simple symmetry. Most particularly, cAMP was found to be present in nerve and muscle as well as in liver, adrenal and parotid. Conversely, calcium was found to serve a coupling function in the hormonal activation of renal gluconeogenesis, the initiation of insulin secretion, the action of ACTH on the adrenal gland, and in a host of other hormone mediated responses. These facts led me to present[1] the thesis that in many different cell types calcium ions and cAMP served interrelated second messenger functions. At that time this thesis was based largely on physiological data. Discoveries in the past 10 years have expanded its base into the biochemical and molecular domains, and lead to a more complete, more comprehensive, and more interesting model of their relationships. Many of these discoveries are discussed in detail by authors of other contributions to this volume, and in recent reviews.[2,3] However, it is worthwhile summarizing some of them briefly (Table 1).

Obviously, if two messengers are to serve interrelated functions one would expect that these messengers influence each in one of several ways. This expectation has been realized. Calcium ion regulates cAMP metabolism either by: (1) increasing cAMP synthesis due to a calcium-calmodulin-dependent activation of adenylate cyclase;[4] (2) increasing cAMP hydrolysis by activating either a soluble or membrane-bound phosphodiesterase;[5,6] or (3) by altering the sensitivity of cAMP response elements to regulation by cAMP.[7] Conversely, cAMP regulates calcium metabolism by stimulating or inhibiting calcium entry into the cell across the plasma membrane[8,9] stimulating calcium efflux of calcium across the plasma membrane,[10] stimulating the energy-dependent calcium accumulation by the endoplasmic reticulum,[11] altering the distribution of calcium between cytosol and mitochondrial matrix space (see Rasmussen and Goodman),[3] and by altering in

Table 1. The cellular and molecular interactions between the calcium and cAMP messenger systems

---

I. Calcium-dependent regulation of cAMP metabolism
  A. Calcium activates soluble phosphodiesterase—brain, heart, many other tissues
  B. Calcium activates membrane-bound phosphodiesterase—rat erythrocyte
  C. Calcium produces a biphasic change of adenylate cyclase function-activation at low and inhibition at high concentrations—brain, adrenal medulla, adrenal cortex, others

II. cAMP-dependent regulation of calcium metabolism
  A. cAMP stimulates energy-dependent uptake of calcium by endoplasmic reticulum
  B. cAMP enhances calcium influx across the plasma membrane—nerve endings, heart, possibly others
  C. cAMP inhibits calcium influx across the plasma membrane—mast cell
  D. cAMP stimulates energy-dependent calcium efflux from cell—smooth muscle, cardiac muscle, others
  E. cAMP alters calcium exchange across the mitochondrial membrane—many tissues

III. Sensitivity modulation of one by the other
  A. cAMP-dependent phosphorylation of phosphorylase b kinase causes an increase in its sensitivity to activation by calcium
  B. cAMP-dependent phosphorylation of myosin light chain kinase causes a decrease in its sensitivity to activation by calcium
  C. Calcium-dependent phosphorylation of cardiac microsomes cause an increase in their sensitivity to activation by cAMP
  D. Calcium-dependent phosphorylation of a response element leading to a decrease in sensitivity to activation by cAMP not yet known

---

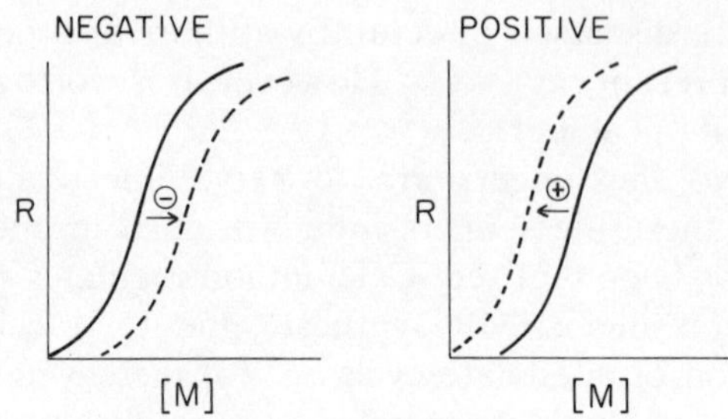

Figure 2. The concept of sensitivity modulation. Response (R) is plotted against messenger concentration [M]. Negative modulation shifts the response to the right, and positive modulation to the left

either a positive or negative way the sensitivity of calcium response elements to activation by calcium ion (Figure 2).

The other expected relationship would be in the realm of regulating the activity of what can best be called cellular response elements. This, too, has been discovered. Calcium and cAMP may function together in one of several ways to control the behaviour of a cellular response element. By this term I mean an integrated unit which is a major site of regulation within the cell, and plays a major role in the cells response to extracellular stimulus. For example, the entire enzyme cascade involved in regulating glycogen metabolism is a prototypical response element. Likewise, the actin–myosin–other regulatory protein complex in a smooth muscle cell is a response element. So too, the luminal membrane of the fly salivary gland, or the sequence of enzymes involved in hepatic or renal gluconeogenesis.

If one considers the way in which the activity of these response elements are controlled by calcium and cAMP, one of several relationships become apparent (Table 2). They may act in a coordinate fashion as in the luminal membrane of the fly salivary gland, one (cAMP) to control K^+ flux, and the other (Ca^{2+}) to control Cl^- permeability and thereby jointly determine the rate of fluid and electrolyte secretion.[12] They may act sequentially as in the control of catecholamine secretion from adrenal medulla.[13] They may act redundantly as in the case of the hormonal control of hepatic gluconeo-genesis.[14] They may act in a hierarchical type of relationship as in the control of insulin from the islets of Langerhans.[15] Finally, they may act in an antagonistic fashion as in control of the state of contraction of many types of smooth muscle.[10]

With this summary as a background, we are now in a position to consider the attributes of synarchic regulation by calcium and cAMP. The concept can be presented most succinctly in terms of the model depicted in Figure 3, When cells are stimulated by extracellular messengers to carry out their specialized function the dual messengers, Ca^{2+} and cAMP, couple stimulus

Table 2. Interrelated control of response element activity
by calcium and cAMP

A. Coordinate
1. Fly salivary gland secretion
2. PTH stimulation of renal gluconeogenesis

B. Sequential
1. Adrenal medullary secretion
2. Cardiac muscle contraction
3. Dopaminergic neurons in CNS

C. Redundant
1. Hepatic gluconeogenesis
2. Skeletal muscle glycogenolysis
3. Aldosterone secretion from *Zona glomerulosa*

D. Hierarchical
1. Exocrine secretion from parotid
2. Insulin secretion from pancreas
3. Muscle contraction in aplysia
4. Synaptic facilitation

E. Antagonistic
1. Smooth muscle contraction
2. Platelet release reaction
3. Histamine release from mast cell

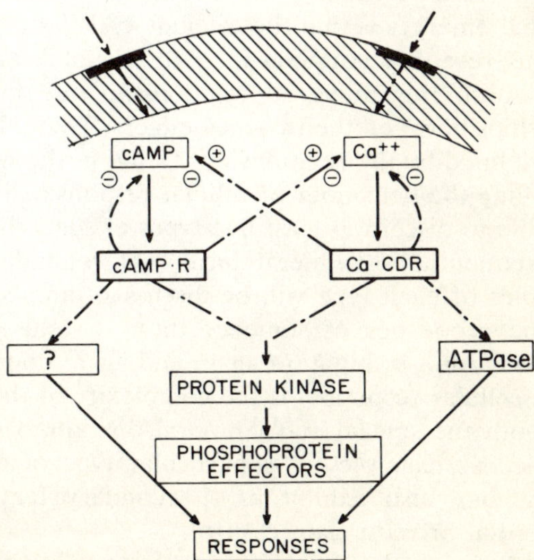

Figure 3. The generalized model of stimulus-response coupling in which synarchic
regulation of cell response is mediated by calcium and cAMP

to response. Within this context, stimulus–response coupling is employed in a more general sense that the usually restricted definition used to define the behaviour of excitable tissues. As will be made evident in the ensuing discussion, when one considers the nature of the systems coupling stimulus to response in either excitable or non-excitable tissue, this universal duality is evident. Hence, it seems both legitimate and appropriate to apply the term stimulus–response coupling to the hormonal activation of aldosterone bio-synthesis as it is to apply it to the neural stimulation of transmitter release.

The features of the model that are universal are: (1) when the totality of extracellular messengers regulating the particular cell's specialized response are concerned than in nearly every case both Ca^{2+} and cAMP act as coupling factors; (2) in every such case Ca^{2+} regulates cAMP metabolism and/or cAMP regulates calcium metabolism in the particular cell; and (3) the two messengers interact in one of a variety of ways (Table 2) to control the activity of cellular response elements.

The system is restricted by definition to extracellular messengers that interact with surface membrane receptors and initiate the specialized re-sponse of that particular cell type. It does not include certain steroid or sterol hormones which may also control the specialized response of a particular cell type but do not do so by interacting with surface receptor. Nor does it include peptide hormones like insulin and somatomedin which also interact with cell surface receptors but elicit a growth or nutritive cellular response rather than a catabolic, specialized work function of a particular cell type. By restricting the definition in this way, the most important truth that emerges is that the calcium–cAMP duality is a common feature of stimulus–response coupling in both excitable and non-excitable cells. However, within this universal relationship one finds a variety of different relationships between the two messengers. These different relation-ships are largely defined by the relationships found in the way in which they interact in controlling the behaviour of cellular response elements (Table 2). Hence, it is possible to discern at least five types of synarchic regulation: (1) coordinate; (2) sequential; (3) hierarchical; (4) redundant; and (5) an-tagonistic. Examples of each type will be discussed, and because of limita-tions of space the properties of some of these systems somewhat over-simplified. However, what is apparent again and again upon deeper analysis of many of these cellular responses is the complexity of the calcium–cAMP relationships in both the spacial and temporal domain. This means that in systems categorized as examples of a particular type of organization, e.g. hierarchical, they may also exhibit as a secondary feature elements of coordinate, sequential or redundant control.

The major conclusion to be drawn from a knowledge of these extensive variations on this universal theme is that the properties of the synarchic system exhibit a high degree of plasticity: this term being used in the sense that Sherrington employed it to describe the control properties of the systems underlying the reflex behaviour of the mammalian nervous system. Its major meaning is that of being the opposite of stereotyped, fixed or formalized, but rather of being open, highly variable and allowing for the expression of a host of operational modes.

2 COORDINATE CONTROL

The concept of coordinate control is represented schematically in Figure 4. A system that demonstrates this type of control is the stimulation of fluid and electrolyte secretion by the fly salivary gland in response to 5-hydroxy-tryptamine (5HT). When 5HT acts upon this cell it causes an increase in the cAMP content of the cell, and also independently stimulates the entry of calcium into the cell across its basolateral membranes.[16] The biochemical correlate of this change in calcium flux appears to be an increase in the turnover of phosphatidylinositol in the membrane.[17] This membrane response is induced by 5HT but by neither the calcium ionophore A23187 or exogenous cAMP even though each of the latter is able to stimulate fluid secretion by the gland. The fact that 5HT, but not exogenous cAMP, is capable of enhancing both PI turnover and calcium uptake argue strongly that 5HT has two independent effects on the plasma membrane being able to simultaneously activate adenylate cyclase and a receptor-mediated calcium channel in this membrane. As illustrated in Figure 4, these two effects are depicted as being the consequence of the interaction of 5HT with two separate types of receptors. This appears likely on the basis of the general properties of receptor coupling in other tissues, but it is a feature of this particular cellular system which remains to be validated.

As depicted in Figure 4, the simultaneous increase in the concentrations of both messengers leads to the coordination of the response events. In this particular cell type, the major response element is the luminal membrane and cAMP appears to activate a K^+ pump in this membrane, and Ca^{2+} to open a Cl^- channel in it. Their combined effects lead to an increase in KCl

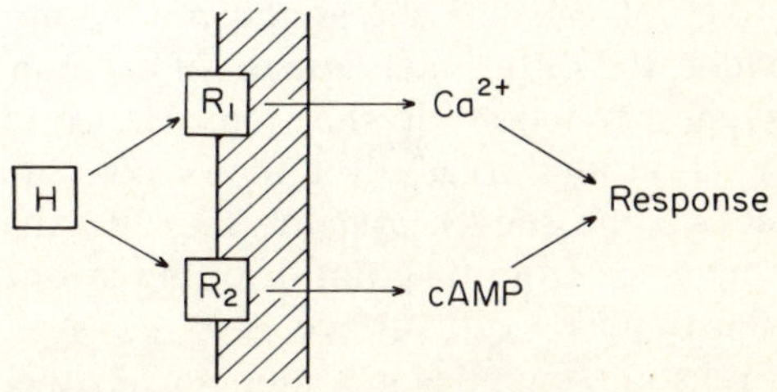

Figure 4. A model of coordinate control of cell function by calcium and cAMP

secretion and consequently to an increase in fluid secretion. Each is necessary for the coordinate response of the cell. In this case it is foolish to talk of either messenger as being of greater importance, as both are of equal significance in determining cellular response.

Not depicted in Figure 4 but an important component of their relationship is that Ca^{2+} and cAMP regulate each other's concentration in this tissue. A rise in calcium acts as a negative feedback regulator of cAMP concentration either by stimulating the activity of phosphodiesterase and/or inhibiting the activity of adenylate cyclase. On the other hand, cAMP acts to increase the mobilization of calcium from an intracellular pool, most likely the mitochondrial pool, and thus extends the spacial domain of the calcium message.

This effect of cAMP on mitochondrial calcium exchange although clearly evident when events are analysed at an intact cellular, physiological level[16] remains a matter of controversy because it has proven difficult to demonstrate an effect of cAMP on calcium exchange in isolated mitochondria from this or a number of other tissues in which a similar cAMP-mediated increase in calcium efflux is observed (see Rasmussen and Goodman[3]). Nonetheless, the physiological data from systems like the fly salivary gland, the mammalian liver and kidney are so consistent that one must continue to believe that directly or indirectly cAMP regulates calcium exchange across the inner mitochondrial membrane.

The potential importance of this effect of cAMP on mitochondrial calcium exchange becomes readily apparent when one considers the situation in a polar secretory cell such as found in the salivary gland. According to our present information, the messenger calcium enters the cell across its basolateral membrane but acts to alter the Cl^- permeability of the luminal membrane. Yet interposed between the two membranes are numerous mitochondria with an enormous capacity to sequester calcium, and thereby being able to greatly restrict the domain of calcium diffusion within the cell. This capacity of mitochondria to restrict the diffusional domain of calcium has been demonstrated directly in other tissues.[18] The paradox one faces is that the diffusion of this calcium is essential if it is to perform its messenger function, yet the cell is normally organized to restrict its diffusion. However, if in conjunction with an increased entry of calcium, an additional messenger acts to alter the fluxes of calcium across the mitochondrial membrane in such a way as to extend the diffusional domain of calcium, it would serve to promote the messenger function of this ion. Hence, in the coordinate regulation of cell function in a polar cell a major consequence between the components of the two messenger systems may be that of insuring the transmission of the message from its point of origin at one cell surface to its point of action at the opposite cell surface.

A similar type of reasoning applies in a number of other systems in which calcium and cAMP interact in such a way that cAMP appears to extend

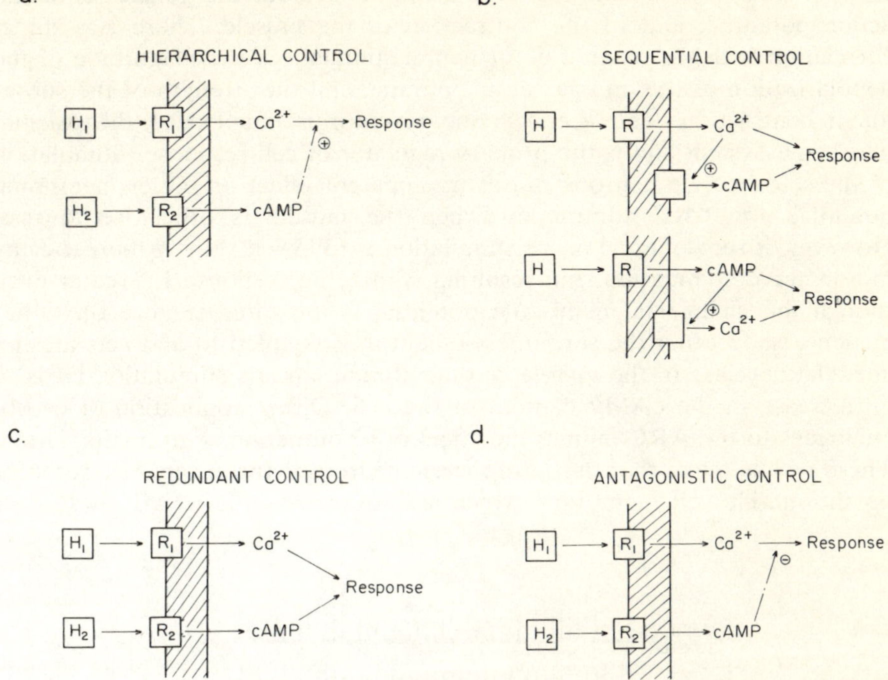

Figure 5. The various models of synarchic regulation of cell function by calcium and cAMP: (a) hierarchical control; (b) sequential control; (c) redundant control; and (d) antagonistic control

either the temporal and/or spacial domains of the calcium message. Clearly, one of the important unresolved issues in the cAMP–calcium relationship is this problem of how cAMP influences mitochondrial calcium exchange.

If we use the coordinate model (Figure 4) as the prototype of the synarchic system, we can easily understand conceptually the other types of control modes illustrated in Figure 5.

3 HIERARCHICAL CONTROL

Two examples of hierarchical control (Figure 5a) will be discussed because in addition to illustrating the properties of this type of control model, each in its own way addresses the question of excitable *vs.* non-excitable cells. The first of these is the neuronal control of contractile activity of the accessory radula closer muscle (ARC) of *Aplysia*;[19] the second the regulation of insulin release from the beta cells of the Islands of Langerhans.

The ARC receives a dual innervation; the first cholinergic motor neurons, the second serotonergic neuron. Release of acetylcholine at the motor nerve endings leads to a membrane depolarization without the generation of an action potential, and to the contraction of the muscle. There is a direct correlation with the magnitude of neural stimulation, the magnitude of the depolarization of the muscle cell membrane, and the strength of the subsequent contraction. The acetylcholine receptor is coupled to the calcium messenger system and is the primary regulator of cell response. Stimulation of the serotonergic neurons causes no apparent effect on either membrane potential nor ARC contraction when the muscle is otherwise at rest. However, if serotonergic nerve stimulation is followed shortly thereafter by motor nerve stimulation, the resulting contractile response is greater even though the change in membrane potential is the same (Figure 6). Other evidence shows that the serotonergic neuron is coupled to and activates an adenylate cyclase in the muscle cell membrane, and its stimulation leads to an increase in the cAMP content of the cell. Direct application of cAMP analogues to the ARC mimics the effect of serotonergic neuron stimulation. These results mean that the serotonergic neurons activate separate receptor on the muscle cell membrane which are coupled to the cAMP messenger

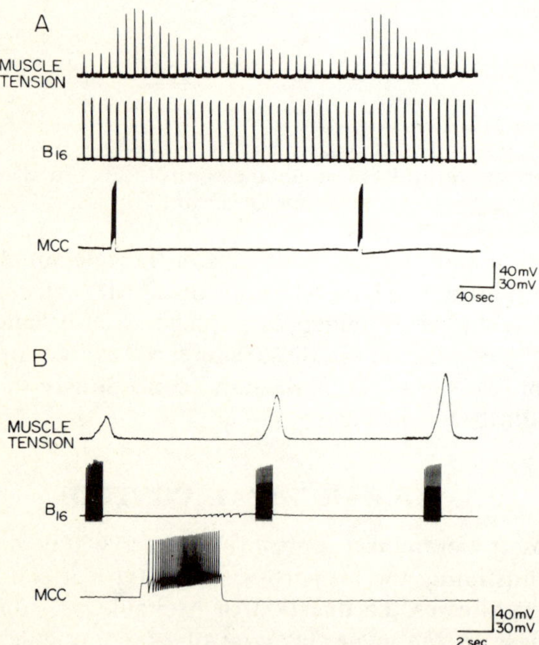

Figure 6. Potentiation of ARC muscle contraction by prior stimulation of the metacerebral cell (MCC). (a) Recording of muscle tension (top) intracellular recording of membrane potential of motor neuron (BIG) and metacerebral cell. (b) Expanded record of second MCC stimulation period in (a). From Kupermann *et al.*[19]

system. It is not yet known how the rise in cAMP enhances the contractile response but by analogy with other systems one of two or both of two possibilities exist. On the one hand, cAMP may extend either the spatial and/or temporal domain of the calcium messenger by altering calcium exchange across one or more subcellular membranes. On the other, cAMP by catalysing the phosphorylation of a protein component of the protein complex regulating actin–myosin interactions may alter the sensitivity of this cellular response element to calcium: the same calcium signal may induce a greater contractile response because the response element is more highly sensitive to this calcium message.

The characteristics of this ARC system raise a point of considerable interest concerning the distinction between excitable and non-excitable tissues. By the classic definition this type of muscle is non-excitable because neural stimulation does not initiate action potential in the muscle cell membrane. Nonetheless, in nearly all other characteristics it is comparable in properties to the excitable tissue, mammalian skeletal muscle. Both are caused to contract when acetylcholine interacts with surface receptors. This interaction leads to a change in the cation permeability properties of the cell membrane which in turn activates the messenger system. The resulting increase in calcium ion concentration couples stimulus to response. In short, the organization similarities of stimulus–response coupling in these two muscle types far outweigh their differences, and tend to call into question the utility of classifying them differently.

The case of insulin release in response to glucose illustrates the opposite situation. When glucose initiates insulin release from the Islets of Langerhans, it induces a train of action potentials in the plasma membrane of the beta cell.[20] Hence, by the classic definition, the beta cell is an excitable cell. However, in this case excitation is induced not by a neurotransmitter but by a blood-borne chemical agent, glucose. Many other endocrine secretory cells which respond to blood-borne messengers liberate their specific product without displaying action potentials, yet many of them employ the calcium messenger system as the primary means of coupling stimulus to response just as occurs in pancreatic islets. Again, the basic operational similarities of stimulus response coupling in these diverse tissues outweighs the fact that one displays action potentials upon activation whereas several others do not.

The factors mediating the release of insulin from the beta cell are also arranged in a hierarchical way (Figure 5). The primary stimulus to insulin release is glucose and this acts predominantly to activate the calcium messenger system. A host of secondary extracellular messengers modulate the response to glucose by influencing the cAMP system.[15] A rise in cAMP enhances and a fall diminishes secretory response to a standard glucose stimulus. Glucagon is an example of an intracellular messenger which potentiates the effect of glucose. Glucagon receptors on the beta cell membrane interacts with specific adenylate cyclase. Epinephrine is an example of an extracellular messenger that suppresses the effect of glucose. It interacts with specific α-receptors on the surface membrane which are also coupled to adenylate cyclase, but interaction of epinephrine with these receptors leads to an inhibition of adenylate cyclase activity. The mechanism by which cAMP influence the glucose-mediated secretory response has not been completely established. It may both increase the intracellular domains of calcium, and alter the sensitivity of the calcium response elements to calcium. These aspects of cAMP–calcium relationships in this tissue remain to be worked out.

In addition to this clearly definable hierarchical type of control employing separate extracellular messengers to regulate the activities respectively of the calcium and cAMP messenger systems, this tissue displays characteristics of sequential control. Incubation of the tissue with low concentrations of glucose initiates secretion by enhancing calcium uptake into the cell without altering the cAMP concentrations. Incubation with higher glucose concentrations causes both a net influx of calcium, and an increase in cAMP content.[21] The glucose-induced increase in cAMP content of the tissue is not due to a direct coupling of a glucose receptor to the adenylate cyclase, but rather appears to involve a calcium dependent increase in the calcium–calmodulin content of the cell, and this complex activates the cyclase. In doing so it acts as a positive feedforward signal to enhance the magnitude of the secretory response to the greater initial glucose signal.

As is evident from the effect of a α-adrenergic activation upon glucose-mediated insulin release, the control systems operating in this tissue also display a component of antagonistic control in that α-adrenergic activation suppresses the effect of a standard glucose signal. In contrast to the other antagonistic systems to be discussed subsequently, the antagonistic feature of this insulin control system is operative at the site of intracellular signal generation and not at the site of intracellular messenger action.

The possession of elements of hierarchical, antagonistic and sequential control by this tissue is but one of numerous examples of the highly plastic capabilities possessed by systems exhibiting synarchic regulation. Using but a few elements (messenger generation and messenger termination) and two messengers, an infinite variety of interactions are possible.

4 SEQUENTIAL CONTROL

A system representing a relatively pure example of sequential regulation (Figure 5b) is the neuronal stimulation of catecholamine secretion from the adrenal medulla.[13] The extracellular stimulus is acetylcholine which acting upon a surface receptor induces a depolarization of the plasma membrane.

This leads to an increased influx of calcium into the cell *via* a voltage-dependent calcium channel. The rise in cytosolic calcium concentration initiates an immediate secretory response, but if the stimulus is sustained, a rise in cAMP concentration occurs. This rise in cAMP is considered a consequence of a calcium–calmodulin mediated activation of adenylate cyclase. The two messages Ca^{2+} and cAMP then acting sequentially appear to be responsible for regulating the activity of preexisting enzyme (calcium) and the content of the enzyme (cAMP), tyrosine hydroxylase, the rate limiting enzyme in catecholamine biosynthesis. So the two messengers arise sequentially and then act coordinately at two different cellular sites to increase the rate of hydroxytyrosine.

As the original extracellular messenger (cholinergic nerve stimulation) continues, the initially elevated cAMP content of the cell falls back to prestimulus values. This is a consequence of a combination of a cAMP-mediated sequestration of calcium within the cell, and by a sequential activation of cytosolic cyclic nucleotide phosphodiesterase by calcium–calmodulin.

Thus, by the sequential activation of the enzyme for cAMP synthesis and the one for cAMP hydrolysis by sequential increases in the calcium ion content of the subsurface cellular domain, and then the more central cell cystole, a transient, but nonetheless essential, rise in cAMP content serves a messenger function in the response of this cell to an extracellular stimulus. It is of interest that in this tissue, as well as in the pancreatic islets, sequential control involves an initial increase in the calcium message which is in turn responsible for evoking the cAMP message. There are examples, for instance, in the response of the mammalian heart to catecholamines in which the opposite appears to hold: the activation of a beta receptor leads to a rise in the cAMP content of the heart which in turn modifies the properties of a voltage-dependent calcium channel in the membrane so as to prolong the duration of calcium flux into the cell.[8]

5 REDUNDANT CONTROL

The hormonal control of hepatic glucose by glucagon and epinephrine was the first system in which the messenger function of cAMP was discovered.[22] From the characteristics of this system, Sutherland and coworkers developed a set of criteria by which to identify cAMP as the second messenger in the action of a peptide or amine hormone. These included: (1) a rise in cAMP concentration in response to the hormone; (2) a hormone-responsive cyclase in the particulate fraction of the cell; (3) an enhanced hormonal effect in the presence of a phosphodiesterase inhibitors; and (4) an ability of exogenous cAMP to mimic the hormonal effect. A fifth criterion, that of the presence of a cAMP-dependent protein kinase, was added later.

The action of glucagon upon hepatic glycogenolysis and gluconeogenesis fulfil all of these criteria. However, even in this case one effect of the cAMP message is an enhanced mobilization of calcium from an intracellular pool,[23] so that the control system mediating glucagon action may well function in a sequential mode similar to that discussed above.

Even though glucagon appears to act via the cAMP messenger system, a number of other agents including catecholamines, angiotensin and vasopressin stimulate hepatic glucose production *via* a calcium-dependent cAMP-independent mechanism.[24,14] The system appears to operate as a redundant one (Figure 5c) in which either intracellular messenger system can activate the final response elements within the cell. Stimulation of these cells with either an α-adrenergic agent or with glucagon leads to a phosphorylation of the same proteins[25] (all of which are probably regulated enzymes of either the glycogenolytic or gluconeogenic sequence).

The concepts that emerge from these findings and those in other systems[26] are: (1) the same regulated protein may be a substrate for either a cAMP-dependent or a calcium-dependent protein kinase; (2) the two different kinases commonly catalyse the phosphorylation of different sites on the same protein molecule; and (3) phosphorylations of these different sites causes qualitatively similar, but quantitatively different changes in the function of the protein (Figure 7). Also, joint phosphorylation by both types of kinases often leads to a greater change in protein function than seen after phosphorylation by either alone. This means that the separate intracellular messenger systems act both in a redundant and a hierarchical fashion to determine cellular response.

A similar redundant type of control is seen in the hormonal regulation of aldosterone secretion from the zona glomerulosa of the adrenal cortex. Three different extracellular messengers are of physiological importance in

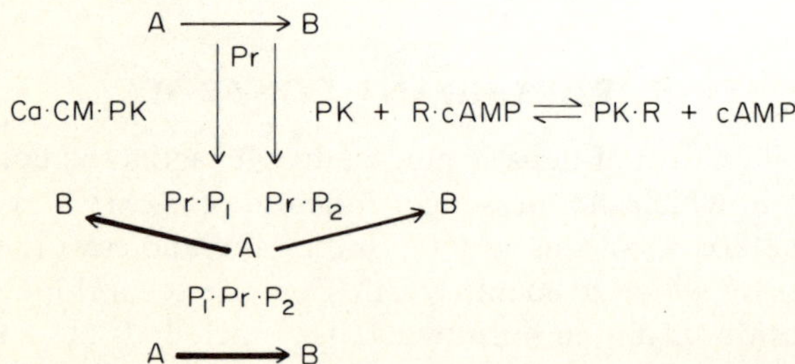

Figure 7. Representation of dual control of the activity of the same enzyme (Pr) by calcium dependent protein kinase (Ca·CM·PK) and cAMP dependent protein kinase (PK). The two protein kinases catalyse phosphorylation of different sites (Pr·P$_1$ and Pr·P$_2$) on the enzyme. Each phosphorylation cause an activation of the enzymes. The doubley phosphorylated enzyme (P$_1$·Pr·P$_2$) is also active

controlling its rate of secretion:[27,28,29] (1) extracellular K^+; (2) angiotension; and (3) adrenocorticotropin (ACTH).

If we contrast the effects of the two peptide hormones, angiotensin and ACTH, we find that each separately stimulates aldosterone secretion, that the effect of angiotensin is highly dependent upon extracellular calcium but that of ACTH less so, that angiotensin stimulates calcium uptake into the gland, and that ACTH but not angiotensin stimulates adenylate cyclase and induces, thereby, a rise in cellular cAMP content. From these data the conclusion has been drawn that angiotensin acts *via* the calcium messenger system, ACTH via the cAMP messenger system. Hence, the situation is operationally similar to that seen in the liver. However, just as in the liver where the action of glucagon, the activator of the cAMP system, probably also involves Ca^{2+}, so in the adrenal glomerulosa, the action of ACTH is also partially Ca^{2+} dependent.

6 ANTAGONISTIC CONTROL

Up until this point in our discussion, the relationships considered have all been ones in which the two messenger systems compliment one another in one fashion or another. However, equally as important from a physiological viewpoint, and in validating the synarchic concept are situations in which the two messenger systems act in an opposing or antagonistic fashion (Figure 5d). This type of control system is of widespread occurrence being found in the case of histamine release from the mast cell[9] in the platelet release reaction,[30] and a large number of different types of smooth muscle.[31] In all of these cases, the primary extracellular stimulus or stimuli initiate response via the calcium messenger system, and a variety of other extracellular messengers acting via the cAMP messenger system exert an inhibitory or suppressive action. An instance in which a primary activating system employs the cAMP messenger, and is antagonized by agents acting via the calcium messenger system has not, to my knowledge, been described but within the conceptual framework of synarchic regulation is likely to be found.

For example, one might conclude from the particular variations of the universal synarchic system depicted in Figure 5 that calcium ion is nearly always the primary messenger involved in initiating the specialized response of the cell, and that, by and large, the cAMP messenger system serves to modulate the behaviour of the calcium messenger system. In many of the examples discussed this appears to be the case. However, the case of glucagon action on the liver is a clear exception. Furthermore, a consideration of the respective roles of two messenger systems play in mediating the secretory response in the two exocrine glands, the parotid and the pancreas, is most instructive.[32,33] To oversimplify the situation slightly, secretion in each of these glands consists of two components: (1) fluid and electrolytes; and (2) digestive enzymes by the process of exocytosis. The accumulated evidence supports the conclusion that in the parotid the cAMP messenger system mediates primarily the exocytotic response and has only a secondary role in stimulating fluid secretion whereas the calcium messenger system has a primary role in regulating electrolyte secretion and plays a secondary role in regulating exocytosis. In contrast, in the exocrine pancreas the calcium messenger system has the primary role in mediating exocytosis, and the cAMP system in controlling fluid and electrolyte output.

These conclusions are understandable in the light of our previous discussion of redundant control systems in which either messenger system can activate the same cellular response elements. Given this further evidence of the plasticity of the control modes available within the concept of synarchy, it is reasonable to propose that eventually cAMP-dominated antagonistic control systems will be found.

Returning to a consideration of what we might call calcium-dominated antagonistic control systems, it is of interest to explore the calcium–cAMP relationships in the regulation of smooth muscle contraction. The summary of these relationships will be based on data from a number of different types of smooth muscle so a composite picture of their relationships in an idealized smooth muscle cell will be presented.[31] In practice, one or more elements of this composite may either be lacking or greatly exaggerated in a specific type of smooth muscle response.

Essential to an understanding of this discussion is a knowledge of how Ca^{2+} mediates the actin–myosin interaction in smooth muscle. In contrast to the situation in skeletal muscle where the calcium receptor protein is a part of a complex of proteins which interact with actin (thin filament regulation) and thereby control the interaction of actin with myosin, in smooth muscle a major means of activation involves a calcium-dependent protein kinase, myosin light chain kinase (MLCK) which catalyses the phosphorylation of the regulatory light chain of myosin. This phosphorylation initiates the actin–myosin interaction (thick filament regulation). This enzyme, MLCK, has as its calcium binding subunit, the calcium receptor protein, calmodulin.

When a smooth muscle is stimulated to contrast by a cholinergic neuron, either there is an increase in calcium entry across the plasma and/or a release of calcium from an intracellular, probably microsomal, pool. The resulting increase in cytosolic calcium leads to the activation of MLCK, the phosphorylation of myosin, and thereby to an increase in force of contraction. Stimulation of adrenergic neurons leads to the interaction of catecholamine is with beta receptors which are coupled to adenylate cyclase. Receptor activation causes an increase in cAMP content of the muscle cell. This rise in cAMP content causes a decrease in contractile force. The decrease is brought about by several different mechanisms: first cAMP

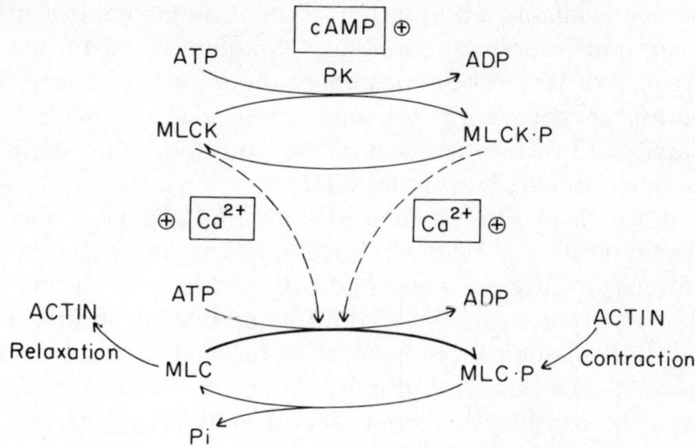

Figure 8. The interrelated roles of cAMP and Ca^{2+} in regulating smooth muscle contraction. The state of contraction is controlled by the balance between the phosphorylated (MLC-P) and non-phosphorylated forms of myosin light chain (MLC). This phosphorylation state is determined by the relative activities of two enzyme; a phosphatase which catalyses the conversion of MLC·P to MLC; and a protein kinase, myosin light chain kinase (MLCK). This MLCK is a calcium-activated, calmodulin-dependent enzyme. However, MLCK can exist either as a free (MLCK) or phosphorylated form (MLCK·P). When in the latter form, MLCK·P, the enzyme is less sensitive to activation by calcium. The state of phosphorylation of MLCK is controlled by a phosphatase and a cAMP-dependent protein kinase

stimulates the energy-dependent accumulation of calcium by the endoplasmic reticulum; second, by one of several possible mechanisms cAMP stimulates the efflux of calcium from the cell; and third, by activating a protein kinase cAMP induces the phosphorylation of MLCK. The phosphorylation of this enzyme by the cAMP-dependent protein kinase decreases its sensitivity to activation by calcium (Figure 8). Since there is a direct correlation between the degree of myosin light chain phosphorylation and the contractile state of the muscle, and the degree of this phosphorylation is a balance between the activity of MLCK and of phosphoprotein phosphatase, a decrease in the magnitude of the activation of MLCK by calcium leads to a decrease in the strength of muscle contraction (Figure 8).

The fact that cAMP induces muscle relaxation by exerting several separate effects on cell function, brings up an important point concerning synarchic regulation in general. As biochemists, we have attempted to define the problem of hormone action in its starkest form by demanding to know the primary or single molecular event by which the hormone initiates cellular response. This demand seemed to have been met in the discovery and elaboration of the second messenger model, but as is clear from the present discussion this model must be significantly modified if it is to account for the presently available information about cellular control mechanisms involved in stimulus–response coupling. Nonetheless, with the apparent simplistic elegance of the second messenger model, the discovery of protein kinase activation as apparently the only molecular mechanism by which cAMP regulates cell function focused further attention onto the philosophic concept of a single unitary hormonal effect.

However, if we drop our biochemist's mantle and view the cell as in endocrine physiologists, the truth that becomes evident is that hormones or neurotransmitters produce an integrated cell response. A variety of cellular response elements are activated in a coordinated fashion so that the changes in the activities of all the different parts of the cell produce an integrated cellular response. The intuitive confirmation of this statement becomes apparent when we consider hormone action at the level of the organism. Just as Sherrington proclaimed the 'integrative action of the nervous system' one can proclaim the integrative action of the endocrine system. If one takes as an example a single hormone such as epinephrine released from the adrenal medulla, it is known that it exerts multiple actions on multiple organs systems all integrated to provide the organism with optimal function to undertake fight or flight. If we consider only its action on energy metabolism, providing both blood glucose, and blood non-esterified fatty acid is an integrative response in terms of providing adequate glucose to the brain and working muscle both by increasing its release into the bloodstream[34] and by providing an alternative energy-yielding substrate for many of the other less immediately essential tissues of the body.

If we carry this concept of integrative action back to the intracellular domain, it then becomes apparent that synarchic regulation is involved above all else in providing the cell with an integrative mechanism by which to couple stimulus to response, and at the same time endow this system with sufficient plasticity to be responsive enough to meet the changing needs of the organism.

One particular feature of the situation in smooth muscle worthy of discussion is the relationship between the calcium-dependent and cAMP-dependent protein kinases. If we contrast their relationship in smooth muscle with that in the phosphorylase system an interesting point emerges. In both cases the phosphorylation of a subunit of a calcium-dependent protein kinase, other than its calcium-binding subunit, leads to a change in the sensitivity of the enzyme to activation by calcium. In the phosphorylase system, cAMP-dependent phosphorylation of the calcium-dependent enzyme increases its sensitivity to activation by calcium; and in the myosin light chain kinase system, it decreases its sensitivity to activation by calcium.

This type of control which can be termed *sensitivity modulation* (Figure 2) is probably a more widespread phenomenon than currently realized. There are examples of it in the cAMP as well as the calcium messenger system.

The point to be understood is that either cAMP or Ca^{2+} may serve a messenger function in a cellular response without their concentrations changing, simply by an alteration in the sensitivity of a response element to their action.

7   THE UNIVERSAL *vs.* THE PARTICULAR

The conclusion which I should like to draw is that the synarchic system underlying stimulus–response coupling in the case where cells are called upon to perform their specialized function is as universal an attribute of cell biochemistry as is the Krebs cycle. The immediate challenge to this conclusion is likely to be that as just discussed this 'universal' control systems has so many variations as to belie the very concept of its universality whereas the Krebs cycle is a very constant and unvarying unit of the biochemical machinery of the cell. However, if we view the activity of the Krebs cycle in its natural setting, the mitochondria, and consider the participation of the enzymes of the cycle in metabolic events in different types of tissue, it becomes apparent that this duality between the universal and the particular is as much a property of the Krebs cycle enzymes and the related transport systems for the Krebs cycle intermediates across the mitochondria membranes as it is of the synarchic system underlying stimulus–response coupling.

In support of this conclusion, one can contrast the behaviour of Krebs cycle enzymes in the mitochondria of the insect flight muscle, and in those of the rat liver. From the physiological point of view the flight muscle is a tissue which when activated has an extremely high rate of energy utilization. This demand is met by mitochondria capable of high rates of oxidative phosphorylation. To function most efficiently to serve this single type of metabolic behaviour, the mitochondria are organized so that other than pyruvate, none of the Krebs cycle intermediates cross the mitochondria membrane to any significant extent. Hence, in a way the Krebs cycle enzymes are essentially isolated within the mitochondrial matrix compartment and serve therein only their stereotyped cyclic function of rapidly and efficiently converting acetyl Co to CO_2 and H_2O and generating thereby ATP.

In contrast, the mitochondria of the hepatocyte are the site of a number of crossroads in energy metabolism. The shuttling of Krebs cycle intermediates across the mitochondrial membrane and the utilization of only a few of the cycle enzymes are an essential feature of the urea cycle, of the process of fatty acid biosynthesis, and of gluconeogenesis. Quite often the flux of carbon skeletons via a segment of the Krebs cycle is far greater than the flux through the entire cycle. This fantastic plasticity exhibited by the participation of these enzymes in the metabolic functions of the liver mitochondria stands in sharp contrast to the very stereotyped metabolic activity of the mitochondria from the insect flight muscle. The difference in the organization complexity of this system in the two tissues is as great as the difference in complexity between the rather stereotyped coupling of stimulus to response in the fly salivary gland, and the highly plastic multifactorial coupling of stimuli to response in the beta cell of the islets of Langerhans.

8 CONCLUDING REMARKS

A summary of the major conclusions drawn from the preceding discussion is presented in Table 3. The purpose of this discussion has been to place a consideration of the phenomenon of stimulus–response coupling within a broader conceptual framework.

The value of this new conceptual framework within which to consider the mechanisms by which stimuli are coupled to responses in animal cells, lies in forcing us to reconsider our prejudices, and to formulate new questions or reformulate old ones. The particular value of the present formulation of a universal dual second messenger system lies in the obvious conclusion to which it leads. Namely, only by considering the roles of both calcium and cAMP in the regulation of any specialized cellular response are we likely to develop a complete understanding of the means by which this regulation is achieved.

Table 3. Summary of the features of stimulus–response coupling by synarchic messengers

---

1. Synarchic regulation by calcium and cAMP is a universal mechanism by which stimulus–response coupling is achieved in differentiated cells when these are activated by extracellular messengers interacting with surface receptors.

2. This type of regulation is involved when a differentiated cell is called upon to perform its specialized function.

3. Although a universal system, synarchic control exhibits a number of specific variations.

4. These variations can be defined by the nature of the interactions between calcium and cAMP and include coordinate, hierarchical, redundant, sequential and antagonistic patterns.

5. In some cellular systems elements of more than one of these patterns operate.

6. Viewed within the context of stimulus–response coupling, there is no difference between so-called excitable and non-excitable cells.

7. A change in a messenger-mediated response can be achieved either by amplitude or sensitivity modulation.

---

9 ACKNOWLEDGEMENT

The work in my laboratory which has formed a part of the basis of this study was supported by a grant (AM-19813) from the National Institutes of Health. I am indebted to Mrs. Nancy Canetti and Ms. Anita See for secretarial assistance.

REFERENCES

1. Rasmussen, H. (1970). Cell communication, calcium ion, and cyclic adenosine monophosphate. *Science*, **170**, 404–412.
2. Berridge, M. J. (1976). The interaction of cyclic nucleotides and calcium in the control of cellular activity. *Advan. Cyclic Nucleotide Res.*, **6**, 1–96.
3. Rasmussen, H., and Goodman, M. A. (1977). Effect of repeated treatment of rabbit platelets with low concentrations of thrombin on their function, metabolism and survival. *Brit. J. Haematol.*, **25**, 675–689.
4. Brostrom, C. O., Huang, Y.-C., Breckenridge, B. McL., and Wolff, O. J. (1975). Identification of a calcium-binding protein as a calcium dependent regulator of brain adenylate cyclase. *Proc. Natl. Acad. Sci. U.S.A.*, **72**, 64–68.
5. Wang, J. H., Teo, T. S., Ho, H. C., and Stevens, F. C. (1975). Bovine heart protein activator of cyclic nucleotide phosphodiesterase. *Advan. Cyclic Nucleotide Res.*, **5**, 179–194.
6. Clayberger, C., Goodman, D. B. P., and Rasmussen, H. (1981). Regulation of cyclic AMP metabolism in the rat erythrocyte during chronic β-adrenergic stimulation. Evidence for calmodulin mediated alteration of membrane-bound phosphodiesterase activity. *J. Memb. Biol.*, **58**, 191–201.
7. LePeuch, C. J., Harech, J., and Demaille, J. G. (1979). Concerted regulation of cardiac sarcoplasmic reticulum calcium transport by cyclic adenosine monophosphate dependent and calcium–calmodulin dependent phosphorylations. *Biochemistry*, **18**, 5150–5157.
8. Tsien, R. W. (1977). Cyclic AMP and contractile activity in the heart. *Advan. Cyclic Nucleotide Res.*, **8**, 364–420.
9. Foreman, J. C., Garland, L. G., and Mongar, S. L. (1976). The role of calcium in secretory processes: model studies in mast cells. In *SEB Symposium XXX*, pp. 193–218.
10. Andersson, R. K., Nilsson, J., Wikberg, S. Jo-Hannson, and Lundholm, L. (1975). Cyclic nucleotides and the contractions of smooth muscle. *Advan. Cyclic Nucleotide Res.*, **5**, 491–518.
11. Katz, A. M., Tada, M., and Kirchberger, M. (1975). Control of calcium transport in the myocardium by the cyclic AMP–protein kinase system. *Advan. Cyclic Nucleotide Res.*, **5**, 453–472.
12. Berridge, M. J., and Prince, W. T. (1972). The role of cyclic AMP and calcium in hormone action. *Advan. Insect Physiol.*, **9**, 1–49.
13. Guidotti, A., Hanbauer, I., and Cosla, E. (1975). Role of cyclic nucleotides in the induction of tyrosine hydroxylase. *Advan. Cyclic Nucleotide Res.*, **5**, 619–640.
14. Exton, J. H., and Harper, S. C. (1975). Role of cyclic AMP in the actions of catecholamines on hepatic carbohydrate metabolism. *Advan. Cyclic Nucleotide Res.*, **5**, 519–532.

15. Malaisse, W. J. (1973). Insulin secretion: multifactorial regulation for a single process of release. *Diabetologia,* **9,** 167–173.

16. Prince, W. T., Berridge, M. J., and Rasmussen, H. (1972). Role of calcium and adenosine 3′:5′-cyclic monophosphate in controlling fly salivary gland secretion. *Proc. Natl. Acad. Sci. U.S.A.,* **69,** 553–557.

17. Fain, J. M., and Berridge, M. J. (1979). Relationship between hormonal activation of phosphotidylinositol hydrolysis fluid secretion and calcium flux in blowfly salivary glands. *Biochem. J.,* **178,** 45–58.

18. Baker, P. F. (1976). The regulation of intracellular calcium. In *Calcium in Biological Systems, SEB Symp.,* **30,** 67–88.

19. Kupfermann, I., Cohen, J. L., Mandelbaum, D. E., Schonberg, M., Susswein, A. J., and Weiss, K. R. (1979). Functional role of serotonergic neuromodulation in *Aplysia. Fed. Proc.,* **38,** 2095–2102.

20. Matthews, E. K. (1975). Calcium and stimulus–secretion coupling in pancreatic islet cells. In *Contraction and Secretion* (Eds. Carafoli, E., Clementi, F., Drabikowsky, W., and Margreth, A.), pp. 203–210, Amsterdam, North-Holland.

21. Grill, V., and Cerasi, E. (1974). Stimulation by D-glucose of cyclic adenosine 3′:5′-monophosphate accumulation and insulin release in isolated pancreatic islets of the rats. *J. Biol. Chem.,* **249,** 4196–4201.

22. Sutherland, E. W., and Rall, T. W. (1958). Formation of cyclic adenine ribonucleotide by tissue particles. *J. Biol. Chem.,* **232,** 1065–1076.

23. Friedmann, N. (1974). Effects of glucagon and cyclic AMP on ion fluxes in the perfused liver. *Biochim. Biophys. Acta,* **274,** 214–225.

24. Tolbert, M. E. M., and Fain, J. N. (1974). Studies on the regulation of gluconeogenesis in isolated rat liver cells by epinephrine and glucagon. *J. Biol. Chem.,* **249,** 1162–1166.

25. Garrison, J. C., Borland, M. K., Florio, V. A., and Twible, D. A. (1979). The role of calcium as a mediator of the effects of angiotensin II, catecholamines and vasopressin on the phosphorylation and activity of enzymes in isolated hepatocytes. *J. Biol. Chem.,* **254,** 7147–7156.

26. Cohen, P. (1978). The role of cyclic AMP-dependent protein kinase in the regulation of glycogen metabolism in mammalian skeletal muscle. *Curr. Top. in Cell Reg.,* **14,** 117–196.

27. Albano, J. D. M., Brown, B. L., Ebens, R. P., Tart, S. A. S., and Tait, J. F. (1974). The effect of potassium, 5-hydroxytryptamine, adrenocorticotrophin and angiotensin II on the concentration of adenosine 3′:5′-cyclic monophosphate in suspensions of dispersed rat adrenal zona glomerulosa and zona fasiculata cells. *Biochem. J.,* **142,** 391–400.

28. Mackie, C., Warren, R. L., and Simpson, E. R. (1978). Investigation into the role of calcium ions in the control of steroid production by isolated adrenal zona glomerulosa cells of the rat. *J. Endocrinol.,* **77,** 119–127.

29. Fakunding, J. L., Chow, R., and Catt, K. J. (1979). The role of calcium in the stimulation of aldosterone production by adrenocorticotropin, angiotensin II, and potassium in isolated glomerulosa cell. *Endocrinol.,* **105,** 327–333.

30. Feinstein, M. B., and Fraser, C. (1974). Human platelet secretion and aggregation induced by calcium ionophores. Inhibition by PGE, and dibutyryl cyclic AMP. *J. Gen. Physiol.,* **66,** 561–581.

31. Adelstein, R. S., and Hathaway, D. R. (1979). Role of calcium and cyclic adenosine 3′:5′-monophosphate in regulating smooth muscle contraction. *Am. J. Cardiology,* **44,** 783–787.

32. Butcher, F. R. (1978). Calcium and cyclic nucleotides in the regulation of

secretion from the rat parotid by autonomic agonists. *Advan. Cyclic Nucleotide Res.*, **9**, 707–721.

33. Schramm, M., and Selinger, Z. (1975). The functions of cyclic AMP and calcium as alternative second messenger in parotid gland and pancreas. *J. Cyclic Nucleotide Res.*, **1**, 181–192.

34. White, J. G., Rao, G. H. R., and Gerrard, J. M. (1974). Effects of the ionophore A23187 on blood platelets. *Am. J. Pathol.*, **77**, 135–149.

10 Receptor mediated modulation of plasma membrane calcium transport

D. L. Gill

1 INTRODUCTION

The importance of Ca^{2+} as an intracellular mediator of response mechanisms elicited by external stimuli is well recognized and has been emphasized in other chapters of this volume. Although considerable information is available regarding the locations and mechanisms of the Ca^{2+} regulatory machinery of cells, there is often confusion regarding which of these components is influenced to bring about a particular Ca^{2+} mediated event. This is a consequence of the complex and poorly understood relationship between the several different categories of Ca^{2+} translocating mechanisms situated at discrete sites within the cell.

Against an extracellular free Ca^{2+} concentration in the millimolar range, cytosolic free Ca^{2+} is maintained at the submicromolar level by three fundamental membrane-mediated mechanisms: (1) enzymic transmembrane Ca^{2+} transport directly utilizing energy from ATP hydrolysis—almost exclusively the $(Ca^{2+}+Mg^{2+})$-dependent ATPase; (2) mediated exchange indirectly coupled to the energy dependent transport of another ion—frequently Na^{+}–Ca^{2+} or H^{+}–Ca^{2+} exchange; (3) immobilization of Ca^{2+} ions either by complexation with acidic residues of phospholipids or proteins, or by deposition as insoluble salts within Ca^{2+}-accumulating compartments with selective permeability towards appropriate anions. In addition to mechanisms serving to limit the concentration of free Ca^{2+} in the cytosol, the cell contains specific mechanisms to permit entry of Ca^{2+} down its electrochemical gradient. Passive diffusion across the membrane is an unlikely entry mechanism due to the high impermeability of membranes to ionic Ca^{2+}. Instead, cells contain facilitated Ca^{2+} influx mechanisms controlled by, for example, membrane potential. However, the exact nature of these influx mechanisms remains largely obscure.

The mechanisms of Ca^{2+} efflux from the cytosol exist primarily at three locations within the cell: the mitochondria, the endoplasmic or sarcoplasmic reticulum, and the plasma membrane. Controlled Ca^{2+} influx sites are believed to exist primarily at the plasma membrane, but similar mechanisms may mediate influx from the internal Ca^{2+}-accumulating organelles. Therefore, cellular events mediated by an increase in cytosolic Ca^{2+} may arise from deactivation or activation of one or more of the efflux or influx mechanisms, respectively.

It is the purpose of this chapter to discuss how agents that interact with plasma membrane receptor sites may *directly* effect changes in the activity of Ca^{2+} translocating mechanisms within the same membrane, that is, without invoking the action of any non-membrane-associated intermediary. An agent that interacts initially with cell surface receptors but affects Ca^{2+} translocation of an internal organelle cannot be classified as 'directly' acting and will not be included in this discussion. Some external stimulators may indirectly affect plasma membrane Ca^{2+} translocation through the release of an intermediary agent. Although the primary stimulator cannot be classified as acting directly, the secondary agent clearly can. Only brief mention of the mechanisms of such intracellular effectors (in particular calmodulin and cyclic AMP) will be included since their actions are considered in more detail elsewhere in this volume. The discussion opens with a brief consideration of the nature of the Ca^{2+} translocating mechanisms of the plasma membrane. For more complete coverage of such mechanisms, the reader is urged to consult the excellent reviews available.[1,2,3,4]

2 PLASMA MEMBRANE CALCIUM TRANSPORT MECHANISMS

2.1 Plasma membrane calcium influx

In describing the role of the plasma membrane in mediating an increase in cytosolic Ca^{2+}, the term 'influx' is not necessarily accurate. Thus, although much evidence suggests the existence of one or more classes of specific channels to allow the influx of Ca^{2+} from the external medium, there are at least two other possible mechanisms by which the plasma membrane may facilitate an increase in cytosolic Ca^{2+} without invoking an increase in the rate of translocation of extracellular Ca^{2+} across the plasma membrane. Firstly, the plasma membrane may have a pool of releasable surface-bound Ca^{2+} associated, for example, with acidic phospholipids, the content and turnover of which appear to be intimately associated with a great number of agents known to affect intracellular Ca^{2+} levels.[5] Secondly, the rate of plasma membrane Ca^{2+} efflux by either of the mechanisms described in Section 2.2 may be attenuated allowing internal Ca^{2+} levels to rise as a result of constant Ca^{2+} entry from either the extracellular fluid or internal Ca^{2+}-containing organelles.

Specific channels for the inward movement of Ca^{2+} through the plasma membrane are believed to operate in most secretory or excitatory cells where they are an important mechanism through which Ca^{2+} enters the cytosol to mediate coupling between external stimulation and cellular secretion. The operation of such channels has been considerably documented in excitatory cells where evidence suggests that the plasma membrane contains a voltage sensitive Ca^{2+} channel, the function of which resembles but is distinct from the inward Na^+ channel (for reviews see Baker,[6] and Mullins[7]). Thus, depolarizing impulses applied to many nerve or muscle cells elicit not only inward Na^+ movement (in most cases a major component of the action potential) but also a significant influx of Ca^{2+}. Although the activated Na^+ channel does conduct Ca^{2+}, albeit with low efficiency (approximately 1% relative to Na^+), the distinction between the two channels has clearly been demonstrated by Baker *et al.*,[8] who observed two components of Ca^{2+} entry into the squid axon in response to depolarizing pulses of varying length (Figure 1). In this experiment Ca^{2+} accumulation was monitored by measuring light emission from aequorin introduced within the axon. Short pulses applied to the axon produce a small increase in internal Ca^{2+} which is inhibited by tetrodotoxin (TTX, a potent Na^+-channel blocker) and represents Ca^{2+} entry through the Na^+ channel. However, longer pulses cause an additional increase in Ca^{2+} accumulation which is TTX-independent. This influx is termed the 'slow' Ca^{2+}-channel and has been shown to have distinct specificity for Ca^{2+} [9] and the so-called Ca^{2+}-channel antagonists, verapamil and D600.[10] Although the current carried by the Ca^{2+}-channel comprises a small proportion (approximately 0.1 per cent) of the total inward current of

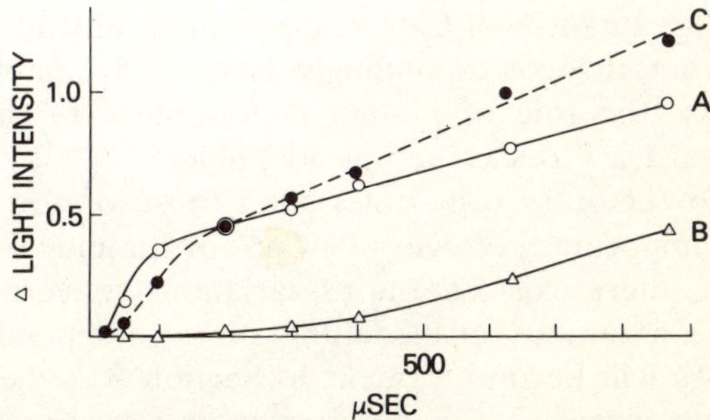

Figure 1. Evidence for two phases of Ca^{2+} entry in the squid giant axon. Curves show the relationship between the duration in μsec. of applied voltage-clamp pulses of 80 mV (abscissa) and the increase in aequorin light emission in response to Ca^{2+} entry, per pulse. Measurements were made before addition of TTX (a), in the presence of TTX (b), or after removal of TTX (c). (Reproduced from Baker *et al.*,[8] by permission of the *Journal of Physiology*, London)

the squid axon action potential, in some tissues, for example, certain crustacean muscle fibres, an inward Ca^{2+} current almost totally replaces the inward Na^+ current in the generation of action potentials.[11] A similar Ca^{2+} current predominates in the action potential of many vertebrate smooth muscles, and both Na^+ and Ca^{2+} contribute to the action potential of cardiac tissue.

In general, the entry of Ca^{2+} under depolarizing conditions is important to initiate stimulus-secretion or stimulus-contraction coupling in excitatory cells. Thus the squid giant synapse contains a much greater density of Ca^{2+} channels than the axon such that Katz and Miledi,[12] during their original studies on the role of Ca^{2+} in transmitter release, were able to measure the Ca^{2+} current at the nerve terminal though noted a considerable diminution in Ca^{2+} conductance away from this area. In fact the function of voltage controlled Ca^{2+} influx sites appears to be a universal property of neuro-secretory nerve terminals, permitting Ca^{2+} entry in response to arrival of the depolarizing action potential, and thus initiating events that lead to movement, fusion and release of neurotransmitter storage vesicles. In muscle cells, the function of sarcolemmal Ca^{2+} influx sites triggered by surface depolarization is not necessarily so intimately related with the response (contraction) even though the latter is absolutely dependent upon raised cytosolic free Ca^{2+}. Thus transsarcolemmal Ca^{2+} influx during the action potential may be insufficient to cause contraction of myofilaments in cardiac muscle[12A] although it is widely reported that Ca^{2+} entering through the plasma membrane triggers the larger release of Ca^{2+} from sacroplasmic reticulum.[13] Contraction of smooth muscle is perhaps more closely related to entry of external Ca^{2+}, although it is not always clear by what mechanism stimulatory agents may induce Ca^{2+} influx and how membrane potential is related to stimulation[14] (see Section 3.2 for further discussion). In skeletal muscle the very rapid and large quantity of Ca^{2+} released internally appears to originate almost totally from the correspondingly large and complex sarcoplasmic reticulum. In fact, the role of plasma membrane Ca^{2+} influx, even as a trigger for internal Ca^{2+} release, is questionable.[15]

With regard to secretory cells other than those of the nervous system, even though the importance of cytosolic Ca^{2+} in stimulus-secretion coupling is generally clear, there exist enormous variations concerning the reported role of external Ca^{2+} influx in mediating stimulus-dependent increases in cytosolic Ca^{2+}. As will become evident in Section 3.2, the actual mechanisms of plasma membrane Ca^{2+} influx in response to stimulators are mostly ill-defined in these tissues. In many cases voltage sensitive Ca^{2+} influx has been observed but the connection between secretogogue action and changes in membrane potential is by no means universal (see chapter by Grollman in this volume for an analysis of the influence of various membrane acting ligands upon membrane potential).

2.2 Plasma membrane calcium efflux

The plasma membrane of vertebrate cells contains two major mechanisms for extruding Ca^{2+} from the cytoplasm against its electrochemical gradient—an enzymic ATP-dependent transport mechanism and a mediated Na^+–Ca^{2+} antiporter. A H^+–Ca^{2+} antiporter mechanism is reported to function in some prokaryotes to limit internal Ca^{2+} levels, for example, the plasma membrane of *E. coli*,[16] and recent evidence suggests that a similar mechanism operates in the eukaryotic fungus, Neurospora.[17] The following discussion is limited to the two major efflux mechanisms observed in eukaryotic cell plasma membranes.

ATP-dependent Ca^{2+} transport occurs through the operation of the plasma membrane $(Ca^{2+} + Mg^{2+})$-dependent ATPase. The Ca^{2+} transporting function of this enzyme has been widely investigated in the plasma membrane of erythrocytes since the initial demonstration of the formation of an outward Ca^{2+} gradient in these cells.[18] Erythrocytes offer the advantage of being readily prepared without loss of function and being easily lysed by treatment with hypotonic media to form osmotically intact 'ghosts' or, under suitable conditions, inverted plasma membrane vesicles.[19] It is likely that ATP-dependent Ca^{2+} transport is a universal property of plasma membranes, similar activity being reported in a variety of cells including L-cells,[20] platelets,[21] kidney cells[22] and possibly liver cells.[23] Excitable cells, which contain an alternative Ca^{2+} extrusion mechanism (Na^+–Ca^{2+} exchange), are not exceptions, ATP-dependent Ca^{2+} transport having been recently observed in the plasma membrane of squid axons,[24] cardiac sarcolemma,[25] and in mammalian nerve terminals,[26] as described in detail below.

As with the widely investigated plasma membrane ion transporter, $(Na^+ + K^+)$-ATPase, the Ca^{2+} translocating enzyme exhibits asymmetry. Thus Ca^{2+}, Mg^{2+} and ATP are all required inside the intact cell to obtain outward Ca^{2+} extrusion.[27] However, these constituents applied externally to inverted erythrocyte plasma membrane vesicles result in inward Ca^{2+} accumulation.[28] The Ca^{2+} transporting activity of the enzyme in intact cells is not affected by the external concentration of Ca^{2+} or Mg^{2+} or any other divalent or monovalent cation.[29] Therefore, unlike $(Na^+ + K^+)$-ATPase, the activity of itself does not involve the mediated co-transport of two ions and must therefore be presumed to be electrogenic. However, in the erythrocyte, passive movement of Cl^- ions may accompany Ca^{2+} transport and thus preserve electroneutrality. The Ca^{2+} transport activity resides in a membrane protein of M_r 140,000 as determined by isolation of the phosphorylated intermediate produced initially after ATP hydrolysis.[30] This protein represents approximately 0.02 per cent of the total erythrocyte membrane protein.

In nerve and muscle cell membranes, outward Ca^{2+} extrusion occurs via the Na^+–Ca^{2+} antiport system which is driven by the influx of Na^+ moving

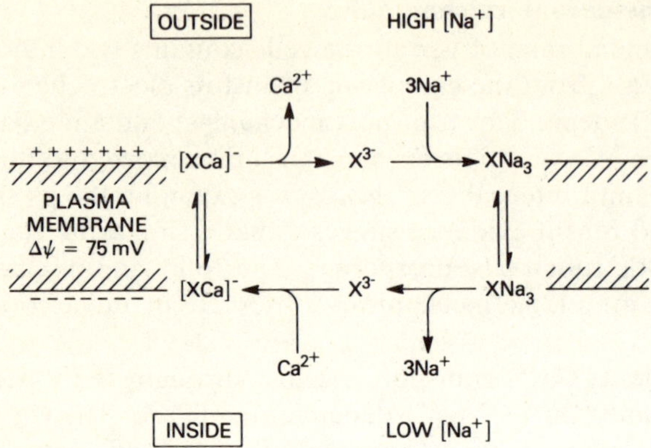

Figure 2. Hypothetical model for the operation of Na^+–Ca^{2+} exchange in the plasma membrane. See Sections 2.2 and 3.2 in the text for explanatory details

down its electrochemical gradient formed by operation of the direct energy utilizing $(Na^+ + K^+)$-ATPase. The concept of such an exchange was originally proposed by Reuter and Seitz.[31] Soon after, Blaustein and Hodgkin[32] demonstrated the dependence of Ca^{2+} efflux from squid axons upon external Na^+. Baker *et al.*,[33] showed that an outward Na^+ gradient (opposite to physiological) produces an influx of Ca^{2+}, suggesting the exchange process is reversible. It was also estimated that 3 or more Na^+ ions are transported per Ca^{2+} ion. This inherent electrogenicity was also suggested by observations on the effects of membrane potential on the exchange mechanism[34] and recently, direct isotopic studies have confirmed the stoichiometry of transport as 3 Na^+ per Ca^{2+} ion.[35]

Na^+–Ca^{2+} exchange is believed to operate by means of a membrane protein which has been solubilized with cholate from cardiac sarcolemma and reconstituted into phospholipid vesicles which then exchange Na^+ for Ca^{2+} in a mode similar to the native membranes.[36] It is unlikely that a selective channel operates to allow the co-movement of the two ions. Rather, it may be envisaged that the protein is an ion carrier, in the simplest case binding either one Ca^{2+} ion or three Na^+ ions as depicted in Figure 2. In this case, under physiological conditions, the inward Na^+ gradient (approximately 10-fold) and the interior-negative membrane potential (approximately 75 mV) are the driving forces that produce the large inward Ca^{2+} gradient. It has been calculated that the actual Ca^{2+} gradient measured across the plasma membrane may be larger than can be maximally attained under physiological conditions[37] and it may therefore be possible that other factors within this system, or additional Ca^{2+} extrusion mechanisms, operate to increase Ca^{2+} efflux (see also Section 3.2).

Na^+–Ca^{2+} exchange functions not only in axons but also in the synaptosomal plasma membrane[38,26] and the sarcolemmal membrane.[39,35] Some cells, for example most erythrocytes, are apparently devoid of this mechanism. However, the dog erythrocyte which apparently contains no Na^+ pump may contain a mechanism that extrudes Na^+ at the expense of Ca^{2+} influx via a Na^+–Ca^{2+} exchange mechanism[40] utilizing the Ca^{2+} gradient formed by ATP-dependent efflux.[41] The occurrence of Na^+–Ca^{2+} exchange in the plasma membranes of secretory cells has in a few cases been conclusively investigated.

Until recently, the existence of both the Ca^+ efflux mechanisms described above in the plasma membrane of a single cell had not been directly reported. Thus, it appeared that cells such as erythrocytes were capable of maintaining a Ca^{2+} gradient across the membrane by the operation of $(Ca^{2+}+Mg^{2+})$-ATPase alone. On the other hand, excitable cells were viewed as utilizing only the Na^+–Ca^{2+} exchange mechanism to extrude Ca^{2+} into the extracellular space. Therefore, it was possible that cells contained only one mechanism or the other. Although it was well established that broken cell preparations from nerve and muscle tissue contained $(Ca^{2+}+Mg^{2+})$-ATPase and non-mitochondrial ATP-dependent Ca^{2+} transport, it was assumed that such activity originated from the well characterized sarcoplasmic reticulum of muscle, or by analogy, the endoplasmic reticulum of mammalian nerve cells.[42,43] However, in the case of nerve tissue, it

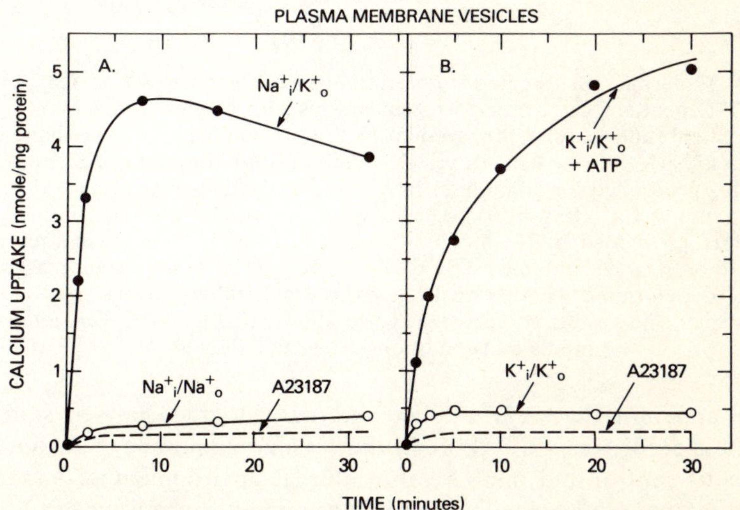

Figure 3. Time course of accumulation of Ca^{2+} within synaptosomal membrane vesicles via either Na^+–Ca^{2+} exchange (a) or ATP-dependent Ca^{2+} uptake (b). In (a) vesicles were pre-equilibrated in 100 mM Na^+ medium and Ca^{2+} (with ^{45}Ca) uptake took place in K^+ medium (●) or Na^+ medium (○). In (b) pre-equilibration was in K^+ medium and Ca^{2+} uptake took place in K^+ medium with (●) or without (○) 1 mM ATP. See Gill *et al.*[26] for details of methodology used

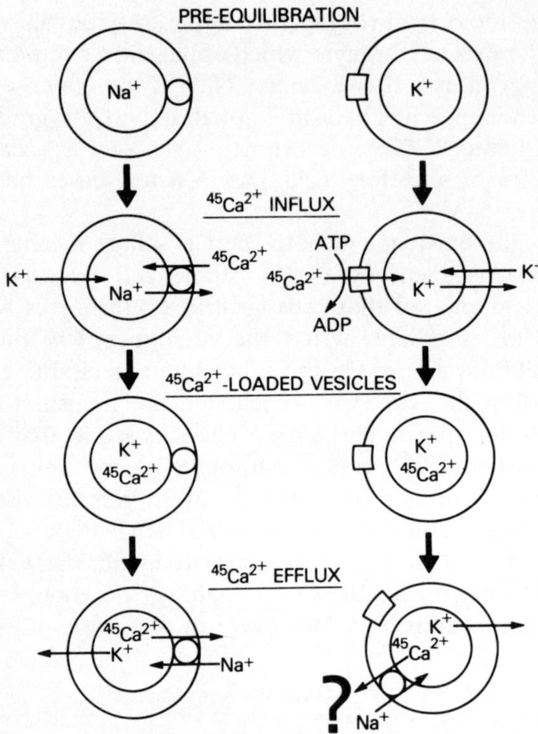

Figure 4. Experimental scheme for determining the existence of Na^+–Ca^{2+} exchange and ATP-dependent Ca^{2+} transport mechanisms in a single or in separate populations of membrane vesicles prepared from brain synaptosomes. Left-hand scheme (loading via Na^+–Ca^{2+} exchange): vesicles were pre-equilibrated in Na^+ medium and Ca^{2+} influx proceeded for 20 minutes in K^+ medium. Right-hand scheme (loading *via* ATP-dependent Ca^{2+} transport): vesicles were pre-equilibrated in K^+ medium and Ca^{2+} influx proceeded in K^+ medium with 0.3 mm ATP. Note that after loading, vesicles in both cases contained only Ca^{2+} and K^+. Vesicles were then placed in Na^+ medium to determine whether in both cases Ca^{2+} efflux via Na^+–Ca^{2+} exchange would occur. The results of such efflux are shown in Figure 5. For details of the media and conditions used see Gill *et al.*[26]

became apparent that ATP applied internally to the dialysed squid axon caused a rapid loss of Ca^{2+} through the plasma membrane.[24] Although it is possible to control and measure the internal environment of large axons, such techniques are not applicable to determining mechanisms of Ca^{2+} efflux in intact mammalian nerves or nerve terminals. Thus the location of non-mitochondrial ATP-dependent Ca^{2+} transport observed in broken synaptosomal preparations remained largely conjecture.

Recently, Gill *et al.*[26] observed both Na^+–Ca^{2+} exchange and ATP-dependent Ca^{2+} transport in a preparation of membrane vesicles isolated from guinea pig brain synaptosomes. The time dependence of Ca^{2+} accumulation *via* the two mechanisms together with a description of conditions for observing uptake by each mechanism is given in Figure 3. Since Na^+–Ca^{2+} exchange is well documented in the synaptic plasma membrane[44,38] it was important to determine whether the ATP-dependent Ca^{2+} transport activity also existed in vesicles from the plasma membrane or whether two sub-populations of vesicles were present, derived from different organelles. By utilizing the reversible nature of the Na^+–Ca^{2+} exchange mechanism it was possible to distinguish between these two possibilities. This reversibility could be demonstrated using the synaptosomal membrane vesicles in an experiment shown diagrammatically in Figure 4 (left-hand scheme). In this case Ca^{2+} loaded within the vesicles in exchange for an outward flux of Na^+ ions, could be effluxed by reversing the Na^+ flux. The resulting Ca^{2+} efflux was both as complete and rapid as that mediated by the Ca^{2+} ionophore A23187 in the absence of Na^+ (Figure 5(a)). Ca^{2+} could alternatively be loaded in vesicles via ATP-dependent Ca^{2+} transport as opposed to Na^+–Ca^{2+} exchange, as shown in Figure 4 (right-hand scheme). In this case, vesicles pre-equilibrated in K^+ medium, accumulated Ca^{2+} in the presence

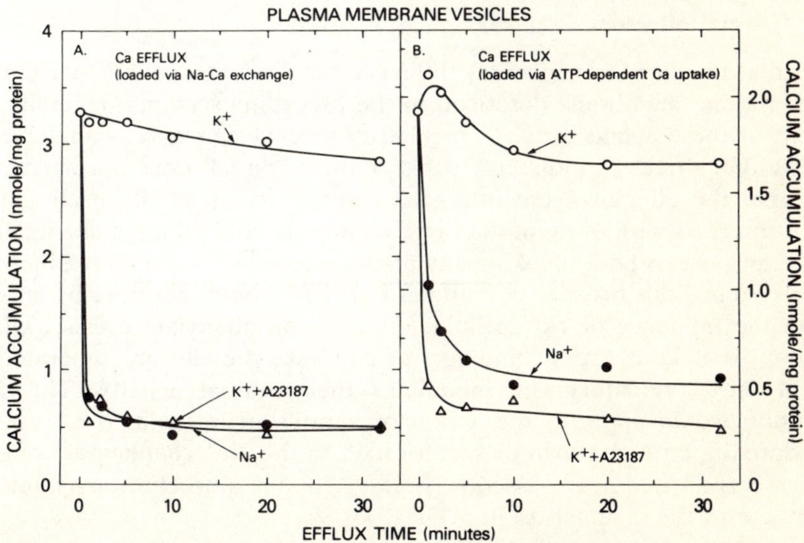

Figure 5. Time course of Ca^{2+} efflux from synaptosomal membrane vesicles loaded with Ca^{2+} (containing ^{45}Ca) via either Na^+–Ca^{2+} exchange (a) or ATP-dependent Ca^{2+} transport (B) as described in Figure 4. Efflux proceeded in each case in either Na^+ medium (●) or K^+ medium with (△) or without (○) 10 μM A23187. Efflux media contained the same concentration of Ca^{2+} (10 μM) and ^{45}Ca as influx media.

Details of media and experimental conditions are described by Gill *et al.*[26]

of K$^+$ medium containing ATP so that after the same period (20 minutes) of Ca^{2+} influx the vesicles contained Ca^{2+} and K$^+$ as in the previous case for Na$^+$–Ca^{2+} exchange-loaded vesicles (see Figure 3). Placing these vesicles now into Na$^+$ medium also produced a rapid and almost complete efflux of internal Ca^{2+} (Figure 5(b)). The somewhat slower rate of efflux in this case may result from residual ATP remaining under the conditions of efflux. This direct result, together with other indirect evidence[26] strongly suggests that both of the Ca^{2+} transport mechanisms are operating in inverted plasma membrane vesicles in the above experiments and most likely function together in the intact nerve terminal to extrude Ca^{2+} from the cell. The operation of two distinct Ca^{2+} transport mechanisms within the same plasma membrane has been useful in the determination of the specificity of various agents which may affect Ca^{2+} efflux. Such studies together with examples of a number of other effector-related changes in plasma membrane Ca^{2+} flux, form the substance of the next section of this article.

3 CONTROL OF PLASMA MEMBRANE CALCIUM REGULATORY MECHANISMS

3.1 'Direct' effectors

A great number of agents may influence the Ca^{2+} controlling mechanisms of the plasma membrane described in the preceding section. Generally, the actions of these agents on Ca^{2+} regulatory mechanisms may be classified as being either direct or indirect. In this context, 'direct' does not necessarily infer that the effector agent interacts with the transport site itself, rather, that it interacts with a site in the same membrane and effects a change upon Ca^{2+} transport without invoking any non-membrane-associated mediator. In other systems, this broader definition of 'direct' would therefore be applicable to the influence of extracellular effectors on adenylate cyclase, where, not only is it known that binding site and catalytic site are separate, but that a third regulatory site modulates their mutual activity. The same definition would apply to the action of various neurotoxins (for example, tetrodotoxin, batrachotoxin or veratridine) on the Na$^+$ channel, although in this case the directness of action is less tenuous since they may interact directly with the channel itself.

Many external agents that influence intracellular Ca^{2+} levels do so only indirectly, that is, they may interact with a receptor at the plasma membrane but effect changes in Ca^{2+} transport through release of another intracellular regulator, for example, cyclic AMP. Such indirect mediated activation must account for any agent that releases Ca^{2+} from intracellular compartments such as mitochondria or endoplasmic reticulum. However, it should be noted that Ca^{2+} entering the cell may itself be such a mediator as, for example, in the case of Ca^{2+}-mediated release of Ca^{2+} from the sarcoplasmic reticulum in muscle.[45] Although indirect activators of plasma membrane Ca^{2+} transport per se are not considered in the following discussion, it should be noted that the action of any intracellular mediator upon a Ca^{2+} regulatory process, constitutes a direct effect and therefore will be considered.

The distinctions between the above classifications of effectors of Ca^{2+} transport are depicted in Figure 6. In this scheme, X represents a direct external effector of plasma membrane Ca^{2+} transport (either influx or efflux). The designation of such an effector may be suggested by various characteristics of its effect on Ca^{2+} transport including the rapidity of its action, the persistence of its effect on purified plasma membranes (or vesicles thereof), and the absence of any effect upon Ca^{2+} transport mechanisms of internal organelles. In Figure 6, Y represents an indirect external effector of Ca^{2+} transport mediated by release of intracellular effector, Z. Frequently the effect of Y on Ca^{2+} flux or Ca^{2+}-mediated events depends on the integrity of the cytosolic compartment necessitating the use of whole cells in order to observe any effects on Ca^{2+}. In the case that Z effects release of Ca^{2+} from an internal organelle, then the events elicited by Y should occur independently of external Ca^{2+}. Although this is infrequently the case, Borle[46] suggested that the dependence of effector action upon external Ca^{2+} does not necessarily permit the inference that the effector mediates entry of Ca^{2+} through the plasma membrane. Furthermore, even an increased cellular uptake of labelled Ca^{2+} in response to an effector, according to Borle, 'does not necessarily mean an increased transport' but may 'reflect an increased intracellular exchangeable pool'. However, distinction between these two possibilities might be obtained by analysis of the kinetics of Ca^{2+} uptake in response to effector. Thus, from Figure 6, if the concentrations of Ca^{2+} in the extracellular, cytosolic and intraorganelle compartments are designated $[Ca]_e$, $[Ca]_c$ and $[Ca]_i$, respectively, and the rate constants for Ca^{2+} movement between the compartments are designated as:

$$[Ca]_e \underset{k_{-1}}{\overset{k_1}{\rightleftharpoons}} [Ca]_c \underset{k_{-2}}{\overset{k_2}{\rightleftharpoons}} [Ca]_i$$

then the rate of change of cytosolic Ca^{2+} concentration is determined by the equation:

$$\frac{d[Ca]_c}{dt} = k_1[Ca]_e + k_{-2}[Ca]_i - (k_{-1} + k_2)[Ca]_c \tag{1}$$

At equilibrium when this rate of change is equal to zero, then:

$$[Ca]_c = \frac{k_1[Ca]_e + k_{-2}[Ca]_i}{k_{-1} + k_2}$$

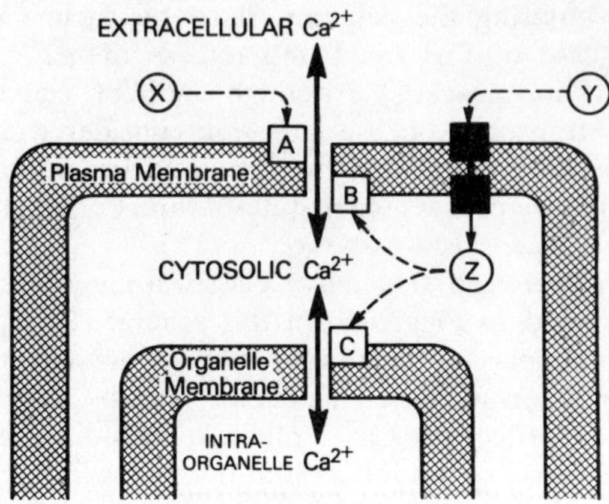

Figure 6. Representation of distinct classes of effectors of Ca^{2+} regulatory mechanisms and their sites of action within the cell. A and B are regulatory sites on the external and internal surface of the plasma membrane, respectively, which may modify Ca^{2+} flux across the membrane in response to effector agents. C is a similar regulatory site on the outer surface of intracellular Ca^{2+}-sequestering organelles such as mitochondria or sarcoplasmic reticulum. X represents a direct external effector of plasma membrane Ca^{2+} transport. Y represents an indirect external effector of Ca^{2+} transport that operates through the release of the internal direct effector, Z, which acts at an internal site to elicit changes in Ca^{2+} flux. Explanatory details are given in the text

that is, the equilibrium cytosolic Ca^{2+} level is dependent on a change in each of the rate constants governing transport across either membrane and therefore is not an indicator of through which membrane (or which Ca^{2+} flux mechanism) a particular agent may elicit an increase in cytosolic Ca^{2+}. However, the initial rate of uptake of labelled Ca^{2+} ($^{*}Ca$) is dependent only upon the rate constant for influx across the plasma membrane (k_1). This is apparent from equation (1) since at the time of label addition to the extracellular medium, both $[^{*}Ca]_c$ and $[^{*}Ca]_i$ are equal to zero, and the equation for the initial rate of increase in $[^{*}Ca]_c$ simplifies to:

$$\frac{d[^{*}Ca]_c}{dt} = k_1[^{*}Ca]_e$$

Thus, by careful examination of the initial rate of labelled Ca^{2+} accumulation in whole cells after treatment with a particular agent, distinction between an effect on either extracellular Ca^{2+} influx or intraorganelle Ca^{2+} efflux should be possible since only in the former case will a change in initial rate occur.

As suggested above, agents directly affecting plasma membrane Ca^{2+} regulatory mechanisms may do so either externally or internally. The following parts of this section describe examples of each type of effector.

Those agents that interact with sites on the external surface of the plasma membrane and directly affect Ca^{2+} influx or efflux in the same membrane will be considered first (Section 3.2). Secondly, agents that act within the cell upon plasma membrane Ca^{2+} transport are considered (Section 3.3) though this discussion is brief since such mechanisms (concerning primarily the interrelationship between cyclic AMP or calmodulin and Ca^{2+} transport) are considered in more detail by other authors in this volume (see Chapters by Rasmusson, Fain, Gardner and Jensen, and Klee).

3.2 Extracellular effectors

3.2.1 *Influence on calcium influx mechanisms*

Of the extracellular agents that may effect changes in plasma membrane Ca^{2+} regulatory mechanisms, most are reported to act on Ca^{2+} influx rather than efflux mechanisms. Although little information is available concerning the precise mechanisms of plasma membrane mediated entry of Ca^{2+} (see Section 2.1), there appear to be three major categories of Ca^{2+} influx under the influence of a variety of external stimuli. These are (a) the voltage sensitive 'slow' inward Ca^{2+} channel, (b) putative voltage-insensitive Ca^{2+} entry mechanisms, and (c) possible release of Ca^{2+} bound to components of the inner membrane surface. In general, effects on such processes have been observed in excitatory or secretory cells where it is well established that cytosolic Ca^{2+} is an essential mediator between stimulus (electrical or hormonal) and response (secretion, contraction or changes in membrane permeability).

Obviously, the number of tissues in which Ca^{2+} influx is under external control and the number of agents that are reported to effect such control are enormous and cannot be comprehensively covered here. Instead, below is an outline of some recent reports concerning systems which illustrate the possible distinction between the above Ca^{2+} entry mechanisms and, in particular, systems where the plasma membrane has been strongly implicated in mediating stimulus-dependent increases in cytosolic Ca^{2+}. As stated above (Section 3.1), in most systems considerable controversy exists regarding whether Ca^{2+} for stimulus-secretion coupling originates from outside the cell (plasma membrane mediated) or from within (released from cellular organelles). This uncertainty is well illustrated in several systems, for example, the actions of those hormones which stimulate exocrine secretion from the pancreas through elevated cytosolic Ca^{2+} (see Schulz and Ullrich[47] and Gardner and Jensen's Chapter, this volume), the Ca^{2+}–dependent action of hypothalamic releasing hormones on adenohypophysial hormone secretion,[48] or the action of α-adrenergic stimulators on liver metabolism via intracellular Ca^{2+}.[49] However, in the case of α-adrenergic agents, these appear to have more defined effects upon plasma membrane Ca^{2+} transport in other tissues, as detailed below. It is possible that a particular effector agent may influence different Ca^{2+} regulatory mechanisms in different cells, but it should be noted that each tissue necessitates its own particular experimental approach, resulting in differing contributions of relevant organelles to overall Ca^{2+} movement either as an implicit property of the cell or as a consequence of the experimental conditions.

Recent evidence implicates the α_1-adrenergic receptor in modifying Ca^{2+} levels in cells of several different tissues.[50] In smooth muscle cells α-adrenergic stimulators may cause increased Ca^{2+} influx and/or release of Ca^{2+} into the cytosol.[51,52] However, Ca^{2+} entry in response to α-adrenergic agonists does not appear to be through the same channel as depolarization-dependent Ca^{2+} influx. Thus, D600 which blocks voltage sensitive Ca^{2+} channels (see below) inhibits only K^+-stimulated uptake of Ca^{2+} in smooth muscle and not that resulting from low, non-depolarizing concentrations of norepinephrine.[53] A further characteristic of Ca^{2+} influx mediated by α-adrenergic stimulation of smooth muscle[52] and rat lacrimal gland[54] is an apparent transient release of Ca^{2+} into the cell in the absence of extracellular Ca^{2+}, whereas a sustained Ca^{2+} entry is observed in the presence of extracellular Ca^{2+}. An ostensibly opposite effect of α-adrenergic stimulation has been observed in rat sympathetic neurons[55] where norepinephrine inhibits Ca^{2+} dependent potentials suggesting that the α-adrenergic receptor may antagonize voltage sensitive Ca^{2+} channels. However, a mechanism reconciling these differing actions was put forward[56] whereby α-adrenergic stimulation of sympathetic neurons may indeed open a specific Ca^{2+} channel (or cause Ca^{2+} release from the membrane), as suggested above. In this case, the observed inhibition of Ca^{2+} influx through the voltage-dependent Ca^{2+} channel may be a consequence of increased cytosolic Ca^{2+} entering through a distinct α-adrenergic sensitive Ca^{2+} channel.

Another major group of hormone effectors that may directly influence plasma membrane mediated Ca^{2+} entry are the cholinergic receptor agonists. However, in this case it is clear that different routes of Ca^{2+} entry are associated with the operation of different classes and locations of cholinergic receptor. Thus the passage of Ca^{2+} ions through Na^+/K^+ channels induced by acetylcholine, may account for increased Ca^{2+} influx through the post-synaptic membrane at the neuromuscular junction.[57] This leakage of Ca^{2+} was also suggested to occur by Ritchie[58] in the rat pheochromocytoma cell line, PC12, though in this case, Ca^{2+} also entered through voltage sensitive Ca^{2+} channels opening as a result of depolarization caused by the induced changes in Na^+ and K^+ flux. On the other hand, the muscarinic cholinergic receptor in smooth muscle or exocrine tissues appears to function to directly regulate a specific Ca^{2+} entry mechanism. This action, which resembles closely the effect of α-adrenergic agonists described above, has been demonstrated in isolated parotid cells[59,60] in which binding of cholinergic agonist, increased Ca^{2+} influx and Ca^{2+}-activated K^+ efflux are all inter-relatable in accordance with the stimulus-secretion coupling model.[61] In fact, it appears that the rat parotid gland contains at least three separate classes of receptor, each specifically binding either α-adrenergic, muscarinic cholinergic, or peptide (Substance P) agonists, all of which function to allow entry of Ca^{2+} into the cytosol upon ligand interaction.[62,63] The non-additivity and similarity of action of these receptors on Ca^{2+} influx suggests that each acts upon a commn Ca^{2+} entry system characterized by an initial phase of influx insensitive to external Ca^{2+}, followed by a sustained phase dependent on external Ca^{2+}. From these studies, Putney[63] proposed that receptor occupancy causes first a rapid release of internal Ca^{2+} (possibly bound to the internal surface of the plasma membrane) which is then replenished by Ca^{2+} transport from the outside across the plasma membrane.

Some important recent observations appear to relate this type of mechanism with the well described effect by which a great number of effector agents that influence cellular Ca^{2+} also increase turnover of plasma membrane phosphatidate and phosphatidyl inositol in their respective target tissues.[64] The latter effect has been implicated in controlling plasma membrane Ca^{2+} gating by Michell and co-workers, although the precise mechanism involved was unknown. It has now been shown by Putney and co-workers that phosphatidic acid applied to the parotid gland can cause a similar Ca^{2+}-dependent response (increased K^+ efflux) as observed with α-adrenergic or muscarinic agonists.[65] Unlike the latter agents, however, phosphatidate did not elicit a transient response in the absence of Ca^{2+}, in other words, its effect was absolutely dependent on external Ca^{2+}. In a separate study, Salmon and Honeyman[66] observed that carbachol promoted a very rapid incorporation of $^{32}PO_4$ into phosphatidate in isolated smooth muscle cells, and that phosphatidate caused a similarly rapid contraction of these cells. The authors predicted that phosphatidate is acting as a Ca^{2+}-binding ionophore and possibly allowing transport of Ca^{2+} across the membrane in the form of inverted micellar complexes. The rapidity of both the carbachol induced labelling of phosphatidate and the phosphatidate induced Ca^{2+}-dependent responses described above is strong evidence in support of such a model. Furthermore, since the phosphatidyl inositol/phosphatidate turnover effect is so fundamentally associated with the action of agents or stimuli that effect Ca^{2+} levels in cells (see Michell,[5] and Fain's Chapter, this volume), it is tempting to believe that the scheme is widely applicable.

In many endocrine secretory cells Ca^{2+} plays an essential role in mediating hormone release in response to secretogogue. In some cases there is evidence that the voltage-dependent Ca^{2+} channel mediates Ca^{2+} entry upon stimulation. In other cases, there is an assumed or poorly characterized link between increased cytosolic Ca^{2+} and the operation of this channel, often such a link being inferred from the inhibitory action of D600 or related Ca^{2+} channel 'blockers'. However, as discussed below, the activity of these agents is not necessarily restricted to blockade of a specific Ca^{2+} channel. For glucose-induced secretion of insulin from pancreatic β-cells, evidence suggests that Ca^{2+} entry does occur through the operation of voltage-sensitive channels.[67] Thus, glucose induces depolarization of the β-bell membrane potential and the initiation of bursts of Ca^{2+}-dependent spike activity which may represent transient Ca^{2+} conductances.[68] The absence of effect of tetrodotoxin suggests that operation of Na^+ channels does not contribute to this process.[69] Depolarization of the membrane has been speculated to result from an observed glucose-dependent decrease in K^+-efflux.[70] However, caution is required in accepting this mechanism since it was recently reported that Ca^{2+} movement in response to glucose is not inhibited by high concentrations of the K^+ ionophore, valinomycin, which abolishes glucose-dependent K^+ conductance.[71] Unlike the effect of glucose on pancreatic β-cells, direct evidence militates against the passage of Ca^{2+} ions through voltage sensitive channels in the response of adenohypophysial cells to thyrotropin releasing hormone (TRH). Thus, direct recordings from pituitary cells have revealed Ca^{2+}-dependent action potentials but no change in the resting membrane potential in response to TRH.[72] Furthermore, Gershengorn[73] recently demonstrated that whereas application of either high extracellular K^+ or TRH to cultured pituitary cells in both cases caused similar effects on Ca^{2+} movement and prolactin release, only high K^+ caused efflux of the lipophilic cation, triphenylmethylphosphonium (TPMP$^+$), which is evidence for depolarization. The absence of a TRH effect on TPMP$^+$ movement indicates that TRH influences Ca^{2+} transport via a mechanism not involving a change in membrane potential and therefore unlikely to be through the voltage-sensitive Ca^{2+} channel.

Considering how widely applicable the role of Ca^{2+} influx appears to be in mediating stimulus-contraction or stimulus-secretion coupling in a great many tissues, it is perhaps surprising that few if any specific antagonists of Ca^{2+} entry have been clearly identified. By contrast, the flux of Na^+ through the Na^+ channel of excitatory tissues is specifically antagonized or promoted by a number of agents. The lack of specific antagonists of the one or more types of Ca^{2+} channel that may exist has been a significant factor in our considerable lack of knowledge regarding mechanism, structure, and in many cases, even the existence of the processes that mediate increased cytosolic Ca^{2+} in response to external stimuli. It has been widely accepted that the organic Ca^{2+} antagonists, most notably verapamil, D600 (methoxy-verapamil) and nifedipine, are specific inhibitors of voltage-dependent Ca^{2+} channels. This was originally proposed by Fleckenstein and co-workers who demonstrated that such agents inhibited only the slow Ca^{2+}-dependent component of the action potential in cardiac muscle without altering the fast Na^+ current.[74] The various effects and likely actions of these agents has been extensively reviewed.[1,10,75] Although these agents are apparently specific blockers of a particular Ca^{2+} channel in mammalian cardiac fibres, their potency and specificity among other tissues and species is variable.[76] For example, they are considerably less potent in the squid axon[77] and may alter other membrane-mediated ion movements, such as outward K^+ flux, independently of any inhibition of Ca^{2+} entry.[78] Whether these differences in activity are due to differences in the structure of Ca^{2+} channels, or, are an inherent consequence of the lack of specificity of the drugs, is unknown. In any case, great caution is required in interpreting the significance of an inhibitory effect of these agents on a given Ca^{2+}-mediated response. As an example, Study *et al.*[79] demonstrated an inhibitory effect of D600 on both voltage sensitive and carbachol-induced Ca^{2+}-dependent cyclic GMP production in neuroblastoma cells (via the muscarinic cholinergic receptor). In fact, the authors concluded from additional data that the muscarinic receptor mediates Ca^{2+} influx via a mechanism distinct from the voltage-sensitive Ca^{2+} channel. The effect of D600 on muscarinic activation may stem from a recently described action of D600 whereby it inhibits binding of muscarinic agonists with their receptor.[80] In this study D600 also inhibited α-adrenergic receptor binding. These effects were shown to be similar to those of local anaesthetics on receptor binding and probably result from 'nonexchange.[83]

Lastly, a recent observation on the influence of black widow spider toxin upon Ca^{2+} uptake into PC12 pheochromocytoma cells suggests that the toxin may specifically influence a Ca^{2+} influx mechanism.[81] Thus the toxin causes a large and rapid increase in Ca^{2+} uptake with a concomitant release of catecholamine from the cell which is dependent on external Ca^{2+}. High K^+ in the medium causes the same Ca^{2+} uptake and catecholamine release but these responses are greatly more sensitive to verapamil than those elicited by toxin. A further difference between the two stimulatory conditions is a much greater sensitivity of the spider toxin induced Ca^{2+} influx or catecholamine release to concanavalin A. Should black widow spider toxin prove to be a specific agonist of a Ca^{2+} influx mechanism possibly distinct from the voltage-sensitive Ca^{2+} channel, its importance in elucidating information about such a site in other tissues (for example, those described above) is obvious.

3.2.2 *Influence on calcium efflux mechanisms*

Two major mechanisms of Ca^{2+} efflux exist in eukaryotic plasma membranes as discussed in Section 2.2. Thus Ca^{2+} is extruded either by the Ca^{2+}

translocating $(Ca^{2+}+Mg^{2+})$-ATPase or the Na^+–Ca^{2+} antiport mechanisms of the plasma membrane. There are remarkably few specific extracellular effectors, whether physiological or pharmacological, which have been shown to directly influence either of these mechanisms according to the definition of 'direct' described in Section 3.1. It is strange that so few effectors have been established as causing a change in the activity of these clearly defined efflux mechanisms, when so many effectors are reported to influence the relatively poorly defined mechanisms of Ca^{2+} influx.

In the case of ATP-dependent Ca^{2+} efflux, the lack of extracellular effectors may be due to the sidedness of the mechanism. ATPase activity occurs exclusively on the inner surface of the plasma membrane and it is becoming apparent that this activity is under allosteric control from intracellular regulators (see Section 3.3) and may contain specific regulatory sites. The Na^+–Ca^{2+} exchange process does not appear to be sided, although thus far, physiological regulation of this activity, other than by levels of the transported ions themselves or possibly membrane potential, has not been conclusively reported. Below, however, is an account of recent observations concerning the influence of tetanus toxin on Ca^{2+} efflux[82] and the possibility that its action may be upon a site on the outside of the plasma membrane regulating the activity of Na^+–Ca^{2+} exchange.[83] The specificity of the toxin's action upon this Ca^{2+} efflux mechanism has been revealed by observing its influence upon the two Ca^{2+} efflux systems operation of which has been clearly defined and characterized in the synaptosomal membrane vesicles described in Section 2.2.[26]

Before considering these experiments, below is an outline of some reported observations on the actions of tetanus toxin relevant to an understanding of any proposed effect it may have. The site and mechanism of action of tetanus toxin have remained unknown even though its physiological effect of 'dysinhibition', involving blockade of inhibitory neurons of the central nervous system, is fairly conclusive.[84] Tetanus toxin appears to inhibit neurotransmitter release, in particular, the release of glycine or gamma-aminobutyric acid (GABA) from inhibitory presynaptic nerve terminals. Thus, in the presence of tetanus toxin, impulses continue to be generated in efferent motor neurons independently of stimulation of impinging inhibitory neurons.[85] However, inhibition is restored in the presence of the toxin upon local administration of glycine or GABA indicating that tetanus toxin blocks the presynaptic release rather than the postsynaptic effect of neurotransmitter. A lack of observed change in the levels of neurotransmitters in nerves from toxin treated animals suggests there is no change in neurotransmitter synthesis.[86] Labelled tetanus toxin is known to bind to the outer surface of nerve terminals[87,88] although binding sites may not be exclusive to nerve endings since toxin binds to the entire surface of neuronal cells in culture.[89] van Heyningen[90] observed that the toxin binds strongly to specific gangliosides found in the plasma membrane (especially types G_{D1b} or G_{T1}) and it is likely that a ganglioside constitutes the receptor or a part thereof.[91] Tetanus toxin is able to enter neurons after binding to peripheral nerve terminals and travel internally *via* retrograde transport to the centrally located cell body and thence across the synapse to an adjacent presynaptic terminal where it may effect inhibition of neurotransmitter release.[92] It appears that binding to ganglioside is a sufficient criterion to permit retrograde transport since other ligands which interact with gangliosides, for example, cholera toxin or thyrotropin, are similarly internalized and transported in neurons.[93] Since after internalization and transport the toxin is able to reenter the extracellular space (in order to cross the synapse), it is possible that the molecule does not enter the cytosolic compartment but remains attached to the inner surface of membrane vesicles within the cell which are either pinocytosed or exocytosed in order to effect toxin entry or exit at the peripheral or central extremities of the neuron, respectively. This hypothetical scheme would permit the observed passage of the intact two-subunit molecule through neurons[94] without invoking transmembrane transport of the toxin as envisaged for some other two-subunit toxins whose intracellular action may, in addition, require subunit cleavage.[95] The interaction of tetanus toxin with an external plasma membrane component may be sufficient not only for its transport within but also for its cellular response upon neurons, as suggested by its effects on plasma membrane ion transport described below.

Since an increase in internal free Ca^{2+} is the triggering mechanism for neurotransmitter release, an obvious action of tetanus toxin would be upon Ca^{2+} flux—either directly on the Ca^{2+} transport mechanisms themselves, or indirectly through, for example, a change in membrane potential. Early experiments on the latter indicated that tetanus toxin, upon binding to brain synaptosomes, caused rapid effects on the uptake of the permeant lipophilic cation, $[^3H]$-tetraphenylphosphonium (TPP^+), indicating that a change in membrane potential is associated with the interaction of the toxin with the cell membrane.[96] However, both the extent and direction of these changes in membrane potential appeared to be influenced by the activity of internal organelles, particularly mitochondria, which due to their own membrane potential also accumulate the lipophilic cation and therefore have a considerable modifying effect upon total label accumulation. Furthermore, Ca^{2+} in the external medium was another factor which influenced membrane potential changes in response to toxin. Infact from the observed effect of tetanus toxin on Ca^{2+} transport (see below) it is possible that changes in Ca^{2+} flux were a cause rather than an effect of observed toxin-dependent changes in membrane potential.

Attempts to identify a direct effect of tetanus toxin on Ca^{2+} accumulation within brain synaptosomes have been more consistent, although using this

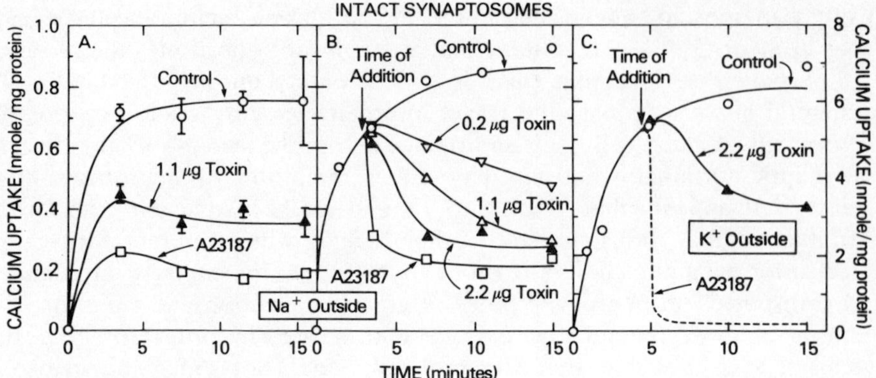

Figure 7. Effect of tetanus toxin on the accumulation of Ca^{2+} by brain synaptosomes. Ca^{2+} uptake was measured after addition of synaptosomes to Krebs–Ringer
phosphate buffer containing ^{45}Ca with either 180 mM NaCl (A and B) or KCl (c) in a
total volume of 100 μl. Tetanus toxin (quantities as indicated), A23187 (final
concentration 10 μM), or control buffer were added either from the start of incubation (A) or at the indicated times after uptake commenced (b) and (c). After the
appropriate time, Ca^{2+} uptake was terminated by rapid filtration. Results in A are
the means of triplicate determinations ± standard deviation

preparation, problems have arisen regarding identification of the exact Ca^{2+}
flux mechanism involved. The accumulation of Ca^{2+} by synaptosomes
(pinched off, resealed presynaptic nerve terminals) has been investigated at
length and found to be influenced both by membrane potential[97] and
external Na^+ [38] through mechanisms involving depolarization-dependent Ca^{2+}
influx and Ca^+–Ca^{2+} exchange, respectively. In recent studies, Gill *et
al.*,[82,83] observed that tetanus toxin inhibits the accumulation of Ca^{2+} within
synaptosomes (Figure 7(a)). The toxin was shown to reverse entry of Ca^{2+}
in a dose dependent manner causing an efflux of up to 90 per cent of the
accumulated free Ca^{2+} (Figure 7(b)). The effect was not observed with
boiled toxin or tetanus toxoid, and was inhibited by pre-treatment of the
toxin with anti-tetanus antiserum. When Na^+ in the medium was replaced
with K^+ (so-called depolarizing conditions) Ca^{2+} uptake was considerably
enhanced and, although tetanus toxin still reversed uptake, the effect was a
smaller proportion of the total free accumulated Ca^{2+} (Figure 7(c)). One
interpretation of these effects is that tetanus toxin inhibits Ca^{2+} entry
through voltage-sensitive Ca^{2+} channels by, for example, causing hyperpolarization of the membrane. However, there are two objections to this
argument. Firstly, direct measurement using $[^3H]$-TPP$^+$ indicates that the
toxin, under the conditions utilized in Figure 7(a) and (b), results in a
depolarizing rather than hyperpolarizing effect.[96] Secondly, if toxin had
directly caused hyperpolarization it would have been expected to exert a

greater influence under depolarizing conditions rather than under the non-depolarizing conditions as observed.

Indeed, several factors indicate that at least a portion of the 'depolarization-dependent' increase in Ca^{2+} uptake observed with synaptosomes in the presence of high external K^+ is due not to a voltage-regulated Ca^{2+} influx but rather to an inhibition of Na^+–Ca^{2+} exchange-dependent efflux. Thus in the presence of high external Na^+, Ca^{2+} efflux via this mechanism is increased resulting in lowered Ca^{2+} accumulation, whereas with high external K^+, Na^+-dependent Ca^{2+} efflux is obviously reduced. This view is supported by a simple experiment measuring Ca^{2+} uptake and $[^3H]$-TPP^+ accumulation in synaptosomes in the presence of external Na^+, K^+ or choline chlorides. The greater permeability of the membrane to K^+ as opposed to Na^+ or Ch^+ ions results in hyperpolarization (interior negative) when the external medium contains either of the latter ions as indicated by increased $[^3H]$-TPP^+ uptake (Figure 8(b)). However, although external K^+ causes increased Ca^{2+} uptake relative to Na^+, so too does external Ch^+ (Figure 8(a)) even though Ch^+ does not depolarize relative to Na^+. Similar

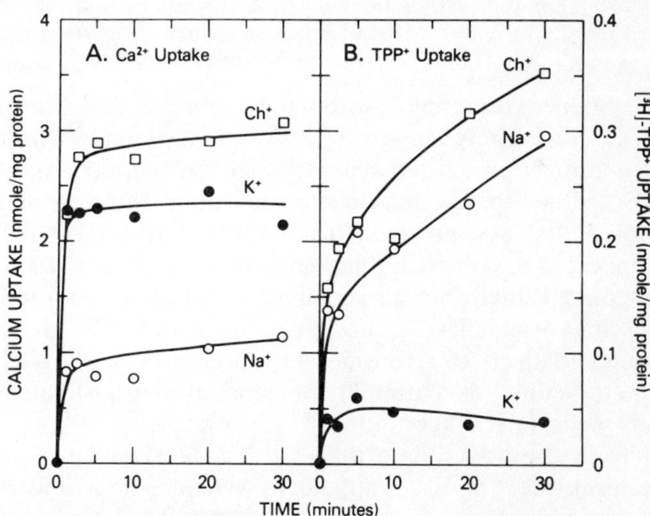

Figure 8. Influence of external ionic conditions on the uptake of Ca^{2+} or TPP^+ by brain synaptosomes. Synaptosomes were pre-equilibrated for 15 minutes in 180 mM KCl containing 1 mM $MgSO_4$, 5 mM Hepes-Tris buffer, pH 7.4. Ca^{2+} uptake (a) or TPP^+ uptake (b) were measured after addition of 5 μl of synaptosomes to 100 μl of the above medium (●) or this medium with KCl replaced by 180 mM NaCl (○) or choline chloride (□) each containing in addition either 10 μM $CaCl_2$ with ^{45}Ca (a) or 5 μM $[H^3]$-TPP^+ (b). After the appropriate times, uptake was determined by rapid filtration

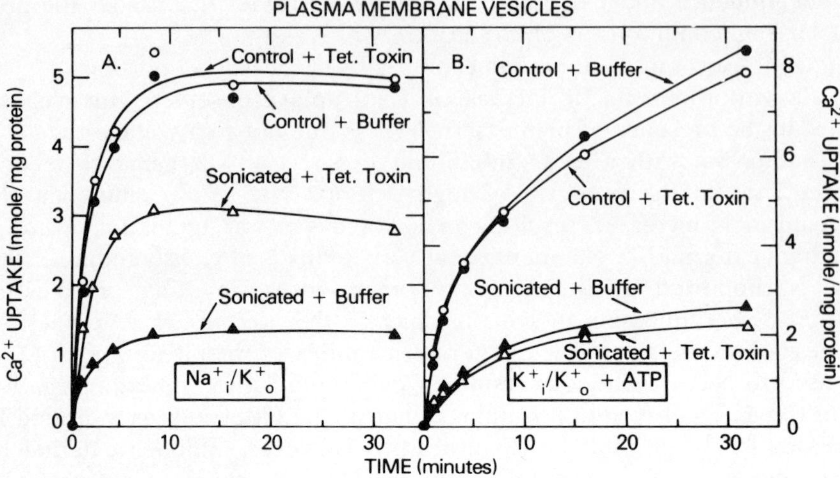

Figure 9. Influence of tetanus toxin on Na^+–Ca^{2+} exchange (a) or ATP-dependent Ca^{2+} uptake (b) in synaptosomal membrane vesicles. Ca^{2+} accumulation via either mechanism was as described in Figure 3. However, after pre-equilibration, tetanus toxin at 20 μg/mg membrane protein (O, $\triangle$) or control buffer ($\bullet$, $\blacktriangle$) were added to the vesicles which were then either treated ($\triangle$, $\blacktriangle$) or not treated ($\bullet$, O) by gentle sonication for 1 minute at 22 °C. After a further 15 minute incubation at 22 °C, Ca^{2+} uptake proceeded as described in Figure 3

results were obtained using synaptosomal membrane vesicles.[26] However, these results do not entirely agree with those of Blaustein[97] who reported a more direct relationship between synaptosomal Ca^{2+} uptake and depolarizing conditions, although the reason for this discrepancy is not clear. It should be noted that synaptosomal Ca^{2+} uptake (like [^{3}H]-TPP$^+$ uptake described above) is a complex phenomenon reflecting not only plasma membrane mediated fluxes but also a large contribution from intraterminal Ca^{2+}-sequestering organelles. It has been observed that their presence (especially mitochondria) can considerably alter total Ca^{2+} accumulation, moreover, their activity is extremely dependent upon slight changes in preparation or experimental conditions.[83]

It is clear from the above data that Na^+–Ca^{2+} exchange is at least one major mechanism affecting Ca^{2+} uptake by synaptosomes and it was thus possible that tetanus toxin may have influenced this mechanism to effect a change in Ca^{2+} accumulation. To investigate this further in a more defined system, synaptosomal plasma membrane vesicles were utilized to detect any specific change in either of the two Ca^{2+} transport mechanisms (Na^+–Ca^{2+} exchange and ATP-dependent Ca^{2+} transport) characterized in this preparation,[26] as described in Section 2.2. In fact, upon addition of tetanus toxin little significant change was observed on Ca^{2+} accumulation via either of these mechanisms (see Figure 9(a) and (b)). However, an important conclusion drawn from studies utilizing these vesicles was that a large proportion function as inverted plasma membrane sacs,[26] although the exact proportion relative to non-inverted structures was not ascertained. If the effect of tetanus toxin is mediated by its association with a specific outer cell surface component (as suggested above), an effect on Ca^{2+} transport may be precluded by non-exposure of this component in a preparation of inverted plasma membrane vesicles. Therefore, to observe a possible specific effect upon one of the Ca^{2+} transport mechanisms of these vesicles, it appeared that it might be necessary to allow access of the toxin to their inner surface. This was achieved by gentle sonication of the vesicles in the presence of tetanus toxin. Using ^{125}I-labelled tetanus toxin, the sonication process was observed to allow toxin access to a saturable binding site as determined by displacement with unlabelled toxin (Table 1). This observation is assumed to indicate the binding of toxin to a specific site that originally existed on the outside surface of the synaptosomal plasma membrane but now exists on the inner surface of the inverted vesicles. The process appears to temporarily open vesicles to the external medium after which they apparently close with the same orientation as before since presonicated vesicles do not bind more toxin than untreated vesicles (Table 1). Uptake of Ca^{2+} by the sonicated vesicles, via either Na^+–Ca^{2+} exchange or ATP-dependent Ca^{2+} transport, is qualitatively comparable with unsonicated vesicles though there is a similar quantitative decrease in uptake via each mechanism suggesting a decrease in

Table 1. Binding of [^{125}I]-tetanus toxin to synaptosomal membrane vesicles (SMV)

[^{125}I]-tetanus toxin with or without an excess of cold tetanus toxin (0.1 mg/ml) was added to untreated vesicles (A, B) or vesicles that had been pre-treated by gentle sonication for 1 minute (C, D). The same sonication procedure was also applied to untreated vesicles *after* addition of toxin (E, F). Results are means of 9 determinations taken from 5 minutes to 1 hour afer addition of toxin±standard deviation. Maximal binding occurred within the first 5 minutes and did not significantly change during the 1 hour incubation. Procedures all took place at 22 °C. Specific binding is the difference between total (binding in the absence of cold toxin) and non-specific (binding in the presence of cold toxin)

Present in binding incubation	[^{125}I]-tetanus toxin bound (pmole/mg membrane protein)	
	Total	Specific
A. Untreated SMV + [^{125}I]-toxin	7.04±0.74	
B. Untreated SMV + [^{125}I]-toxin + cold toxin	6.43±0.36	0.61
C. Sonicated SMV + [^{125}I]-toxin	5.58±0.51	
D. Sonicated SMV + [^{125}I]-toxin + cold toxin	5.39±0.53	0.19
E. Sonicated SMV + [^{125}I]-toxin	8.13±0.86	
F. Sonicated SMV + [^{125}I]-toxin + cold toxin	5.03±0.27	3.10

either size or number of intact vesicles (see Figures 9 and 10). Most significantly, vesicles sonicated in the presence of tetanus toxin accumulated more than 2-fold the amount of Ca^{2+} via Na^+–Ca^{2+} exchange compared with vesicles sonicated without toxin (Figure 9(a)). By contrast, tetanus toxin whether added before sonication or added to untreated vesicles, had no effect upon ATP-dependent Ca^{2+} accumulation within the vesicles (Figure 9(b)). This provides a good indication of the specificity of the tetanus toxin effect which cannot therefore be due to non-specific changes in, for example, Ca^{2+} binding or internal volume of the vesicles. Just as prior sonication of vesicles was not a sufficient treatment to permit tetanus toxin access to its binding site (Table 1), so toxin added after sonication was without any effect on Na^+–Ca^{2+} exchange in contrast to that added before sonication (Figure 10).

The effects of tetanus toxin upon Ca^{2+} accumulation in the intact synaptosomes and synaptosomal plasma membrane vesicles are opposite, though consistent, in light of the reversed orientation of membranes in the two preparations. A scheme depicting the possible mechanism of tetanus toxin in

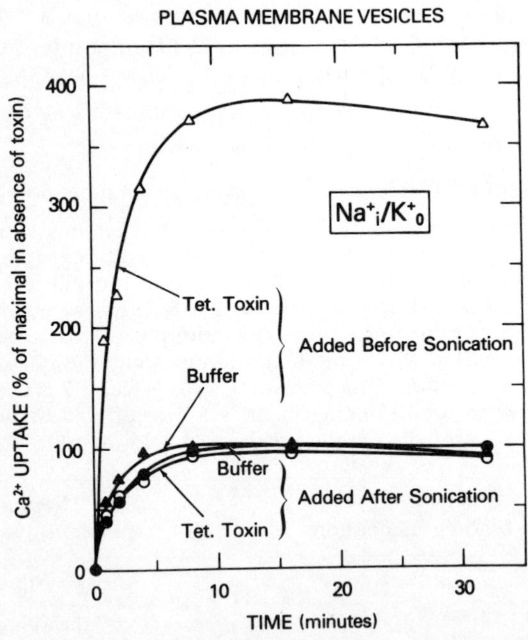

Figure 10. Ca^{2+} accumulation by Na^+–Ca^{2+} exchange in synaptosomal membrane vesicles sonicated before or after addition of tetanus toxin. Ca^{2+} uptake was measured as described for Figure 3(a). Tetanus toxin (20 μg/mg membrane protein) (○ △) or buffer (● ▲) were added either before (△ ▲) or after (○ ●) sonication which was as described in Figure 9

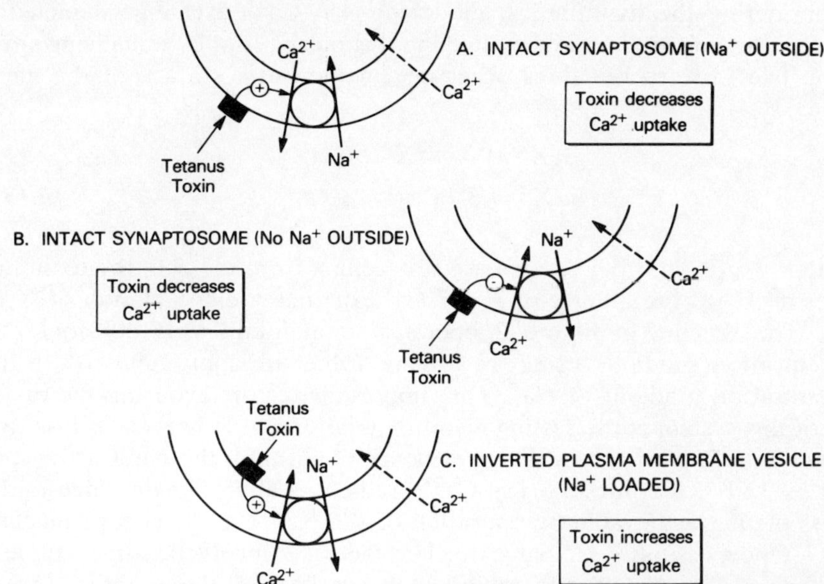

Figure 11. Schematic model for the possible mode of action of tetanus toxin on Na^+–Ca^{2+} exchange in the plasma membrane of intact synaptosomes (a) and (b) or in synaptosomal membrane vesicles (c). The schemes depicted in (a) (b), and (c), are derived from the data shown in Figure 7(a) and (b), Figure 7(c), and Figures 9(a) and 10, respectively. Details of these proposed mechanisms are described in the text

these preparations is shown diagrammatically in Figure 11. Thus, for the intact synaptosome accumulating Ca^{2+} with external Na^+, Ca^{2+} may enter (for example by passive or mediated diffusion) though will be subjected to efflux in exchange for entering Na^+. Tetanus toxin, which lowers Ca^{2+} accumulation under these conditions, may enhance the outward movement of Ca^{2+} occurring specifically via Na^+–Ca^{2+} exchange (Figure 11(a)). When intact synaptosomes accumulate Ca^{2+} in the absence of external Na^+ (for example with external K^+), Na^+-dependent efflux of Ca^{2+} is inhibited. Furthermore, since synaptosomes as used in these experiments have a significant Na^+ content,[98,99] an outward Na^+ movement may significantly contribute to Ca^{2+} accumulation by exchanging for external Ca^{2+}. By favoring the outward movement of Ca^{2+} via this mechanism, tetanus toxin may inhibit Ca^{2+} influx and therefore result in lowered Ca^{2+} accumulation (Figure 11(b)), though, since this may not be the major Ca^{2+} influx mechanism operating, with a reduced efficiency relative to its effect in the presence of external Na^+. Lastly, in the case of the membrane vesicles, when toxin gains access to their internal surface, it enhances Na^+ efflux-dependent Ca^{2+} influx (equivalent to its effect on Na^+–Ca^{2+} exchange in Figure 11(a)) and therefore results in increased Ca^{2+} accumulation (Figure 11(c)).

Considering the hypothetical model of Na^+–Ca^{2+} exchange depicted in Figure 2, tetanus toxin may be affecting this mechanism by enhancing either of the two forward reactions occurring on the outer surface of the membrane:

$$[XCa]^- \rightleftharpoons X^{3-} + Ca^{2+} \tag{3.1}$$

$$X^{3-} + 3Na^+ \rightleftharpoons XNa_3 \tag{3.2}$$

In other words, in order to increase Ca^{2+} efflux from the cell, the toxin may either decrease the affinity of X for Ca^{2+} or increase the affinity of X for Na^+. The scheme in Figure 2 operates to maintain a 10,000-fold Ca^{2+} concentration gradient using as driving force an approximately 10-fold concentration gradient of Na^+. One important factor favouring the operation of this system is the resting membrane potential. However, as discussed in Section 2.2, with recent direct evidence confirming the Na:Ca transport ratio as $3:1$,[35] the physiological Ca^{2+} gradient is approximately five-fold in excess of that achievable by operation of the Na^+–Ca^{2+} exchange mechanism.[37] One possibility is that a further factor is involved, for example, a mechanism that specifically inhibits the stability of the putative $[XCa]^-$ complex on the outer surface of the membrane and thus promotes Ca^{2+} dissociation even in the presence of the high external Ca^{2+} concentration. Such a mechanism might serve not only to increase the efficiency of the process but also as a possible regulatory site on the outside of the cell. It is possible that tetanus toxin, either acting directly or through such a site, may enhance the breakdown of the $[XCa]^-$ complex.

So far, this discussion has considered one possible direct external effector of Na^+–Ca^{2+} exchange. For ATP-dependent Ca^{2+} efflux, there is a similar scarcity of direct external effectors even though many membrane acting agents including several hormones may indirectly affect this mechanism through release of intracellular regulators. However, recent evidence suggests that insulin may have a direct influence upon ATP-dependent Ca^{2+} efflux. The action of insulin on, for example, glucose uptake into adipocytes[100] or muscle cells,[101] is known to be dependent upon external Ca^{2+}. Thus its effect may be mimicked by the calcium ionophore A23187.[102] Such results indicate that increased intracellular Ca^{2+} may mediate the effects of insulin. However, verapamil is apparently ineffective in inhibiting glucose uptake in adipocytes[100] suggesting that the mechanism by which insulin may increase internal Ca^{2+} is other than by an increase in Ca^{2+} influx through Ca^{2+} channels. Recently, Pershadsingh and McDonald[103] observed a direct inhibition by insulin of the $(Ca^{2+} + Mg^{2+})$-ATPase activity in isolated adipocyte plasma membranes, suggesting that a possible inhibition of the ATP-dependent Ca^{2+} pump is the mechanism by which insulin elevates internal Ca^{2+}. To date, studies directly observing Ca^{2+} fluxes in response to insulin have not been easy to interpret. Thus, insulin apparently increases labelled Ca^{2+} efflux from preloaded adipocytes[104] though it is possible that this is only a reflection of increased cytosolic Ca^{2+} and not a true effect on efflux. It is possible that any direct action of insulin on plasma membrane Ca^{2+} efflux may only be discernable using isolated vesicles of plasma membranes in experiments similar to those described above for tetanus toxin. It is also possible that insulin is affecting levels of an intracellular regulator of Ca^{2+} fluxes. The existence of such regulators is discussed in the remaining part of this section.

3.3 Intracellular effectors

Plasma membrane Ca^{2+} regulatory mechanisms are influenced by a diverse range of intracellular effectors though the physiological significance of many of these effects is obscure. Each of the transport mechanisms considered in Section 2 is influenced by modification of the internal ionic environment and/or substrate levels. Obviously the internal free Ca^{2+} concentration affects all of these mechanisms. Internal concentrations of monovalent and other divalent cations also influence Ca^{2+} transport, either directly, or by causing a modification of the membrane potential. Efflux of Ca^{2+} via Na^+–Ca^{2+} exchange and ATP-dependent Ca^{2+} transport will be directly influenced by internal levels of Na^+ and ATP, respectively. In addition, Na^+ in the absence of any net Na^+ gradient inhibits the K_m for ATP of ATP-dependent Ca^{2+} transport in synaptosomal membrane vesicles by almost 10-fold.[26,105] Also, ATP at high concentrations decreases the efficiency of Na^+–Ca^{2+} exchange. This suggests a complex coordinate control of the activity of the two plasma membrane efflux mechanisms exists mediated via internal levels of their substances.[105]

Although it is possible that all the plasma membrane Ca^{2+} transport mechanisms are internally regulated, most reports concerning specific cytosolic regulators of Ca^{2+} fluxes relate to effects upon ATP-dependent Ca^{2+} transport. This may be expected in view of the inherent asymmetry of this transport mechanism. However, as discussed in Section 3.1, determination of the influence of a particular internal effector upon plasma membrane as opposed to other intracellular sites of Ca^{2+} transport, may present a problem. $(Ca^{2+} + Mg^{2+})$–ATPase activity involved in Ca^{2+} transport is located on the cytosolic face of both the plasma membrane and the endoplasmic reticulum. In addition, mitochondria exert an ATP-dependent accumulation of Ca^{2+}. Therefore careful distinction between these sites is required in order to localize any effect to the plasma membrane.

ATP-dependent Ca^{2+} transport is known to be under the control of two major intracellular regulatory mechanisms—the cyclic AMP/protein kinase and the Ca^{2+}/calmodulin systems. The problem of determining site of action is evident in considering, for example, the properties of cyclic AMP-dependent modulation of ATP-dependent Ca^{2+} transport in muscle tissue. Thus ATP-dependent Ca^{2+} accumulation by sacroplasmic reticulum is known to be activated by cyclic AMP or cyclic AMP-dependent protein kinase[106] in conjunction with the phosphorylation of a specific protein within the membrane.[107] Cardiac sarcolemmal $(Ca^{2+} + Mg^{2+})$–ATPase activity also appears to be increased in response to cyclic AMP-dependent protein kinase-mediated phosphorylation.[108] However, the relative contribution of cyclic AMP-dependent modification of ATPase activity at the two locations, upon overall control of intracellular Ca^{2+}, is still unclear. Moreover, the effect of cyclic AMP upon sarcolemmal Ca^{2+} transport is further complicated by its apparent modulation of the slow inward Ca^{2+} channel of this membrane.[109] It is likely that a similarly complex view pertains to the role of cyclic AMP in regulating Ca^{2+} transport across the synaptic plasma membrane which is known to be cyclic AMP-dependent.[110] Thus an intricate relationship between cyclic AMP mediated events and changes in the activity of different Ca^{2+} translocating mechanisms is likely to be important in most excitatory and secretory tissues. The interrelationship between the cyclic AMP and Ca^{2+} regulatory systems exists on different levels with particular physiological relevance to a particular cell. Various categories of this interrelationship and their significance to cellular function are considered in detail by Rasmusson in this volume.

The activity of ATP-dependent Ca^{2+} transport in the plasma membrane also appears to be directly modulated internally by the Ca^{2+}-dependent regulatory protein, calmodulin. This has been repeatedly observed using relatively pure plasma membrane preparations from red blood cells and is therefore unlikely to reflect modulation of any internal ATP-dependent Ca^{2+} flux or any other plasma membrane Ca^{2+} transport mechanism (voltage-dependent or Na-dependent Ca^{2+} fluxes are not observed in most blood cell membranes). The presence of a non-dialysable, trypsin sensitive stimulator of plasma membrane $(Ca^{2+}-Mg^{2+})$-ATPase activity was known some time ago to exist in red blood cell haemolysates.[111] This factor, which increased the affinity of the enzyme for Ca^{2+} in EDTA-treated erythrocyte plasma membranes[112] was partially purified by Luthra *et al.*[113] Jarrett and Penniston[114] further purified and characterized the factor as a small, acidic, temperature stable protein remarkably similar to the Ca^{2+}-dependent regulator protein from brain known to affect adenylate cyclase and phosphodiesterase activity. Their results together with those of Gopinath and Vincenzi[115] suggested that both proteins were similarly effective at stimulating erythrocyte plasma membrane $(Ca^{2+}-Mg^{2+})$-ATPase activity. Recently, Larsen and Vincenzi[116] have shown that both the red blood cell activator and purified brain calmodulin directly activate ATP-dependent Ca^{2+} transport in inverted plasma membrane vesicles with a dependency similar to that of their

$(Ca^{2+}+Mg^{2+})$-ATPase activation. It appears that calmodulin directly interacts with solubilized $(Ca^{2+}+Mg^{2+})$-ATPase activity from the erythrocyte plasma membrane via a Ca^{2+}-dependent mechanism.[117] Furthermore, this interaction has recently been used to purify the enzyme activity using a calmodulin affinity column.[118] It is unknown how universal the effect of calmodulin may be upon ATP-dependent Ca^{2+} transport amongst tissues. Kuo *et al.*[119] reported effects of calmodulin upon $(Ca^{2+}+Mg^{2+})$-ATPase and ATP-dependent Ca^{2+} transport in synaptic plasma membranes, although preliminary experiments by the author (unpublished observations) indicated little or no effect of calmodulin on ATP-dependent Ca^{2+} transport in the synaptosomal plasma membrane vesicles described in Section 2.2. However, at least in the red blood cell, the activation of Ca^{2+} efflux via a Ca^{2+}-dependent mechanism mediated by calmodulin may represent an important internal feedback response following elevation of Ca^{2+} levels as a result of an external stimulus.[120]

There have been a number of reports concerning the action of certain agents which may directly modify or block ATP-dependent Ca^{2+} transport within cells. Such agents include certain mercurials[121,122] some local anesthetics including tetracaine[121,123] and the orthovanadate ion[124] which is also a potent inhibitor of $(Na^{+}+K^{+})$-ATPase.[125] A useful means for evaluating the specificity of such agents was by examining their effects upon the two Ca^{2+} transport mechanisms that co-reside in synaptosomal membrane vesicles.[26] Of these agents only vanadate appeared to specifically block ATP-dependent Ca^{2+} transport (Table 2). The other agents caused large effects

Table 2. Effect of putative modifiers of Ca^{2+} flux on Na^{+}–Ca^{2+} exchange and ATP-dependent Ca^{2+} uptake in synaptosomal membrane vesicles

Ca^{2+} uptake via the two mechanisms was as described in Figure 3. In all cases, Ca^{2+} proceeded for 15 minutes at room temperature. Details of media and conditions are as described by Gill *et al.*[26]

Addition	Na–Ca exchange		ATP-dependent Ca uptake	
	nmole Ca/mg protein	Percentage of control	nmole Ca/mg protein	Percentage of control
None	2.43 ± 0.15	100	2.70 ± 0.26	100
A23187 (10 μM)	0.17 ± 0.06	7	0.19 ± 0.02	7
Orthovanadate (1 mM)	1.97 ± 0.03	81	0.33 ± 0.03	12
Mersalyl (1 mM)	0.93 ± 0.06	38	0.22 ± 0.02	8
Tetracaine HCl (3 mM)	0.95 ± 0.03	39	1.87 ± 0.20	69
AMP–PNP (2 mM)	1.47 ± 0.19	60	0.49 ± 0.02	18
D600 (0.2 mM)	2.54 ± 0.13	105	2.69 ± 0.10	100

on both Ca^{2+} transport mechanisms suggesting their actions are non-specific. For example, the action of tetracaine probably arises from a more general change in membrane properties as a result of lipid modulation[126] rather than any specific effect upon Ca^{2+} transport. The absence of an effect of D600 upon either Ca^{2+} transport mechanism is consistent with its likely action upon a separate Ca^{2+} channel. Lastly, although AMP–PNP (a non-hydrolysable analogue of ATP) is expected to block ATP-dependent Ca^{2+} uptake as observed, its partial effect upon Na^{+}–Ca^{2+} exchange is evidence of the possible modifying effect of ATP alluded to earlier in this section.[26,105]

4 SUMMARY AND CONCLUSIONS

This chapter has attempted to describe how certain plasma membrane effector molecules may directly modulate the Ca^{2+} regulatory mechanisms of the plasma membrane. Although this is not a comprehensive review, it is hoped that the examples of effectors described provide an account of the variety of mechanisms by which such modulation may occur. It should be emphasized that the criteria by which effectors are determined as influencing plasma membrane as opposed to intracellular Ca^{2+} regulatory mechanisms, are not always conclusive. Even in the cases where such a plasma membrane effect is clear it is likely that in the whole cell any effect on Ca^{2+} accumulation is greatly influenced by the activity of internal Ca^{2+}-sequestering organelles. This is because the control of cytosolic Ca^{2+} levels appears to be a subtle balance between the Ca^{2+} transporting activities of the relevant internal organelles and the plasma membrane. The reader may attempt to determine which examples cited in this chapter most convincingly demonstrate direct receptor-mediated changes in plasma membrane Ca^{2+} regulatory activity. Clearly, it is the author's belief that a defined Ca^{2+} transport activity in the plasma membrane, separate and distinguishable from internal Ca^{2+} transport activity, is required in order to show such a direct effect. Hence the discussion contains details of the author's characterization and use of a plasma membrane vesicular preparation from brain synaptosomes where such activity is discernible. Plasma membrane preparations with similarly defined Ca^{2+} transport activity have also been isolated from erythrocytes and muscle as described in Section 2.2.

Whereas the Ca^{2+} efflux mechanisms of the plasma membrane have been well defined using isolated plasma membrane preparations, the mechanisms of Ca^{2+} influx have not been clearly observed except in whole cells where the problems alluded to above are apparent. Although there may be methods for determining the operation of plasma membrane Ca^{2+} influx in whole cells (described in Section 3.1), influx mechanisms involving passive facilitated diffusion are by definition difficult to observe against the operation of highly efficient active Ca^{2+} efflux mechanisms. Nonetheless, there is good evidence for the operation of Ca^{2+} entry channels sensitive to membrane potential, mainly from experiments on giant axons, the internal and external environments of which are easily controllable. However, such channels do not operate in all cells. Furthermore, with regard to agents effecting increased Ca^{2+} influx, as suggested in Section 3.2.1, there is increasing evidence that such effectors may operate via Ca^{2+} entry mechanisms distinct from the voltage sensitive Ca^{2+} channel. Such mechanisms may be related to the widely described effects upon plasma membrane phosphatidate and phosphatidyl inositol turnover which appear to be associated with effector-related changes in Ca^{2+} transport. The recent predictions by Salmon and Honeyman[66] and Putney et al.[65] that phosphatidate may act as an endogenous Ca^{2+} ionophore are most intriguing in this regard. However, apart from this possible mechanism, little is known concerning the mechanisms controlling other categories of Ca^{2+} transport across the plasma membrane. The elucidation of mechanisms within the plasma membrane that directly couple receptor-binding with modification of Ca^{2+} transport activity may in the future reveal complex regulatory components similar in function to the components that regulate receptor-mediated adenylate cyclase activity.

5 ACKNOWLEDGEMENTS

I am indebted to Dr. Leonard D. Kohn and Dr. Evelyn F. Grollman for their invaluable contribution to the work and ideas described in this chapter. I also thank Dr. Allen P. Minton and Dr. Michael Epstein for useful discussions, and Mrs. Bobbi Armstrong for preparation of the manuscript.

REFERENCES

1. Rosenberger, L., and Triggle, D. J. (1978). Calcium, calcium translocation, and specific calcium antagonists. In: *Calcium in Drug Action* (Weiss, G. B., ed.), pp. 3–31, Plenum Press, New York.
2. Carafoli, E., and Crompton, M. (1978). The regulation of intracellular calcium. In: *Current Topics in Membranes and Transport* (Bronner, F., and Kleinzeller, A., eds.), Vol. 10, pp. 151–216, Academic Press, New York.
3. Erulkar, S. D., and Fine, A. (1979). Calcium in the nervous system. In: *Reviews of Neuroscience* (Schneider, D. M., ed.) Vol. 4, pp. 179–232, Raven Press, New York.
4. Racker, E. (1979). Fluxes of Ca^{2+} ions and concepts. *Fed. Proc.*, **39**, 2422–2426.
5. Michell, R. H. (1979). Inositol phospholipids in membrane function. *Trends Biochem. Sci.* **4**, 128–131.
6. Baker, P. F. (1977). Calcium and the control of neurosecretion. In: *Biochemistry of Membrane Transport* (Semenza, G., and Carafoli, E., eds.), FEBS-symposium No. 42, pp. 430–441, Springer-Verlag, Berlin.

7. Mullins, L. J. (1978). The mechanisms of Ca transport in squid axons. In: *Membrane Transport Processes* (Tosteson, D. C., Ovchinnikov, Y. A., and Latorre, R., eds.) Vol. 2, pp. 371–381, Raven Press, New York.

8. Baker, P. F., Hodgkin, A. L., and Ridgway, E. B. (1971). Depolarization and calcium entry in squid giant axons. *J. Physiol.*, **218,** 709–755.

9. Reuter, H., and Scholz, H. (1977). A study of the ion selectivity and the kinetic properties of the calcium dependent slow inward current in mammalian cardiac muscle. *J. Physiol.*, **264,** 17–47.

10. Fleckenstein, A. (1971). Specific inhibitors and promoters of calcium action in the excitation-contraction coupling of heart muscle and their role in the prevention or production of myocardial lesions. In: *Calcium and the Heart* (Harris, P., and Opie, L. H., eds.), pp. 135–188, Academic Press, London.

11. Hagiwara, S. (1975). Ca-dependent action potential. In: *Dynamic Properties of Lipid Bilayers* (Eisenman, G., ed.), pp. 359–381, Dekker, New York.

12. Katz, B., and Miledi, R. (1969). Tetrodotoxin-resistant electric activity in presynaptic terminals. *J. Physiol.*, **203,** 459–487.

12[A]. Reuter, H. (1974). Exchange of calcium ions in the mammalian myocardium: mechanisms and physiological significance. *Circ. Res.*, **34,** 599–605.

13. Fabiato, A., and Fabiato, F. (1977). Calcium release from the sarcoplasmic reticulum. *Circ. Res.*, **40,** 119–129.

14. Hurwitz, L., and McGuffee, M. (1978). Calcium in smooth muscle function. In: *Calcium in Drug action* (Weiss, G. B., ed.), pp. 75–93, Plenum Press, New York.

15. Stephenson, E. W., and Podolsky, R. J. (1978). The regulation of calcium in skeletal muscle. *Ann. N.Y. Acad. Sci.*, **307,** 462–476.

16. Bray, R. N., and Rosen, B. P. (1979). Cation/proton antiport systems in Escherichia coli. *J. Biol. Chem.*, **254,** 1957–1963.

17. Stroobant, P., Dame, J. B., and Scarborough, G. A. (1980). The neurospora plasma membrane Ca^{2+} pump. *Fed. Proc.*, **39,** 2437–2441.

18. Schatzmann, H. J. (1966). ATP-dependent Ca^{2+} extrusion from human red cell. *Experientia*, **22,** 364–365.

19. Steck, T. L., and Kant, J. A. (1974). Preparation of impermeable ghosts and inside-out vesicles from human erythrocyte membranes. *Methods Enzymol.*, **31,** 172–180.

20. Lamb, J. F., and Lindsay, R. (1971). Effect of Na, metabolic inhibitors and ATP on Ca movements in L cells. *J. Physiol.*, **218,** 691–708.

21. Robblee, L. S., Shepro, D., and Belamarich, F. A. (1973). Calcium uptake and associated adenosine triphosphatase activity of isolated platelet membranes. *J. Gen. Physiol.*, **61,** 462–481.

22. Moore, L., Fitzpatrick, D. F., Chen, T. S., and Landon, E. J. (1974). Calcium pump activity of the renal plasma membrane and renal microsomes. *Biochim. Biophys. Acta.*, **345,** 405–418.

23. Van Rossum, G. D. V. (1970). Net movements of Ca^{2+} and Mg^{2+} in slices of rat liver. *J. Gen. Physiol.*, **55,** 18–32.

24. DiPolo, R. (1978). Ca pump driven by ATP in squid axons. *Nature*, **274,** 390–392.

25. Caroni, P., and Carafoli, E. (1980). An ATP-dependent Ca^{2+}-pumping system in dog heart sarcolemma. *Nature*, **283,** 765–767.

26. Gill, D. L., Grollman, E. F., and Kohn, L. D. (1981). Calcium transport mechanisms in membrane vesicles from guinea pig brain synaptosomes. *J. Biol. Chem.*, **256,** 184–192.

27. Schatzmann, H. J., and Vincenzi, F. F. (1969). Calcium movements across the membrane of human red cells. *J. Physiol.*, **201**, 369–395.
28. Weiner, M. L., and Lee, K. S. (1972). Active calcium ion uptake by inside-out and right side-out vesicles of red blood cell membranes. *J. Gen. Physiol.*, **59**, 462–475.
29. Sarkadi, B., and Tosteson, D. C. (1979). Active cation transport in human red cells. In: *Membrane Transport in Biology* (Giebisch, G., Tosteson, D. C., and Ussing, H. H., eds.) Vol. 2, pp. 117–160, Springer-Verlag, Berlin.
30. Knauf, P. A., Proverbio, F., and Hoffman, J. F. (1974). Electrophoretic separation of different phosphoproteins associated with Ca-ATPase and Na, K-ATPase in human red cell ghosts. *J. Gen. Physiol.*, **63**, 324–336.
31. Reuter, H., and Seitz, N. (1968). The dependence of calcium efflux from cardiac muscle on temperature and external ion composition. *J. Physiol.*, **195**, 451–470.
32. Blaustein, M. P., and Hodgkin, A. L. (1969). The effect of cyanide on the efflux of calcium from squid axons. *J. Physiol.*, **200**, 497–527.
33. Baker, P. F., Blaustein, M. P., Hodgkin, A. L., and Steinhardt, R. A. (1969). The influence of calcium on sodium efflux in squid axons. *J. Physiol.*, **200**, 431–458.
34. Mullins, L. J., and Brinley, F. J. (1975). The sensitivity of calcium efflux from squid axons to changes in membrane potential. *J. Gen. Physiol.*, **65**, 135–152.
35. Pitts, B. J. R. (1979). Stoichiometry of sodium-calcium exchange in cardiac sarcolemmal vesicles. *J. Biol. Chem.*, **254**, 6232–6235.
36. Miyamoto, H., and Racker, E. (1980). Solubilization and partial purification of the Ca^{2+}/Na^+ antiporter from the plasma membrane of bovine heart. *J. Biol. Chem.*, **255**, 2656–2658.
37. Mullins, L. J. (1979). The generation of electric currents in cardiac fibers by Na/Ca exchange. *Am. J. Physiol.*, **236**, C103–C110.
38. Blaustein, M. P., and Oborn, C. J. (1975). The influence of sodium on calcium fluxes in pinched-off nerve terminals in vitro. *J. Physiol.*, **247**, 657–686.
39. Reeves, J. P., and Sutko, J. L. (1979). Sodium-calcium ion exchange in cardiac membrane vesicles. *Proc. Natl. Acad. Sci. U.S.A.*, **76**, 590–594.
40. Parker, J. C., Gitelman, H. J., Glosson, P. S., and Leonard, D. L. (1975). Role of calcium in volume regulation by dog red blood cells. *J. Gen. Physiol.*, **65**, 84–96.
41. Brown, A. M. (1979). Evidence for a magnesium- and ATP-dependent calcium extrusion pump in dog erythrocytes. *Biochim. Biophys. Acta*, **554**, 195–203.
42. Kendrick, N. C., Blaustein, M. P., Fried, R. C., and Ratzlaff, R. W. (1977). ATP-dependent calcium storage in presynaptic nerve terminals. *Nature*, **265**, 246–248.
43. Blaustein, M. P., Ratzlaff, R. W. and Kendrick, N. C. (1978). The regulation of intracellular calcium in presynaptic nerve terminals. *Ann. N.Y. Acad. Sci.*, **307**, 195–212.
44. Swanson, P. D., Anderson, L., and Stahl, W. L. (1974). Uptake of calcium ions by synaptosomes from rat brain. *Biochim. Biophys. Acta*, **356**, 174–183.
45. Fabiato, A., and Fabiato, F. (1975). Contractions induced by a calcium-triggered release of calcium from the sarcoplasmic reticulum of single skinned cardiac cells. *J. Physiol.*, **249**, 469–495.
46. Borle, A. B. (1978). On the difficulty of assessing the role of extracellular calcium in cell function. *Ann. N.Y. Acad. Sci.*, **307**, 431–432.

47. Schulz, I., and Ullrich, K. J. (1979). Transport processes in the exocrine pancreas. In: *Membrane Transport in Biology* (Giebisch, G., Tosteson, D. C., and Ussing, H. H., eds.) Vol. 4B, pp. 811–852, Springer-Verlag, Berlin.

48. Moriarty, C. M. (1978). Role of calcium in the regulation of adenohypophysial hormone release. *Life Sci.*, **23**, 185–194.

49. Exton, J. H. (1980). Mechanisms involved in α-adrenergic phenomena: role of calcium ions in actions of catecholamines in liver and other tissues. *Am. J. Physiol.*, **238**, E3–E12.

50. Fain, J. N., and García-Sáinz, J. A. (1980). Role of phosphatidylinositol turnover in $alpha_1$ and of adenylate cyclase inhibition in $alpha_2$ effects of catecholamines. *Life Sci.*, **26**, 1183–1194.

51. Deth, R., and van Breemen, C. (1977). Agonist-induced release of intracellular Ca^{2+} in the rabbit aorta. *J. Membr. Biol.*, **30**, 363–380.

52. van Breemen, C., and Siegel, B. (1980). The mechanism of α-adrenergic activation of the dog coronary artery. *Circ. Res.*, **46**, 426–429.

53. Karaki, H., and Weiss, G. B. (1979). Alterations in high and low affinity binding of ^{45}Ca in rabbit aortic smooth muscle by norepinephrine and potassium after exposure to lanthanum and low temperature. *J. Pharmacol. Exp. Ther.*, **211**, 86–92.

54. Parod, R. J., and Putney, R. W. (1978). The role of calcium in the receptor mediated control of potassium permeability in the rat lacrimal gland. *J. Physiol.*, **281**, 371–381.

55. Horn, J. P., and McAfee, D. A. (1979). Norepinephrine inhibits calcium-dependent potentials in rat sympathetic neurons. *Science*, **204**, 1233–1235.

56. Horn, J. P., and McAfee, D. A. (1980). Alpha-adrenergic inhibition of calcium-dependent potentials in rat sympathetic neurones. *J. Physiol.*, **301**, 191–204.

57. Bregestovski, P. D., Miledi, R., and Parker, I. (1979). Calcium conductance of acetylcholine-induced endplate channels. *Nature*, **279**, 638–639.

58. Ritchie, A. K. (1979). Catecholamine secretion in a rat pheochromocytoma cell line: two pathways for calcium entry. *J. Physiol.*, **286**, 541–561.

59. Putney, J. W. (1978). Role of calcium in the actions of agents affecting membrane permeability. In: *Calcium in Drug Action* (Weiss, G. B., ed.) pp. 173–194, Plenum Press, New York.

60. Putney, J. W., Van De Walle, C. M., and Leslie, B. A. (1978). Receptor control of calcium influx in parotid acinar cells. *Mol. Pharmacol.*, **14**, 1046–1053.

61. Putney, J. W., and Van De Walle, C. M. (1980). The relationship between muscarinic receptor binding and ion movements in rat parotid cells. *J. Physiol.*, **299**, 521–531.

62. Marier, S. H., Putney, J. W., and Van De Walle, C. M. (1978). Control of calcium channels by membrane receptors in the rat parotid gland. *J. Physiol.*, **279**, 141–151.

63. Putney, J. W. (1979). Receptor regulation of calcium channels in exocrine gland cells. *Proc. West. Pharmacol. Soc.*, **22**, 295–299.

64. Michell, R. H., Jafferji, S. S., and Jones, L. M. (1977). The possible involvement of phosphatidylinositol breakdown in the mechanism of stimulus-response coupling at receptors which control cell-surface calcium gates. *Adv. Exp. Med. Biol.*, **83**, 447–464.

65. Putney, J. W., Weiss, S. J., Van De Walle, C. M., and Haddas, R. A. (1980). Is phosphatidic acid a calcium ionophore under neurohumoral control? *Nature*, **284**, 345–347.

66. Salmon, D. M., and Honeyman, T. W. (1980). Proposed mechanism of cholinergic action in smooth muscle. *Nature*, **284,** 344–345.
67. Matthews, E. K. (1979). Calcium translocation and control mechanisms for endocrine secretion. *Symp. Soc. Exp. Biol.*, **33,** 225–249.
68. Meissner, H. P., and Schmelz, H. (1974). Membrane potential of beta-cells in pancreatic islets. *Pfluegers Arch.*, **351,** 195–206.
69. Matthews, E. K. (1975). Calcium and stimulus-secretion coupling in pancreatic islet cells. In: *Calcium Transport in Contraction and Secretion* (Carafoli, E., Clementi, F., Drabikowski, W., and Margreth, A., eds.) pp. 203–210, North-Holland, Amsterdam.
70. Henquin, J. C. (1978). D-glucose inhibits potassium efflux from pancreatic islet cells. *Nature*, **271,** 271–273.
71. Boschero, A. C., Kawazu, S., Sener, A., Herchuelz, A., and Malaisse, W. J. (1979). The stimulus-secretion coupling of glucose-induced insulin release: effects of valinomycin upon pancreatic islet function. *Arch. Biochem. Biophys.*, **196,** 54–63.
72. Dufy, B., Vincent, J. D., Fleury, H., Du Pasquier, P., Gourdji, D., and Tixier-Vidal, A. (1979). Membrane effects of thyrotropin-releasing hormone and estrogen shown by intracellular recording from pituitary cells. *Science*, **204,** 509–511.
73. Gershengorn, M. C. (1980). Thyrotropin releasing hormone stimulation of prolactin release: evidence for a membrane potential-independent, Ca^{2+}-dependent mechanism of action. *J. Biol. Chem.*, **255,** 1801–1803.
74. Kohlhardt, M., Bauer, B., Krause, H., and Fleckenstein, A. (1972). Differentiation of the transmembrane Na and Ca channels in mammalian cardiac fibres by use of specific inhibitors. *Pfluegers Arch.*, **335,** 309–320.
75. Fleckenstein, A. (1977). Specific pharmacology of calcium in myocardium, cardiac pacemakers, and vascular smooth muscle. *Ann. Rev. Pharmacol. Toxicol.* **17,** 149–166.
76. Hagiwara, S. (1980). The Ca ion permeability of the cell membrane. *Jap. Circ. J.*, **44,** 239–248.
77. Baker, P. F., Meves, H., and Ridgway, E. B. (1973). Effects of manganese and other agents on the calcium uptake that follows depolarization of squid axons. *J. Physiol.*, **231,** 511–526.
78. Kass, R. S., and Tsien, R. W. (1975). Multiple effects of calcium antagonists on plateau currents in cardiac Purkinje fibers. *J. Gen. Physiol.*, **66,** 169–192.
79. Study, R. E., Breakefield, X. O., Bartfai, T., and Greengard, P. (1978). Voltage-sensitive calcium channels regulate guanosine 3′,5′-cyclic monophosphate levels in neuroblastoma cells. *Proc. Nat. Acad. Sci. U.S.A.*, **75,** 6295–6299.
80. Fairhurst, A. S., Whittaker, M. L., and Ehlert, F. J. (1980). Interactions of D600 (methoxy-verapamil) and local anesthetics with rat brain α-adrenergic and muscarinic receptors. *Biochem. Pharmacol.*, **29,** 155–162.
81. Grasso, A., Alemà, S., Rufini, S., and Senni, M. I. (1980). Black widow spider toxin-induced calcium fluxes and transmitter release in a neurosecretory cell line. *Nature*, **283,** 774–776.
82. Gill, D. L., Dyer, S. A., Kohn, L. D., and Grollman, E. F. (1981) Influence of tetanus toxin on synaptic plasma membrane calcium transport. *Fed. Proc.*, **40,** 1781.
83. Gill, D. L., Dyer, S. A., and Grollman, E. F. (1982) Effect of tetanus toxin on calcium transport across the synaptic plasma membrane. *J. Biol. Chem.* (submitted for publication).

84. Bizzini, B. (1977). Tetanus toxin structure as a basis for elucidating its immunological and neuropharmacological activities. In: *Receptors and Recognition* (Cuatrecasas, P., ed.) Series B, Vol. 1, pp. 175–218, Chapman and Hall, London.

85. Curtis, D. R., and DeGroat, W. C. (1968). Tetanus toxin and spinal inhibition. *Brain Res.*, **10**, 208–212.

86. Osborne, R. H., and Bradford, H. F. (1973). Patterns of amino acid release from nerve-endings isolated from spinal cord and medulla. *J. Neurochem.*, **21**, 407–419.

87. Price, D. L., Griffin, J. W., and Peck, K. (1977). Tetanus toxin: evidence for binding at presynaptic nerve endings. *Brain Res.*, **121**, 379–384.

88. Schwab, M. E., and Thoenen, H. (1978). Selective binding, uptake, and retrograde transport of tetanus toxin by nerve terminals in the rat iris. *J. Cell. Biol.*, **77**, 1–13.

89. Mirsky, R., Wendon, L. M. B., Black, P., Stolkin, C., and Bray, D. (1978). Tetanus toxin: a cell surface marker for neurons in culture. *Brain Res.*, **148**, 251–259.

90. van Heyningen, W. E. (1974). Gangliosides as membrane receptors for tetanus toxin, cholera toxin and serotonin. *Nature*, **249**, 415–417.

91. Kohn, L. D. (1978). Relationships in the structure and function of receptors for glycoprotein hormones, bacterial toxins and interferon. In: *Receptors and Recognition* (Cuatrecasas, P., and Greaves, M. F., eds.) Vol. 5, pp. 133–212, Chapman and Hall, London.

92. Schwab, M. E., and Thoenen, H. (1977). Retrograde axonal and transsynaptic transport of macromolecules: physiological and pathophysiological importance. *Agents and Actions*, **7**, 361–368.

93. Lee, G., Grollman, E. F., Dyer, S. A., Beguinot, F., Kohn, L. D., Habig, W. H., and Hardegree, M. C. (1979). Tetanus toxin and thyrotropin interactions with rat brain membrane preparations. *J. Biol. Chem.*, **254**, 3826–3832.

94. Dumas, M., Schwab, M. E., Baumann, R., and Thoenen, H. (1979). Retrograde transport of tetanus toxin through a chain of two neurons. *Brain Res.*, **165**, 354–357.

95. Neville, D. M., and Chang, T-M. (1978). Receptor-mediated protein transport into cells. Entry mechanisms for toxins, hormones, antibodies, viruses, lysosomal hydrolases, asialoglycoproteins, and carrier proteins. In: *Current Topics in Membranes and Transport* (Bronner, F., and Kleinzeller, A., eds.) Vol. 10, pp. 65–150, Academic Press, New York.

96. Ramos, S., Grollman, E. F., Lazo, P. S., Dyer, S. A., Habig, W. H., Hardegree, M. C., Kaback, H. R., and Kohn, L. D. (1979). Effect of tetanus toxin on the accumulation of the permeant lipophilic cation tetraphenylphosphonium by guinea pig brain synaptosomes. *Proc. Nat. Acad. Sci. U.S.A.*, **76**, 4783–4787.

97. Blaustein, M. P. (1975). Effects of potassium, veratridine, and scorpion venom on calcium accumulation and transmitter release by nerve terminals in vitro. *J. Physiol.*, **247**, 617–655.

98. Campbell, C. W. B. (1976). The Na^+, K^+, and Cl^- contents and derived membrane potentials of presynaptic nerve endings in vitro. *Brain Res.*, **101**, 594–599.

99. Li, P. P., and White, T. D. (1977). Rapid effects of veratridine, tetrodotoxin, gramicidin D, valinomycin and NaCN on the Na^+, K^+ and ATP content of synaptosomes. *J. Neurochem.*, **28**, 967–975.

100. Bonne, D., Belhadj, O., and Cohen, P. (1977). Modulation by calcium of the

insulin action and of the insulin-like effect of oxytocin on isolated rat adipocytes. *Eur. J. Biochem.*, **75**, 101–105.

101. Gould, M. K., and Chaudry, I. H. (1970). The action of insulin on glucose uptake by isolated rat soleus muscle: 1. Effects of cations. *Biochim. Biophys. Acta*, **215**, 247–249.

102. Clausen, T., and Martin, B. R. (1977). The effect of insulin on the washout of (^{45}Ca) calcium from adipocytes and soleus muscle of the rat. *Biochem. J.*, **164**, 251–255.

103. Pershadsingh, H. A., and McDonald, J. M. (1979). Direct addition of insulin inhibits a high affinity Ca^{2+}-ATPase in isolated adipocyte plasma membranes. *Nature*, **281**, 495–497.

104. Hopc-Gill, H., Kissebah, A., Tulloch, B., Clarke, P., Vydelingum, N., and Fraser, T. R. (1975). The effects of insulin on adipocyte calcium flux and the interaction with the effects of dibutyryl cyclic AMP and adrenaline. *Horm. Metab. Res.*, **7**, 195–196.

105. Gill, D. L., Grollman, E. F., and Kohn, L. D. (1980). Interrelationship of calcium transport mechanisms in synaptic membrane vesicles. *Fed. Proc.*, **39**, 2039.

106. Kirchberger, M. A., Tada, M., and Katz, A. M. (1974). Adenosine 3′,5′-monophosphate-dependent protein kinase-catalyzed phosphorylation reaction and its relationship to calcium transport in cardiac sarcoplasmic reticulum. *J. Biol. Chem.*, **249**, 6166–6173.

107. Kirchberger, M. A., and Tada, M. (1976). Effects of adenosine 3′,5′-monophosphate-dependent protein kinase on sarcoplasmic reticulum isolated from cardiac and slow and fast contracting skeletal muscles. *J. Biol. Chem.*, **251**, 725–729.

108. Hui, C. W., Drummond, M., and Drummond, G. I. (1976). Calcium accumulation and cyclic AMP-stimulated phosphorylation in plasma membrane-enriched preparations of myocardium. *Arch. Biochem. Biophys.*, **173**, 415–427.

109. Tsien, R. W. (1977). Cyclic AMP and contractile activity in the heart. *Adv. Cyclic. Nucleotide Res.*, **8**, 363–420.

110. Skirboll, L. R., Baizer, L., and Dretchen, K. L. (1977). Evidence for a cyclic nucleotide-mediated calcium flux in motor nerve terminals. *Nature*, **268**, 352–355.

111. Bond, G. H., and Clough, D. L. (1973). A soluble protein activator of $(Mg^{2+}+Ca^{2+})$-dependent ATPase in human red cell membranes. *Biochim. Biophys. Acta*, **323**, 592–599.

112. Quist, E. E., and Roufogalis, B. D. (1975). Calcium transport in human erythrocytes: separation and reconstitution of high and low Ca affinity (Mg+Ca)-ATPase activities in membranes prepared at low ionic strength. *Arch. Biochem. Biophys.*, **168**, 240–251.

113. Luthra, M. G., Hildenbrandt, G. R., and Hanahan, D. J. (1976). Studies on an activator of the $(Ca^{2+}+Mg^{2+})$-ATPase of human erythrocyte membranes. *Biochim. Biohys. Acta*, **419**, 164–179.

114. Jarrett, H. W., and Penniston, J. T. (1977). Partial purification of the Ca^{2+}–Mg^{2+} ATPase activator from human erythrocytes: its similarity to the activator of 3′:5′-cyclic nucleotide phosphodiesterase. *Biochem. Biophys. Res. Commun.*, **77**, 1210–1216.

115. Gopinath, R. M., and Vincenzi, F. F. (1977). Phosphodiesterase protein activator mimics red blood cell cytoplasmic activator of $(Ca^{2+}–Mg^{2+})$ ATPase. *Biochem. Biophys. Res. Commun.*, **77**, 1203–1209.

116. Larsen, F. L., and Vincenzi, F. F. (1979). Calcium transport across the plasma membrane: stimulation by calmodulin. *Science*, **204**, 306–309.
117. Lynch, T. J., and Cheung, W. Y. (1979). Human erythrocyte Ca^{2+}–Mg^{2+}-ATPase: mechanism of stimulation by Ca^{2+}. *Arch. Biochem. Biophys.*, **194**, 165–170.
118. Niggli, V., Penniston, J. T., and Carafoli, E. (1979). Purification of the (Ca^{2+}–Mg^{2+})-ATPase from human erythrocyte membranes using a calmodulin affinity column. *J. Biol. Chem.*, **254**, 9955–9958.
119. Kuo, C-H., Ichida, S., Matsuda, T., Kakiuchi, S., and Yoshida, H. (1979). Regulation of ATP-dependent Ca-uptake of synaptic plasma membranes by Ca-dependent modulator protein. *Life Sci.*, **25**, 235–239.
120. Vincenzi, F. F. (1979). Calmodulin in the regulation of intracellular calcium. *Proc. West. Pharmacol. Soc.*, **22**, 289–294.
121. Blaustein, M. P., Ratzlaff, R. W., Kendrick, N. C., and Schweitzer, E. S. (1978). Calcium buffering in presynaptic nerve terminals. I: Evidence for involvement of a non-mitochondrial ATP-dependent sequestration mechanism. *J. Gen. Physiol.*, **72**, 15–41.
122. Binah, O., Meiri, U., and Rahamimoff, H. (1978). The effects of $HgCl_2$ and mersalyl on mechanisms regulating intracellular calcium and transmitter release. *Eur. J. Pharmacol.*, **51**, 453–457.
123. Roufogalis, B. D. (1973). Properties of a (Mg^{2+}+Ca^{2+})-dependent ATPase of bovine brain cortex: effects of detergents, freezing, cations and local anesthetics. *Biochim. Biophys. Acta*, **318**, 360–370.
124. DiPolo, R., Rojas, H. R., and Beaugé, L. (1979). Vanadate inhibits uncoupled Ca efflux but not Na-Ca exchange in squid axons. *Nature*, **281**, 228–229.
125. Cantley, L. C., Josephson, L., Warner, R., Yanagisawa, M., Lechene, C., and Guidotti, G. (1977). Vanadate is a potent inhibitor found in ATP derived from muscle. *J. Biol. Chem.*, **252**, 7421–7423.
126. Papahadjopoulos, D. (1972). Studies on the mechanism of action of local anesthetics with phospholipid model membranes. *Biochim. Biophys. Acta*, **265**, 169–186.

Hormone Receptors
Edited by L. D. Kohn
© 1982, John Wiley & Sons, Ltd

11 *Involvement of phosphatidyl-inositol breakdown in elevation of cytosol Ca^{2+} by hormones and relationship to prostaglandin formation*

J. N. Fain

1 INTRODUCTION

Myo-inositol was isolated from biological sources over a hundred years ago but its biological function is still incompletely understood. Inositol is required in the medium for survival of most cultured cell lines as they lack the pathways for synthesis of inositol from glucose or other sugars.[1] The serum level of inositol in the rat is less than 0.1 mM while the concentration of free inositol in tissues is in the range of 5 to 20 mM.[2,3] These findings suggest that inositol is actively transported into and accumulated by many cells but no function for free inositol has yet been established. In most cultured cell lines a concentration of 0.001 mM is required for growth which is higher than is required for essential vitamins.[4] Inositol, like choline, is a lipotropic factor under certain dietary conditions but it has been difficult to produce inositol deficiency in any species under ordinary conditions.[1] Inositol is derived from the diet, synthesized in the gut of animals and some cells are even able to convert glucose to inositol. The fatty liver which occurs in inositol deficient animals has generally been attributed to a reduction in lipoprotein secretion. The triglycerides which would normally combine with phospholipids and proteins to make lipoproteins instead accumulate in the liver and account for the abnormal fat deposition.

The reduced lipoprotein formation is probably related to the inability of hepatocytes to synthesize enough of the phospholipid which contains inositol linked to the phosphate group of phosphatidic acid. The only known function of inositol is as a precursor of phosphatidylinositol a minor phospholipid which accounts for less than 10% of the total phospholipids in most cells. In liver[5,6] platelets[7] and adipose tissue[8] of rats the highest arachidonate content of the different phospholipids is in phosphatidylinositol. The predominant fatty acids in phosphatidylinositol are stearate as the 1-acyl and arachidonate as the 2-acyl group. Arachidonic acid is the precursor from which prostaglandins and thromboxanes are formed. The availability of the arachidonic acid appears to be the major rate-limiting factor in the formation of prostaglandins and recent work suggests that the breakdown of phosphatidyl-inositol in response to stimuli is the major source of arachidonic acid in the exocrine pancreas[9] and platelets.[10]

The role of phosphatidylinositol turnover in hormone action is not well understood. Michell[11] pointed out the association between elevation of cytosol Ca^{2+} and phosphatidylinositol turnover and suggested that the elevation in cytosol Ca^{2+} was secondary to breakdown of phosphatidylinositol in the plasma membrane. The present review is an attempt to interpret the available data, much of which has been reported during the past few years, in light of the hypothesis of Michell.[11] This chapter like most attempts at simplification will probably turn out to be incorrect in some aspects. Future experiments will determine the validity of linking phosphatidylinositol turnover and calcium in such diverse systems as 5-HT induced calcium gating in blowfly salivary glands, ACTH action on the adrenal, TSH action on the thyroid and thrombin activation of prostaglandin formation in platelets.

The remarkable advances which have been made in techniques for the study of nucleic acids and proteins have not been matched by comparable advances in phospholipids. The basic enzymatic reactions involved in phospholipid turnover are incompletely understood because of the difficulty in working with water-insoluble substrates and enzymes.

The proceedings of a symposium devoted to studies on the chemistry, biosynthesis and function of inositol and phosphoinositides was recently published.[12] The proceedings review the status of phosphatidylinositol turnover in response to extracellular stimuli and indicate there was little agreement on the meaning of this response.

2   PATHWAYS FOR PHOSPHATIDYLINOSITOL SYNTHESIS

Phosphatidylinositol is synthesized from phosphatidic acid, inositol and cytidine triphosphate (CTP) as depicted in Figure 1. Phosphatidic acid is formed via diacylglycerol kinase using available diglyceride in the presence of ATP or de novo via successive addition of fatty acyl groups to alpha-glycerophosphate. The diacylglycerol kinase present in both membranes[13,14]

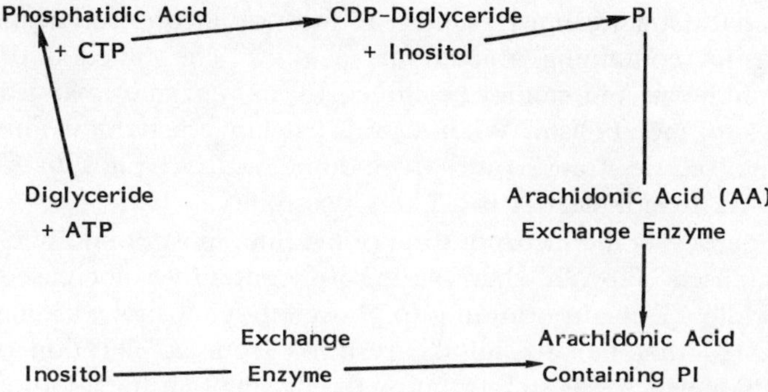

Figure 1. Pathways for formation of phosphatidylinositol (PI)

and cytosol is able to phosphorylate diglyceride to phosphatidic acid. Phosphatidylinositol synthesis involves the reaction of CTP and phosphatidic acid to give CDP-diglyceride and pyrophosphate. The CDP-diglyceride in turn react with free myo-inositol via the enzyme CDP-diglyceride:inositol transferase to form phosphatidylinositol.[15,16] This enzyme is membrane-bound and found in the endoplasmic reticulum of rat liver as well as the plasma membrane.[17] Phosphatidylinositol synthesis in several mammalian tissues[18,19] and blowfly salivary glands[20] is inhibited by Ca^{2+}. This may reflect inhibition by Ca^{2+} of the CDP-diglyceride inositol transferase which has been demonstrated in preparations from rat liver,[21] rabbit lung[22] and the protozoan *Crithidia fasciculata*.[23]

There is another enzyme present in rat liver which can be separated from the CDP-diglyceride inositol transferase[24] and catalyses the exchange of myo-inositol label with phosphatidylinositol. The exchange enzyme is activated by Mn^{2+}, does not require CDP-diglyceride and is inhibited by CTP.[17] Wallace and Fain[21] confirmed these findings and demonstrated that the Mn^{2+}-activated inositol exchange seen in the absence of CDP-diglyceride, CTP or CMP is little affected by Ca^{2+}.

In vivo labeled inositol and ^{32}P are preferentially incorporated into the phosphatidylinositol molecules which are rich in monoenoic plus dienoic fatty acids at the 2 position of glycerol.[5,6,25,26] However, the predominant species of phosphatidylinositol in rat liver contains arachidonic acid (a tetraenoic fatty acid) which serves as the precursor for prostaglandins and thromboxanes) at the 2-acyl position of glycerol.[26] With time, the monoenoic and dienoic fatty acids at the 2-acyl position of newly synthesized phosphatidyl-inositols exchange for arachidonic acid through another enzyme which will be discussed subsequently. Interestingly, the Mn^{2+}-catalysed exchange incorporation of inositol into phosphatidylinositol of isolated hepatic microsomes results in inositol incorporation into phosphatidylinositol containing arachidonic acid.[6,25] The function of this exchange is unknown and cannot be linked to any enzyme involved in phosphatidylinositol metabolism. Whether this exchange activity is linked to the enzyme involved in transfer of arachidonic acid to phosphatidylinositol remains to be elucidated but is a likely possibility.

In rat hepatocytes the incorporation of medium inositol into phosphatidylinositol increased 100-fold when the inositol content was increased from 0.4 to 80 μM while ^{32}P-incorporation into phosphatidylinositol was unaffected.[27] This suggested that inositol uptake resulted from acceleration of the exchange activity which was supported by the finding that the addition of Mn^{2+} (1–10 mM) markedly increased inositol uptake into phosphatidylinositol without affecting that of ^{32}P by isolated rat hepatocytes. Furthermore, the uptake of ^{32}P into phosphatidylinositol was accelerated by alpha catecholamines or vasopressin while that of inositol was unaffected both in the absence or presence of Mn^{2+}.[27] Similar results have been seen in synaptosomes prepared from rat brain where acetylcholine increases incorporation of ^{32}P while Mn^{2+} increases labelled inositol uptake into phosphatidylinositol.[28] The major conclusion from these findings is that in tissues such as rat liver which contains appreciable amounts of the inositol exchange enzyme, the use of labelled inositol as an index of phosphatidylinositol turnover is unreliable. However, the Mn^{2+}-activated exchange enzyme can be used to label phosphatidylinositol in tissues where the turnover of phosphatidylinositol is low. Adipose tissue incubated with Mn^{2+} incorporates appreciable amounts of labelled inositol into phosphatidylinositol.[29] The tissue can be washed free of Mn^{2+} and inositol and utilized for studies on the breakdown of phosphatidylinositol in isolated adipocytes or homogenates. Attempts to label phosphatidylinositol *in vivo* by this procedure have been difficult since an effective dose of Mn^{2+} is toxic to the rats.[30] However, for studies on the turnover of labelled phosphatidylinositol in rat liver an appreciable amount of label is found in phosphatidylinositol 18 hours after the injection of tritiated inositol (50–75 microcuries/rat).

In tissues such as rat hepatocytes and adipocytes where hormones do not accelerate the incorporation of medium inositol into phospholipids, the medium inositol may not equilibrate with the pool of inositol used for phosphatidylinositol synthesis *via* the CDP-diglyceride:inositol transferase. Many tissues have a large pool of cytosol inositol due to active transport and accumulate inositol intracellularly. Examples are rat adipocytes,[31] hepatocytes[27] and kidney.[32] Another factor may be the rate of exchange between the inositol pool of the endoplasmic reticulum and the cytosol. Two or more different intracellular pools for inositol have been described in chick embryo cells.[33] However, in blowfly salivary glands the medium inositol readily equilibrates with a smaller intracellular pool of inositol.[34] Labelled inositol

can be useful for studies on the breakdown of prelabelled phosphatidylinositol but should be used with caution in studies on acceleration of phosphatidylinositol turnover by hormones or other stimuli.

The work of Holub[6,25,26;35,36] suggested that phosphatidylinositol in liver
containing stearic acid at the one position of glycerol and arachidonic acid
at the two position was not formed *de novo* but rather by transacylation.
Irvine and Dawson[37] demonstrated the presence of an enzyme in rat liver
microsomes which would transfer arachidonic acid present in the two
position of phosphatidylcholine to lysophosphatidylinositol. The transfer
appeared to be a direct coenzyme-A-mediated transfer which did not
involve free fatty acid intermediates or phospholipase A_2.

Another way by which arachidonic acid can get into phosphatidylinositol
is by reaction of lysophosphatidic acid with arachidonoyl-CoA. In platelets
there is a rapid breakdown of phosphatidylinositol containing stearate at the
1 position on glycerol and arachidonate at the 2 carbon to diglyceride with
the release of arachidonic acid.[38,10] However, there is no release of stearate
which suggested that the 1-monoacyglycerol formed from diglyceride hydrolysis via diglyceride lipase is converted to lysophosphatidic acid and then
reacylated with arachidonoyl CoA.

3 PHOSPHOLIPID EFFECT IN RESPONSE TO STIMULI

The first description of the so-called phospholipid effect was the report by
Hokin and Hokin[39] of a cholinergic stimulation of amylase secretion in
pigeon pancreas slices accompanied by a large increase in the incorporation
of ^{32}P-orthophosphate into phospholipids. The increase in turnover of
phospholipids in the pancreas was limited to phosphatidic acid and phosphatidylinositol[40] and this has turned out to be the case in other tissues. In
some tissues stimuli increase incorporation of ^{32}P into phosphatidic acid but
in most tissues the primary effect is on phosphatidylinositol formation.

The increased incorporation of ^{32}P into phosphatidylinositol or phosphatidic acid (which is the precursor of the former) is not accompanied by an
equivalent increase in the turnover of fatty acids or the glycerol moieties of
these phospholipids.[11] Stimuli are thought to accelerate breakdown of phosphatidylinositol *via* a phospholipase C type enzyme to give diglyceride
which is subsequently resynthesized to phosphatidic acid in the presence of
ATP containing ^{32}P. The phosphatidic acid is activated by reaction with CTP
to give ^{32}P-labeled CDP-diglyceride which reacts with free inositol to form
phosphatidylinositol. The only other phospholipid formed from CDP diglyceride is phosphatidylglycerol. In rat pineals,[41] isolated rat pancreatic
islets,[42] and blowfly salivary glands,[34] appreciable amounts of phosphatidylglycerol are formed in the absence of added inositol. In most tissues there is
little incorporation of ^{32}P into phosphatidylglycerol. The synthesis of other

phospholipids does not result in incorporation of the phosphate group of phosphatidic acid into the phospholipid. Instead CTP reacts with choline or ethanolamine to give the CDP-base which reacts with diglyceride.

The most serious misunderstanding with regard to the so-called phospholipid effect of increased incorporation of ^{32}P into phosphatidic acid and hence into phosphatidylinositol has been the failure to realize that it is often a secondary event. Initial investigations focused on the synthesis of phosphatidylinositol since it is very easy to measure ^{32}P uptake into phospholipids. However, it is much more difficult to measure phosphatidylinositol breakdown. This review emphasizes studies related to phospholipid breakdown. Michell[11,43] and Michell *et al.*[44] have reviewed the extensive evidence for stimulation of phosphatidylinositol formation in response to hormones or other stimuli which elevate tissue calcium.

4 TISSUES WITH A LARGE DROP IN PHOSPHATIDYL-INOSITOL AND A RISE IN PHOSPHATIDIC ACID: PAROTID AND PANCREAS

The addition of large concentrations of acetylcholine to incubated mouse pancreases or the albatross salt gland resulted in a 40% drop over an hour incubation in total phosphatidylinositol which was accompanied by an equivalent rise in phosphatidic acid.[45,46] It was possible to demonstrate that the phosphatidic acid was derived from diglyceride formed as a result of phosphatidylinositol breakdown. The majority of the phosphatidylinositol in mammalian tissues is 1-stearoyl, 2-arachidonyl-sn-glycerol-3-phosphorylinositol and the newly formed phosphatidic acid contains stearate and arachidonate in amounts equivalent to that lost from phosphatidylinositol.[46] The diglyceride pool derived from phosphatidylinositol breakdown which is used for synthesis of phosphatidic acid does not equilibrate with the total diglyceride pool of the pancreas. Banschbach *et al.*[47] found a large increase in total diglyceride in the mouse pancreas incubated with acetylcholine and the arachidonate content was much less than would be expected of diglyceride derived exclusively from phosphatidylinositol breakdown. In the pancreas and especially in adipose tissue, diglyceride is a intermediate involved in both triacylglycerol synthesis and degradation. Elevation of the diglyceride content of adipocytes with lipolytic agents (which increase triacylglycerol hydrolysis) does not cause any increase in ^{32}P-incorporation into phosphatidate.[48] Apparently the pools of diglyceride used for phospholipid synthesis are separate from those for triacylglycerol formation. In the pancreas and platelets phosphatidylinositol breakdown is associated with a remarkable stimulation of phosphatidic acid formation which results in a large drop in phosphatidylinositol and an increase in phosphatidic acid containing the diglyceride backbone of the phosphatidylinositol and phosphate derived from ATP. These findings support the hypothesis that the primary event in phospholipid turnover initiated by cholinergic stimulation of the mouse pancreas is activation of a phospholipase C which cleaves phosphatidylinositol to give diglyceride and inositol-1,2 cyclic phosphate. Formation of phosphatidic acid by a reversal of the pathway for phosphatidylinositol formation with cleavage of the CDP-diglyceride to diglyceride has been suggested by Hokin-Neaverson[46] but no enzyme is known which would carry out the cleavage of CDP-diglyceride. It is difficult to see accumulation of cyclic inositol phosphate in the pancreas but that probably reflects the high activity of the enzyme which cleaves the cyclic inositol phosphate to give free inositol.

Marshall *et al.*[49] recently suggested that the breakdown of phosphatidylinositol results in release of arachidonic acid through cleavage of the diglyceride by diglyceride lipase. This suggests that the diglyceride lipase and diglyceride kinase compete for diglyceride derived from phosphatidylinositol. The conversion of phosphatidylinositol to phosphatidic acid due to large concentrations of hormone could limit the availability of phosphatidylinositol for arachidonic acid release. In pancreases prelabelled with arachidonic acid the addition of a stimulant for secretion resulted in a 44 per cent loss of arachidonic acid from phosphatidylinositol and about half of this was released as arachidonic acid. There was no loss of arachidonic acid from other phospholipids.[49]

The evidence that arachidonic acid is converted to prostaglandins and that these compounds are involved in stimulus-secretion coupling is as follows: 1. Prostaglandins of the E, F and D series but not thromboxane B_2 stimulated amylase release. 2. Arachidonic acid but not other fatty acids stimulated amylase secretion and this stimulation was blocked by indomethacin. 3. Four different inhibitors of prostaglandin formation inhibited amylase release due to secretagogues and their potencies in this respect paralleled their ability to inhibit cyclooxygenase.[49]

In rat pancreatic islets the stimulation of insulin release by glucose results in a large decrease in loss of labelled inositol from phosphatidylinositol.[50] Farese *et al.*[51] have recently reported a breakdown of phosphatidylinositol of rat pancreatic islets which was equivalent to a 25 per cent decrease in total content and accompanied by an increase in phosphatidic acid and phosphatidylglycerol. The resynthesis of phosphatidylinositol is seen during insulin secretion only if inositol is added to the medium.[42] Insulin release is the same in the absence or presence of inositol which indicates that resynthesis of phosphatidylinositol plays little role in insulin secretion. The increase in phosphatidylglycerol and phosphatidic acid formation along with accumulation of labeled CDP-diglyceride is apparently due to a lack of inositol in islets obtained by collagenase digestion.[42] The link between glucose stimulation of insulin release and phosphatidylinositol degradation may be the elevation of Ca²⁺ seen with glucose.[52,53] Possibly glucose induced calcium gating involves breakdown of phosphatidylinositol.

In rat parotid glands Jones and Michell[54] also saw a 30 per cent drop in total phosphatidylinositol content after incubation for only 5 min with high concentrations of acetylcholine. The breakdown of such large amounts of the total phosphatidylinositol content in the parotid, acinar pancreas and pancreatic islets in response to maximal hormonal stimulation is unlikely to result from degradation of phosphatidylinositol present in the plasma membrane which is involved in calcium gating. In pancreatic islets the degradation of phosphatidylinositol was primarily localized to the secretory granule membranes.[50] In the acinar pancreas Hokin-Neaverson[46] reported that the loss of phosphatidylinositol was primarily from the rough endoplasmic reticulum. These results could be explained if there is transfer of phosphatidylinositol from the endoplasmic reticulum to the plasma membrane or if prostaglandins are activators of the phosphatidylinositol hydrolase.

Calcium is particularly important since secretagogues elevate cytosol calcium[55,56,57] in the acinar pancreas and in the endocrine pancreas in the presence of glucose.[52,58] How do secretagogues increase the influx of extracellular calcium and release bound intracellular Ca? Possibly the effects of stimuli on calcium dynamics are secondary to an increase in phosphatidylinositol degradation in the plasma membrane. The increases in prostaglandins (derived from phosphatidylinositol *via* diglyceride breakdown) may act as a biological cascade. Prostaglandins derived from arachidonic acid may activate phospholipase C in the cytosol which degrades the large amounts of phosphatidylinositol present in intracellular membranes (particularly those of secretory granules). The prostaglandins could act as intracellular calcium ionophores to release calcium from organelles such as mitochondria. Such an effect of prostaglandin E_1 on rat liver mitochondria has been reported.[59] The increases in prostaglandins and cytosol calcium along with the marked change in the phospholipid composition of the secretory membranes (drop in phosphatidylinositol and rise in phosphatidic acid) may facilitate fusion of secretory granules with the plasma membrane.

It should be noted that Chauvelot *et al.*[60] could find no role for prostaglandins in the release of amylase by dispersed rat pancreatic acinar cells incubated with carbamylcholine. In the acinar cells no effects of added arachidonic acid or prostaglandins E_1, E_2, F_2 and I_3 on secretion were noted. Furthermore, the microsomes isolated from the cells were unable to synthesize prostaglandins. Neither imidazole or indomethacin blocked amylase secretion due to carbamylcholine. Quinacrine stimulated amylase secretion but blocked any additional increase due to carbamylcholine or cholecystokinin.[60]

In some tissues the elevation in cytosol Ca^{2+} accelerates phosphatidylinositol turnover. However, there is little evidence for this in the exocrine pancreas or the parotid. While secretion in both glands can be stimulated by addition of the calcium ionophore A 23187, there is no acceleration of phosphatidylinositol turnover in the rat parotid by the ionophore.[54] Similar results have been seen in the exocrine pancreas.[61,62]

Secretion in both the pancreas and parotid can be almost completely abolished by prolonged incubation in Ca^{2+}-free buffers but the ability of secretagogues to accelerate phosphatidylinositol turnover is still present in pigeon pancreas[63] rat pancreas[61,62] and in the rat parotid.[54] These data indicate that entry of extracellular Ca^{2+} is not required for demonstration of phosphatidylinositol turnover. Furthermore, the ability to block secretion in these tissues without abolishing phosphatidylinositol turnover rules out the hypothesis that the increased phospholipid turnover is secondary to some process involved in fusion of the secretory granule membranes with the plasma membrane. In the rat pancreas secretogogues such as carbamylcholine or pancreozymin also stimulate phospholipid turnover when Na$^+$ is replaced by Li$^+$ and in medium containing 90 per cent deuterium oxide (heavy water). In neither condition is secretion observed.[60]

Intracellular release of bound Ca^{2+} is involved in secretion in the exocrine pancreas.[56,57] Thus, phosphatidylinositol breakdown needs to be linked to entry of extracellular Ca^{2+} as well as the release of intracellular bound Ca^{2+}. Release of bound Ca^{2+}, especially from mitochondria in response to muscarinic cholinergic stimulation or other secretagogues needs a second messenger. Diglyceride, cyclic-1,2-inositol phosphate, arachidonic acid or prostaglandins all are possible candidates. Secretin and VIP which elevate cyclic AMP formation in the pancreas and stimulate secretion do not affect phosphatidylinositol turnover.[56] In the parotid beta catecholamines elevate cyclic AMP which seems to be linked to enzyme secretion.[64] Beta catecholamines do not stimulate phosphatidylinositol turnover. In contrast, the secretion of K$^+$, elevation of cytosol Ca^{2+} and phosphatidylinositol breakdown in the parotid are all alpha effects of catecholamines.[54,65,66]

5 PLATELETS AS A SYSTEM IN WHICH PHOSPHATIDYLINOSITOL BREAKDOWN REGULATES PROSTAGLANDIN FORMATION

Phosphatidylinositol breakdown now appears to be the main source of the arachidonic acid released by platelets in response to stimuli such as thrombin. This is depicted in Figure 2. The platelet system differs from that of many other cells since the calcium ionophore A-23187 mimics the increases in phosphatidylinositol degradation due to thrombin. This suggests that thrombin might initially cause an increase in degradation of plasma-membrane-bound phosphatidylinositol which results in elevation of cytosol Ca^{2+}. The diglyceride which results is rapidly cleaved by diglyceride lipase to give arachidonic acid. The arachidonic acid is converted to active cyclic endoperoxides, prostaglandins, prostacylins and thromboxane A$_2$. The

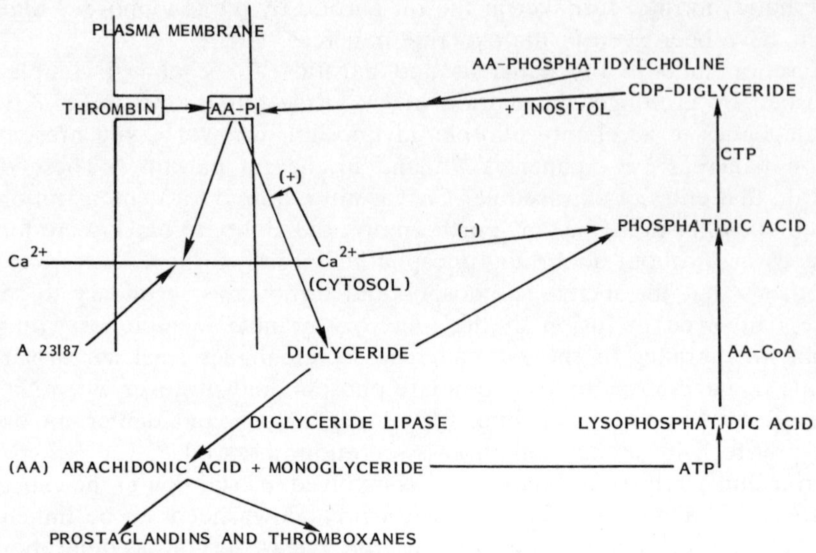

Figure 2. Phosphatidylinositol (PI) turnover in platelets

thromboxane A_2 may release intracellular Ca^{2+} stores in platelets[67] and the Ca^{2+} further activates the phospholipase C which results in additional hydrolysis of phosphatidylinositol.

Until recently it was assumed that arachidonic acid release was due to activation by calcium of a phospholipase A_2 enzyme which degraded phosphatidylcholine. The rise in calcium may indeed activate this enzyme but this is probably a secondary effect. It is not clear whether the oxygenated compounds formed from arachidonic acid by the cyclooxygenase and lipoxygenase enzyme are necessary for platelet aggregation. Lapetina *et al.*[68] reported that while arachidonate metabolism could be blocked by indomethacin and eicosatetraenoic acid the platelet aggregation induced by thrombin or A-23187 was not blocked. These data suggest that the elevation in Ca^{2+} entry due to maximal concentrations of thrombin is sufficient to induce platelet aggregation.

The platelet aggregation seen with agents which elevate Ca^{2+} is blocked by cyclic AMP which indicates a reciprocal relationship between these two messengers in platelets. Prostaglandins of the E series or prostacyclins could act as feedback inhibitors of the Ca^{2+} signal since they are able to elevate cyclic AMP formation in platelets.[69,70,71] The inhibition of platelet aggregation and of phosphatidylinositol breakdown may result from activation by cyclic AMP of calcium pumps which lower cytosol Ca^{2+}. The phosphatidylinositol-specific phospholipase C of platelets[72] is activated by deoxycholate.[73] Even the activation of this enzyme seen after addition of 2.5 mM deoxycholate to intact platelets was blocked by prior incubation of platelets with agents which elevated cyclic AMP.[73] These results are best explained by assuming that cyclic AMP reduced the effective concentration of Ca^{2+}.

Rittenhouse-Simmons[74] found that human platelets generate large amounts of diglyceride within 5 seconds of exposure to thrombin and this process was blocked by high levels of cyclic AMP. This is one of the earliest metabolic events which occurs during activation of platelets by thrombin and seems to be a reflection of an immediate breakdown of phosphatidylinositol. The accumulation of diglyceride was transient and by 2 minutes had returned to control values. Rittenhouse-Simmons[74] concluded that the most likely source of the diglyceride was breakdown of phosphatidylinositol which has since been demonstrated by Bell and Majerus.[10]

Bell and Majerus[10] found a 50 per cent drop in total phosphatidylinositol within 10 seconds after the addition of thrombin or A-23187 to intact human platelets. In dilute suspensions of platelets there was resynthesis of all of the phosphatidylinositol by 14 minutes in cells incubated with thrombin but only a small re-synthesis of phosphatidylinositol in cells exposed to A-23187. The reason for the block in re-synthesis in the presence of A-23187 is not known but may reflect an inhibition of one or more of the enzyme involved by high levels of Ca^{2+}. If the breakdown of phosphatidylinositol by thrombin is involved in calcium gating, the loss of so much phosphatidylinositol may impair further calcium gating.

A novel function of diglyceride has recently been reported. The calcium-dependent protein kinase activity present in many tissues is activated by phospholipids and this effect is greatly potentiated by diglyceride.[75] The net result would be potentiation of calcium-activated phosphorylation of proteins which should act synergistically with Ca^{2+}. Whether this might be involved in platelet aggregation remains to be demonstrated.

Bell *et al.*[38] found that diglyceride lipase is active in human platelets and that the 1-stearoyl-2-arachidonyl diglyceride formed by hydrolysis of phosphatidylinositol is readily cleaved to release arachidonate. There is also diglyceride kinase present in human platelets which competes with diglyceride lipase in the metabolism of diglycerides.[76] A curious finding was that stearate was not released which suggests that breakdown only to the 1-stearoyl glycerol occurs and this monoglyceride is rapidly converted to lysophosphatidic acid (see Figure 2). The lysophosphatidic acid could react with arachidonyl CoA to form phosphatidic acid. Alternatively, there may be some pathway for selective activation of free stearate and reacylation if the diglyceride is completely hydrolyzed to glycerol and stearate.

The incorporation of ³²P into phospholipids occurs at the step in which phosphatidic acid is formed. The phosphatidic acid reacts with CTP to form

CDP-diglyceride which is converted to phosphatidylinositol. The drop in arachidonic acid content of platelet phospholipids seen after prolonged incubation with thrombin was not restricted to phosphatidylinositol but also involved phosphatidylcholine and phosphatidylethanolamine.[7] While these authors attributed the release of arachidonate to a phospholipase A_2 they were unable to find appreciable amounts of this enzyme in platelets. Possibly a large part of the arachidonic acid lost from phosphatidylcholine and phosphatidylethanolamine was through transacylation to phosphatidylinositol via the type of reaction described by Irvine and Dawson.[37]

Probably the inhibition by local anaesthetics, chlorpromazine, quinacrine and propanolol of arachidonic acid release by platelets is due to inhibition of phosphatidylinositol phosphohydrolase rather than phospholipase A_2. In nearly all the papers on fatty acid release from phospholipids of platelets, it has been assumed that phospholipase A_2 was involved without any proof. Inhibitors of arachidonic acid release from phospholipids have commonly been described as phospholipase A_2 inhibitors.[77,78,79] However, most of these drugs alter the function of many other enzymes and probably inhibit phosphatidylinositol phosphohydrolase. Quinacrine, also known as mepacrine, inhibits lipolysis, glucose metabolism and cyclic AMP accumulation in rat adipocytes.[80] Quinacrine has been claimed to also affect platelet metabolism by inhibiting phospholipase A_2[78,79] but this remains to be proved. Quinacrine is a well-known lysomyotropic agent[80] which inhibits the activity of lysosomes due to the increase in lysosomal pH resulting from accumulation of quinacrine which is a lipid soluble weak base like chloroquine.[81]

There is certainly phospholipase A_2 activity in platelets since a homogeneous preparation of phospholipase A_2 has been isolated from human platelets.[82,83] This enzyme is calcium-dependent and inhibited by high concentrations of indomethacin.[83] The addition of calmodulin to human platelet membranes increased the release of arachidonic acid from phosphatidylcholine but the addition of 0.1 mM Ca^{2+} had the same effect as 0.001 mM calmodulin.[84] One curious finding was that calmodulin addition to intact platelets also increased arachidonic acid release. Future studies on release of arachidonic acid from platelet phospholipids should provide data to distinguish between phospholipase A_2 and phosphatidylinositol specific phospholipase C.

In conclusion, recent work has suggested that thrombin activation of phosphatidylinositol breakdown is likely to be related to the mechanism by which thrombin elevates cytosol Ca^{2+}. In this system, the rise in calcium acts to further amplify the signal by activating phospholipase C. The arachidonic acid released during the initial stages of platelet aggregation is due to activation of phosphatidylinositol phosphohyrolase rather than phospholipase A_2.

6 ACTIVATION OF PHOSPHATIDYLINOSITOL HYDROLYSIS IN THE THYROID BY TSH

Freinkel[85] first demonstrated an increase in turnover of phosphatidylinositol in the thyroid after the addition of TSH. The phospholipid effect is unrelated to the ability of TSH to elevate cyclic AMP but rather appears to be linked to a separate effect of TSH on cytosol Ca^{2+} since it is mimicked by muscarinic cholinergic agonists[86] and A-23187.[87]

Figure 3 represents a schematic view of phosphatidylinositol in the thyroid. In the thyroid it appears that Ca^{2+} is an activator of phosphatidylinositol breakdown which might amplify the small increases in breakdown of phosphatidylinositol in the plasma membrane due to hormones. The diglyceride is cleaved by diglyceride lipase to release arachidonic acid which is metabolized in part to thromboxanes, prostacyclins and prostaglandins of the E and F series.[88]

The sole source of arachidonic acid appears to be phosphatidylinositol and Haye *et al.*[89] have reported a remarkable stimulation of its breakdown in thyroid homogenates incubated with TSH. The same laboratory even demonstrated the accumulation of inositol-1,2-cyclic phosphate in a supernatant fraction obtained from thyroid previously incubated with TSH.[90] Previously Jungalwala *et al.*[18] had shown the presence of Ca^{2+}-dependent phospholipase C activity in thyroid which cleaved phosphatidylinositol to give diglyceride and inositol phosphate. The function of accelerated phospholipid turnover in TSH action remains to be established but there is evidence that prostaglandins may be involved in the secretory effect of TSH on the thyroid.[91,92] Since there is little stimulation of thyroid secretion by agents such as muscarinic cholinergic agonists, it appears that the Ca^{2+} and prostaglandins probably potentiate the action of cyclic AMP on secretion.

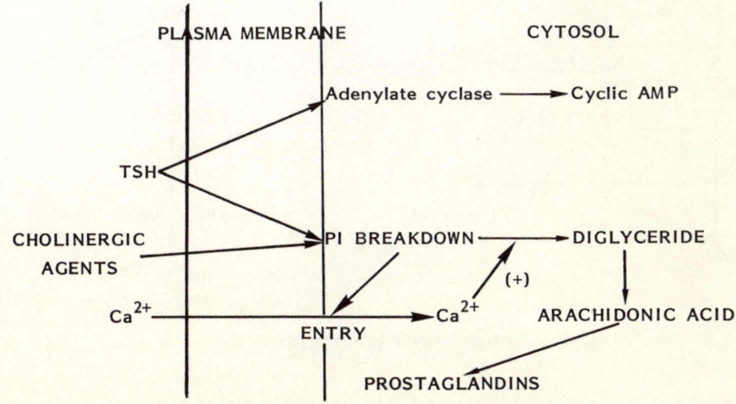

Figure 3. TSH action on thyroid

7 PHOSPHATIDYLINOSITOL TURNOVER AND ACTH ACTIVATION OF STEROIDOGENESIS IN THE ADRENAL

The action of ACTH on the adrenal like that of TSH on the thyroid seems to involve a phospholipid effect in addition to the well-established elevation of cyclic AMP. A speculative hypothesis to explain ACTH action on phospholipid turnover is shown in Figure 4. In the adrenal as in platelets too much emphasis may have been placed on the role of phospholipase A_2 in the release of arachidonic acid and not enough on the role of phosphatidylinositol specific phospholipase C followed by diglyceride lipase.

Schrey and Rubin[93] found that incubation of adrenal cells for two minutes with arachidonic acid in the presence of ACTH resulted in an increased incorporation of arachidonic acid into phosphatidylinositol. This effect was specific for arachidonic acid since the uptake of palmitic, stearic or oleic acids into phospholipids was not affected by ACTH. The increase in incorporation of arachidonic acid was limited to phosphatidylinositol and could be mimicked by A-23187.[93] This suggests that in the adrenal as in the thyroid and platelets there is a phosphatidylinositol specific phospholipase C which is activated by an elevation in cytosol Ca^{2+}. The increase in uptake of arachidonic acid may occur during conversion of lysophosphatidic acid to phosphatidic acid as suggested in Figure 4.

The increase in arachidonic acid turnover in the presence of ACTH suggests that the formation of prostaglandins was accelerated by ACTH. Rubin and Laychock[94] demonstrated an increase in formation of prostaglandins of the E and F series by ACTH. Prostaglandins whose formation is blocked by indomethacin such as postaglandin of the E and F series are not the mediators of steroidogenesis but may play a regulatory function.[94] In the

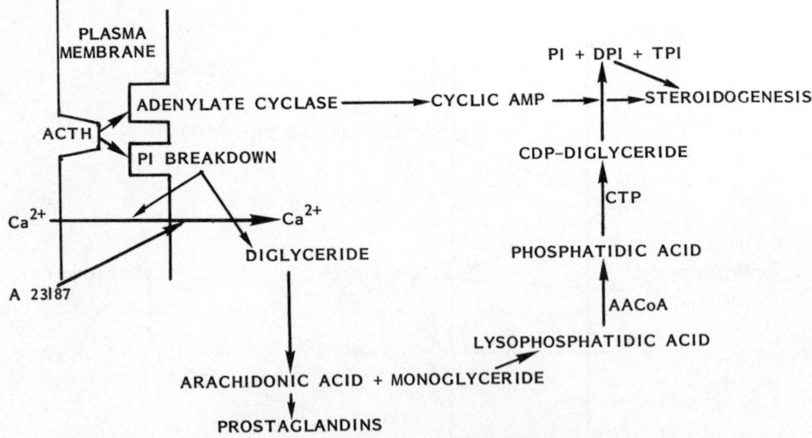

Figure 4. ACTH, PI turnover and steroidogenesis in adrenal glands

adrenal as in the thyroid, agents which only elevate Ca^{2+} such as A-23187 are potent activators of prostaglandin formation but have relatively weak stimulatory effects on steroidogenesis.

The increased turnover of arachidonyl-phosphatidylinositol is also accompanied by an increase in ^{32}P incorporation into phosphatidic acid and phosphatidylinositol in adrenal glands.[95] The increase in ^{32}P incorporation into phosphatidic acid was accompanied by net elevation in phosphatidic acid which was maximal at 5 minutes. This was followed sequentially by increases in phosphatidylinositol and then di- and triphosphatidylinositols. While the release of arachidonic acid and the formation of prostaglandins does not appear to be stimulated by cyclic AMP the net increases in phosphatidyl-inositols are. Farese *et al.*[96] reported a 32 per cent increase in total phosphatidylinositol after 15 minutes incubation with ACTH. Similarly Laychock *et al.*[97] found a 54 per cent increase in total phosphatidylinositol after incubation of adrenal cells for 90 minutes with ACTH.

This is the first example of a phospholipid effect due to hormone which appears to involve net increases rather than decreases in the level of phosphatidylinositols. The total tissue content of diphosphoinositide was about 5 per cent and that of triphosphoinositide about 1 per cent that of phosphatidylinositol. In the presence of ACTH the content of diphosphoinositide increased 2.5-fold and that of triphosphoinositide almost 10-fold.[96] Farese *et al.*[98] suggest that the polyphosphoinositides are involved in mediation of ACTH action on steroidogenesis. Addition of diphosphatidylinositol to adrenal mitochondria accelerated the production of pregnenolone. The possibility that polyphosphoinositides are the third messengers for ACTH action is most exciting. The increase in polyphosphoinositide formation due to cyclic AMP like that of ACTH is blocked by cycloheximide[98] which suggests that protein synthesis is involved in polyphosphoinositide formation.

8 PHOSPHATIDYLINOSITOL TURNOVER, Ca²⁺ AND THE REGULATION OF GLYCOGEN PHOSPHORYLASE IN RAT HEPATOCYTES BY ALPHA CATECHOLAMINES AND VASOPRESSIN

De Torrontegui and Berthet[99] reported that epinephrine but not glucagon increased the uptake of ^{32}P into all phospholipids by slices of rat liver. Subsequent studies have confirmed the increase in uptake of ^{32}P into phosphatidylinositol but were unable to see any stimulation of ^{32}P incorporation into any other phospholipids of intact hepatocytes incubated with epinephrine.[100,27] In contrast to the adrenal cortex no stimulation of ^{32}P incorporation into polyphosphoinositides was noted with epinephrine.[27]

The initial report of phospholipid stimulation by epinephrine but not by glucagon in liver was virtually ignored for the next ten years. The discovery that the activation of hepatic glycogen phosphorylase by both epinephrine and glucagon is associated with an elevation of cyclic AMP[101,102] suggested that the effects of these hormones were secondary to activation of adenylate cyclase. The finding of an effect of epinephrine not shared by glucagon was incompatible with this simple model for hormonal regulation of glycogenolysis.

Sherline *et al.*[103] first reported that in the perfused rat liver alpha catecholamines (phenylephrine) increased glycogen phosphorylase without affecting cyclic AMP. Phentolamine, an alpha-adrenergic antagonist, markedly reduced the increase in phosphorylase due to epinephrine without affecting its stimulation of cyclic AMP accumulation. Sherline *et al.*[103] explained their findings in light of the hypothesis that cyclic AMP is the sole regulator of glycogen phosphorylase by suggesting that vasoconstriction by alpha catecholamines elevated AMP which activated glycogen phosphorylase. The problem of how to explain the findings of De Torrontegui and Berthet[99] was essentially solved by ignoring them. It remained for Tolbert *et al.*[104] to suggest that catecholamines have metabolic effects on rat liver carbohydrate metabolism (in this case gluconeogenesis) which are alpha in nature and unrelated to cyclic AMP. In their experimental system isoproterenol (a pure beta agonist) elevated cyclic AMP to the same extent as epinephrine without having any effect on gluconeogenesis. Obviously, in isolated rat hepatocytes there are no alpha effects of catecholamines on vasoconstriction.

Subsequent studies from my laboratory[105,106,107] and elsewhere[108,109,110] have established that in isolated rat hepatocytes alpha-adrenergic stimulation of glycogen phosphorylase does not involve cyclic AMP. Several reviews on this subject have appeared in recent years which summarize the extensive data suggesting that Ca^{2+} rather than cyclic AMP or cyclic GMP is the mediator of alpha catecholamine and vasopressin activation of glycogen phosphorylase.[111,112,113,114]

Pointer *et al.*[106] found that the divalent cation ionophore A-23187 activated glycogenolysis in the presence but not in the absence of Ca^{2+} from the extracellular buffer. Hutson *et al.*[107] Assimacopoulos-Jeannet *et al.*[115], and Keppens *et al.*[110] demonstrated that A-23187 activated glycogen phosphorylase. They also found that alpha catecholamines increased both influx and efflux of $^{45}Ca^{2+}$ by intact hepatocytes. However, Chen *et al.*[116] found that alpha catecholamines stimulated efflux of Ca^{2+} to a much greater extent than influx and suggested that the major source of the Ca^{2+} released to the medium was the mitochondrial stores. Subsequent studies by Babcock *et al.*;[117] Blackmore *et al.*;[118] and Murphy *et al.*[119] have confirmed that alpha catecholamines and vasopressin cause a substantial loss of Ca^{2+} from the mitochondrial pool. There is little agreement about the effects of catecholamines on hepatic Ca^{2+} flux except for the loss of mitochondrial Ca^{2+}. Foden and Randle[120] and Murphy *et al.*[119] found no net loss of Ca^{2+} from hepatocytes and the latter group claimed that the drop in mitochondrial Ca^{2+} was balanced by uptake of Ca^{2+} into the endoplasmic reticulum. Althaus-Saltzmann *et al.*[121] suggested that the major effect of alpha catecholamines is to release Ca^{2+} from intracellular non-mitochondrial pools. It should be noted that in Ca^{2+} studies the usual procedure (for technical reasons) is to incubate cells in medium containing about 5 per cent of the normal Ca^{2+} content but if studies are done in medium containing 1.3 mM Ca a stimulation of Ca^{2+} influx can be demonstrated with alpha catecholamines.[120,122] The net result is always an elevation of cytosol Ca^{2+}, but whether there is net influx or efflux of Ca^{2+} from hepatocytes depends on experimental conditions.

The lack of a requirement for entry of extracellular Ca^{2+} in order for alpha catecholamines to activate glycogen phosphorylase was first noted by Stubbs *et al.*[123] and confirmed by Blackmore *et al.*[124] and Malbon *et al.*[106] Chan and Exton[124A] also found an activation of glycogen phosphorylase by alpha catecholamines in the absence of extracellular Ca^{2+} which they attributed to an elevation of cyclic AMP. However, Malbon *et al.*[106] found that under similar conditions the alpha catecholamine activation of glycogen phosphorylase was not accompanied by any changes in cyclic AMP. Unpublished studies in my laboratory indicate that only prolonged Ca^{2+} deprivation of hepatocytes which initially had poor viability results in alpha activation of adenylate cyclase.

The finding that entry of extracellular Ca^{2+} is not required for alpha catecholamines to activate glycogen phosphorylase indicates that if Ca^{2+} is the second messenger it can be obtained by release of bound intracellular stores of Ca^{2+} to the cytosol. The large drop in total Ca^{2+} content of rat liver mitochondria after alpha catecholamines suggests the need for a second messenger which transmits a signal from the plasma membrane to mitochondria.

An increase in the breakdown of phosphatidylinositol in the plasma membrane of hepatocytes in response to alpha-1-catecholamine binding to receptors might be involved in generation of a second messenger for this response. The phosphatidylinositol present in rat liver is predominantly 1-stearoyl 2-arachidonyl glycerophosphoryl inositol.[26] Its breakdown by a phospholipase C type enzyme could lead to a net release of arachidonic acid as in platelets and other tissues. Thus, either digylceride, cyclic 1,2-inositol diphosphate, arachidonic acid or other substances generated by the cyclooxygenase or lipoxygenase pathway might be second messengers which release bound Ca^{2+} from organelles such as mitochondria. The initial event would be an increased access of plasma membrane phosphatidylinositol to degradation by either a cytosol enzyme or direct activation of phosphatidylinositol phosphohydrolase (phospholipase C).

Billah and Michell[125] found a 5 per cent decrease in total phosphatidylinositol after incubation of hepatocytes with epinephrine. There was no effect of A-23187 on phosphatidylinositol breakdown in either the absence or presence of vasopressin for 15 minutes. The small drop in total phosphatidylinositol was also seen after vasopressin addition to cells incubated in calcium-free buffer containing 0.2 mm EGTA (a divalent cation chelator with a high affinity for Ca^{2+} and low affinity for Mg^{2+}). We have seen similar results in our laboratory[126] which indicate that in hepatocytes, unlike the pancreas or platelets, there is no appreciable breakdown of the bulk of cellular phosphatidylinositol. However, there was a specific breakdown of phosphatidylinositol in the plasma membrane but not in the endoplasmic reticulum fraction of rat hepatocytes after incubation of hepatocytes for 5 minutes or less with alpha catecholamines.[126]

One difference between hepatocytes and the cells discussed previously was the finding that the ionophore A-23187 had no effect on phosphatidylinositol breakdown. In addition, the Ca^{2+} content of the medium had little effect on either breakdown[125] or resynthesis of phosphatidylinositol.[104,125,127] These data suggest that the breakdown of phosphatidylinositol in hepatocytes is catalysed by enzymes which are relatively insensitive to changes in cytosol Ca^{2+}.

The reason alpha catecholamines and vasopressin have such a small effect on total phosphatidylinositol hydrolysis in hepatocytes but markedly stimulate phosphatidylinositol synthesis remains to be established. It is unlikely to be due to an elevation of diglyceride since the amounts which accumulate in the cytosol are relatively small. Diglyceride is also the common intermediate derived from phosphatidic acid breakdown which is utilized for synthesis of triacylglycerols, phosphatidylcholine and phosphatidylethanolamine. The stimulation of triacylglycerol synthesis by agents such as insulin should increase both phosphatidic acid and diglyceride accumulation but there is no evidence that this results in activation of phosphatidylinositol synthesis. In cultured 3T3 fibroblasts undergoing transformation to lipid-laden cells in the presence of insulin, there are large increases in all the enzymes involved in triacylglycerol synthesis except for phosphatidic acid phosphohydrolase which is present initially at high concentrations and little affected by insulin. The most likely explanation for these findings is that triacylglycerol synthesis occurs in a separate compartment of the cell from that for phosphatidylinositol synthesis. This appears to be the case as triacylgylcerol synthesis occurs in the cytosol and phospholipid synthesis in the endoplasmic reticulum of liver. The major point is that the amount of diglyceride released to the cytosol is probably small in relationship to the pool used for triacylglycerol synthesis and what accumulates should be converted to triacylglycerol. For this reason I consider it unlikely that the stimulus for increased formation of phosphatidylinositol is related in any way to diglyceride or that it is a likely candidate to act as a second messenger.

The reason for the big stimulation of phosphatidylinositol synthesis in liver by vasopressin, angiotensin and alpha-catecholamines is not clear. One feature all these hormones have in common is the release of bound Ca^{2+} from all cellular organelles including the endoplasmic reticulum.[118,121] Possibly the drop in Ca^{2+} content of the endoplasmic reticulum is the signal for increased resynthesis of phosphatidylinositol as depicted in Figure 5. The enzyme involved in reacting CDP-diglyceride with free inositol is inhibited by Ca^{2+} in many tissues which we have confirmed for vesicles derived from the endoplasmic reticulum and plasma membrane of rat hepatocyres.[21] This hypothesis is compatible with the failure of the ionophore A-23187 to increase phosphatidylinositol synthesis in medium containing Ca^{2+} where it would equilibrate the medium Ca^{2+} pool (present at a concentration of 1.4 mM) with intracellular pools. However, if cells are incubated with the ionophore in medium containing a large amount of the calcium-chelator EGTA then the net effect might be a drop in Ca^{2+} in the endoplasmic reticulum. Tolbert *et al.*[27] found that in hepatocytes incubated in buffer containing 2.8 mM EGTA + 1.4 mM Ca^{2+} for 20 minutes prior to the addition of A-23187, phosphatidylinositol synthesis was stimulated by A-23187 to the same extent as by epinephrine. Addition of EGTA either with or 10 minutes before A-23187 was not sufficient for demonstration of a stimulatory effect due to A-23187.

Further evidence for the hypothesis that phosphatidylinositol breakdown in the plasma membrane of hepatocytes is a primary effect rather than a secondary result of elevation of cytosol Ca^{2+} comes from studies with vasopressin. In Ca^{2+}-free buffer vasopressin is unable to elevate intracellular Ca^{2+} enough to activate glycogen phosphorylase but the acceleration of

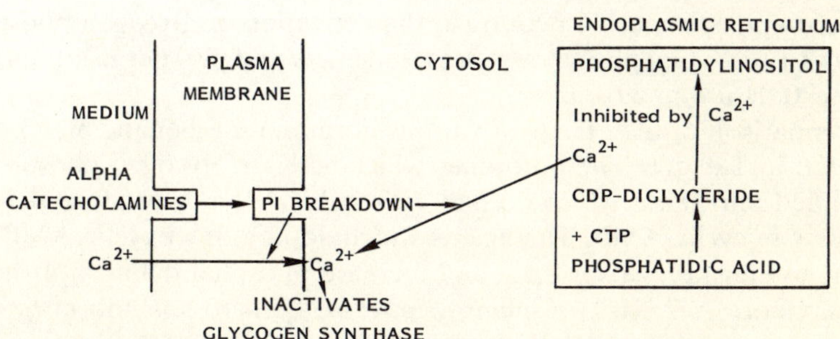

Figure 5. PI turnover in hepatocytes and adipocytes

phosphatidylinositol turnover is still present.[127,125,27] The hypothesis outlined in Figure 5 suggests that vasopressin must cause release of bound Ca^{2+} from the endoplasmic reticulum in amounts sufficient to activate phosphatidylinositol synthesis but not enough to activate glycogen phosphorylase. In contrast, alpha catecholamines are apparently able to release enough Ca^{2+} to both activate glycogen phosphorylase and phosphatidylinositol synthesis. There may be some additional factors involved which account for the difference in Ca^{2+} between vasopressin and alpha catecholamines. However, both hormones have identical effects on phosphatidylinositol turnover.

The receptors involved in alpha catecholamine effects have recently been divided into alpha-1 and alpha-2 receptors based on binding studies with labeled catecholamines agonists and antagonists[128,129,130] pharmacological studies[131,132] and physiological effects.[133] The alpha-1 effects of catecholamines are preferentially blocked by prazosin and methoxamine acts as a weak agonist at alpha-1 sites. In contrast, yohimbine is a preferential antagonist of alpha-2 effects and clonidine is a selective agonist for alpha-2 sites. Phentolamine and phenoxybenzamine are non-selective antagonists which block both types of alpha effects. Jones and Michell[134] originally noted that the increases in phosphatidylinositol turnover seen with catecholamines are characteristic of alpha-1 receptor activation. Wikberg[132] and Sabol and Nirenberg[135] pointed out that the inhibitory effects of alpha catecholamines on adenylate cyclase are characteristics of alpha-2 receptors. Fain and Garcia-Sainz[133] suggested that alpha-1 receptor activation results in both elevation of cytosol Ca^{2+} and increased turnover of phosphatidylinositol. In contrast, alpha-2 effects are independent of Ca^{2+} and due to direct inhibition of adenylate cyclase activation by hormones or even by cholera toxin.

The effects of catecholamines on hepatic gluconeogenesis[134] and glycogenolysis[137,138] appear to be mediated through alpha-1 receptors since they are readily inhibited by prazosin. As little as 0.0001 μM prazosin produced a 50 per cent inhibition of the stimulation in phosphatidylinositol labelling due to 3.3 μM epinephrine.[27] In contrast, the activation of glycogen phosphorylase by 0.1 μM epinephrine was less sensitive and 80 per cent inhibition required 0.01 μM prazosin.[138]

A comparison of the effects of various hormones on hepatic metabolism is presented in Table 1. The hormones which activate glycogen phosphorylase are divided into those which do and those which do not produce detectable elevations in cyclic AMP. The agents which do not affect cyclic AMP are all thought to elevate cytosol Ca^{2+} and increase phosphatidylinositol turnover. The ionophore A-23187 presumably moves Ca^{2+} across membranes without involvement of phosphatidylinositol turnover.

It is indicated in Table 1 that hormones which activate glycogenolysis

Table 1. Comparison of hormone effects on rat hepatocytes

Agent	Cyclic AMP elevated	Effects on glycogen phosphorylase and Ca^{2+} requirement in medium	Effect on phosphatidyl-inositol (PI) turnover
A-23187	No	Activates glycogen phosphorylase only in presence of extra-cellular Ca^{2+}	No effect
Vasopressin V-1 or alpha effects	No	Same as for A-23187	Increases PI turnover even in absence of extracellular Ca^{2+}
Angiotensin	No	Same as for A-23187	Same as for vasopressin
Alpha-1 catecholamines	No	Activates glycogen phosphorylase in absence of extra-cellular Ca^{2+}	Increases PI turnover in absence of extra-cellular Ca^{2+}
Glucagon and beta-catecholamines	Yes	Activates glycogen phosphorylase	No effect. Decreases PE to PC conversion[a]
Alpha-2 catechola-mines	No effects in rat hepatocytes but in other cells inhibition of adenylate cyclase is seen		None in any tissue
Vasopressin V-2 or beta effects	No effects in rat hepatocytes but in other cells activation of adenylate cyclase is seen		None in any tissue

[a] The addition of beta catecholamines, glucagon or even vasopressin and alpha-1 catecho-lamines inhibits the conversion of phosphatidylethanolamine (PE) to phosphatidylcholine (PC) *via* methylation.

actually inhibit phospholipid methylation. Hirata *et al.*[139] suggested that the beta catecholamine isoproterenol had the opposite effect in rat reticulocyte ghosts of stimulating the conversion of phosphatidylethanolamine to phosphatidylcholine. This is usually referred to as phospholipid methylation since the reactions involved result in the successive addition to phosphatidyl-ethanolamine of three methyl groups derived from S-adenosyl-methionine. In reticulocyte ghosts these methylations occur in the plasma membrane. However, in hepatocytes nearly all the detectable methylation of phos-phatidylethanolamine occurs in the endoplasmic reticulum.[140,141]

Klevan and Fain[142] found that in rat hepatocytes isoproterenol actually inhibited phospholipid methylation. There was an inhibition of phospholipid methylation by vasopressin, alpha catecholamines and glucagon. The turn-over of phosphatidylcholine labelled by methylation was also unaffected by isoproterenol. The inhibitory action of isoproterenol on phospholipid methylation was greater than the inhibition of phosphatidylcholine synthesis from choline.[142] These data indicate that the effects of beta-adrenergic agents on phospholipid methylation in hepatocytes are rather different from those on rat reticulocytes[139] and cultured C_6 astrocytoma cells[143] where opposite effects have been reported.

9   ALPHA-1 CATECHOLAMINES AND PHOSPHATIDYL-INOSITOL TURNOVER IN ADIPOCYTES

Stein and Hales[144] first reported that the addition of epinephrine in the presence of propanolol to rat adipocytes increased ^{32}P incorporation into phosphatidylinositol and phosphatidic acid. Previously Perry and Hales[145] had described an alpha-adrenergic stimulation of ^{42}K efflux from rat adipocytes. In other cells ^{42}K efflux has been attributed to a rise in cytosol Ca^{2+} resulting from either alpha-adrenergic stimulation of bound Ca^{2+} release or entry of extracellular Ca^{2+}. However, recognition of the possibility that there are alpha catecholamine effects on adipocyte Ca^{2+} was delayed for some time until a link to metabolism could be demonstrated. In rat adipocytes it has been difficult to detect any effects of A-23187 or alpha catecholamines on lipolysis or cyclic AMP metabolism.[146,147]

There are inhibitory effects of alpha catecholamines on cyclic AMP accumulation by hamster and human adipocytes. However, these are alpha-2 effects which are probably secondary to inhibition of adenylate cyclase.[133,148,149] Fain and Garcia-Sainz[133] suggested that all alpha-2 effects of catecholamines are independent of Ca^{2+} and linked to inhibition of adenylate cyclase while alpha-1 effects are independent of cyclic AMP and linked to elevation of cytosol Ca^{2+} as well as increased turnover of phosphatidylinositol.

The only known effect in rat adipocytes of an elevation in cytosol Ca^{2+} due to alpha catecholamines is an inactivation of glycogen synthase.[150,151] This effect was demonstrated by Garcia-Sainz and Fain[152] to be mediated through alpha-1 receptors. In adipocytes from hypothyroid rats the ability of alpha-1 catecholamines to inactivate glycogen synthase was unaltered while that of beta catecholamines to affect cyclic AMP accumulation was greatly reduced.[152] These results indicate that there is no reciprocal regulation of alpha-1 and beta receptors in rat adipocytes.

The increase in phosphatidylinositol turnover in rat adipocytes due to epinephrine has been shown to be an alpha-1 effect unrelated to lipolysis or cyclic AMP metabolism.[46] The increase in turnover of phosphatidic acid and phosphatidylinositol due to catecholamines was unaffected by incubation in Ca^{2+}-free buffer containing 1 mM EGTA.[46] A stimulation of ^{32}P incorporation into phosphatidylinositol due to either insulin or A-23187 was seen in calcium-free buffer containing 1 mM EGTA. However, if cells were incubated in buffer containing 1 mM EGTA plus 2.5 mM Ca^{2+}, the stimulatory effects of both insulin and A-23187 were abolished and they now inhibited phosphatidylinositol synthesis.[46] These data support the hypothesis mentioned in the section on turnover of phosphatidylinositol in hepatocytes that a loss of Ca^{2+} from the endoplasmic reticulum increases phosphatidylinositol synthesis. McDonald *et al.*[153,154,155] found that in buffer containing Ca^{2+} insulin increased the uptake of Ca^{2+} into the endoplasmic reticulum which correlates with the decrease in phosphatidylinositol turnover due to insulin seen in medium containing 2.5 mM Ca^{2+} and 1 mM EGTA. The ionophore and insulin might decrease Ca^{2+} in the endoplasmic reticulum of adipocytes incubated in the presence of EGTA if they facilitate Ca^{2+} transfer across membranes. In the presence of large amounts of extracellular Ca^{2+} the same agents might move Ca^{2+} in the opposite direction from the medium into the endoplasmic reticulum. Further work is required to determine whether the synthesis of phosphatidylinositol in the endoplasmic reticulum is a reflection of the Ca^{2+} content of that organelle.

In adipocytes it is unlikely that diglyceride formed from breakdown of phosphatidylinositol in the plasma membrane is an intracellular messenger since addition of lipolytic agents should result in a massive accumulation of diglyceride in the cytosol. However, lipolytic agents are inhibitory to phosphatidylinositol synthesis.[46] The same argument applies to arachidonic acid and products formed from it by lipoxygenase or cyclooxygenase pathways. The arachidonic acid content of phosphatidylinositol in adipocytes is higher than that of other phospholipids[8,156] but most of the arachidonic acid is present in triacylglycerols. Triacylglycerol hydrolysis in adipose tissue results in appreciable release of arachidonic acid.[157] These data suggest that neither cytosol diglycerides or arachidonic acid are likely regulators of phosphatidylinositol synthesis in the endoplasmic reticulum. However, separate compartmentation is also a theoretical explanation but in this case as in most cases where it is invoked, it is brought in to explain data which do not conform to existing hypotheses.

An alpha-1 catecholamine effect on uptake of ^{32}P into phosphatidylinositol is readily demonstrable in rat adipocytes but it has been difficult to demonstrate any appreciable increase in breakdown of prelabelled phosphatidylinositol. However, in adipocytes breakdown of the phosphatidylinositol present in the plasma membrane which is involved in Ca^{2+} gating may account for such a small fraction of total breakdown that it cannot be readily detected.

The hypothesis that breakdown of phosphatidylinositol is involved in the elevation of cytosol Ca^{2+} is the best explanation available. However, a second messenger or mechanism for the release of Ca^{2+} from mitochondria and the endoplasmic reticulum is still needed. Cyclic inositol phosphate is the only other product of phosphatidylinositol breakdown released to the cytosol of adipocytes. There is another possibility since it has been suggested that the release of Ca^{2+} bound to the plasma membrane acts as trigger to cause an autocatalytic release of Ca^{2+} from organelles such as the endoplasmic reticulum. It should be noted that the endoplasmic reticulum of adipocytes is analogous to the sacroplasmic reticulum in muscle and is able to bind appreciable amounts of Ca^{2+}.[158]

Deoxycholate activates phosphatidylinositol phosphohydrolase (phospholipase C often used for convenience to describe this enzyme which cleaves phospholipids containing diglyceride linked to inositol phosphate) in many tissues.[159,160,161,162] In homogenates of rat adipocytes the addition of deoxycholate resulted in a marked loss of pre-labeled phosphatidylinositol under conditions in which there was no degradation of phosphatidylcholine or phosphatidylethanolamine.[29] Since deoxycholate activates many membrane bound enzymes involved in both synthesis and degradation of phospholipids, it was somewhat surprising that no effect on breakdown of phosphatidylcholine or ethanolamine was observed. However, these phospholipids are not degraded by phospholipase C but rather phospholipases of the A_2 variety. There appears to be little endogenous A_2 activity in adipocytes[29] even in the presence of large amounts of pure melittin which is a potent activation of these enzymes in many tissues.[163,164] In these studies on adipocyte homogenates the ability of endogenous enzymes to break down prelabeled phospholipids of cellular membranes was measured. In plasma membrane vesicles isolated from prelabeled adipocytes it was possible to demonstrate breakdown of endogenous phospholipids in vesicles isolated by sucrose-density gradient centrifugation.[29] The phospholipase C may be in the cytosol of the vesicles as suggested in brain by Irvine and Dawson.[160] Although the cellular localization of the phospholipase C remains to be established in adipocytes, our data clearly demonstrate that the phosphatidylinositol in the plasma membrane vesicles is readily cleaved by endogenous lipase. The phospholipase C activity in adipocyte homogenates is Ca^{2+}-dependent and little activity can be demonstrated in the absence of added deoxycholate. This is characteristic of mammalian tissues in contrast to the situation in the blowfly salivary gland as discussed in the next section. Why added deoxycholate is often needed to demonstrate activity of this enzyme remains to be demonstrated. Whether hormones increase access of the substrate to the enzyme[20] or directly activate phospholipase C as implied by Michell[11] remains to be demonstrated.

10 PHOSPHATIDYLINOSITOL BREAKDOWN AND Ca^{2+} GATING IN BLOWFLY SALIVARY GLANDS

The fly salivary gland is a good model in which to study salivary secretion since the gland is essentially a tube enclosed at one end made up of a single layer of secretory cells.[165] The gland can be readily removed without damage to its neutral or circulatory inputs since it has none. The signal for salivary secretion is transmitted via a hormone (5-HT also known as serotonin) and both oxygen and nutrients are obtained from the insect haemolymph. Unlike the situation with mammalian salivary glands, secretion can be readily measured *in vitro* by incubating most of the gland in a drop of medium contained in liquid paraffin (also known as mineral oil or liquid petroleum) and pulling the tip of the gland into the paraffin layer. The saliva which is secreted collects as a drop, can be removed at timed intervals and secretion readily measured by sizing the droplets.[165,166]

The study of the hormonal regulation of salivary secretion has followed the fashion of the times with regard to second messengers. Berridge and Patel[165] reported that low concentrations (10^{-10} to 10^{-9} M) of 5-HT stimulated salivary secretion by isolated blowfly salivary glands and this was mimicked by the addition of cyclic AMP. Subsequently, the effects of 5-HT on fluid secretion were all attributed to 5-HT activation of adenylate cyclase.[166] However, in the same year, Rasmussen[167] put forth the hypothesis that Ca^{2+} was a far more important second messenger for hormone action than was cyclic AMP. Rasmussen shortly thereafter went to Cambridge on sabbatical leave and examined the role of Ca^{2+} in blowfly salivary secretion in collaboration with Prince and Berridge.[168] Prince *et al.*[168] soon found effects of 5-HT on Ca^{2+} dynamics which were unrelated to cyclic AMP. The uptake of Ca^{2+} from the medium was stimulated by 5-HT but not by added cyclic AMP. Additionally the transepithelial potential (potential in the lumen with respect to the bathing medium) became more negative in the presence of 5-HT. However, the addition of cyclic AMP had the opposite effect of making the potential across the gland more positive and similar results could be seen with respect to 5-HT if added to glands in medium containing EGTA where no Ca^{2+} uptake can occur.[168]

Hansen-Bay[169] working in the same laboratory subsequentially demonstrated that the secretion of enzymes seen with 5-HT was mimicked by cyclic AMP addition although there was some effect of Ca^{2+} on this process. Studies on the role of Ca^{2+} in salivary secretion were facilitated by the discovery of Berridge and Lipke[170] that transepithelial flux of Ca^{2+} was a good monitor for Ca^{2+} dynamics. They also demonstrated that 5-HT mobilized internal Ca^{2+} and stimulated the entry of external Ca^{2+} through an action independent of cyclic AMP.

In 1977, I arrived for a sabbatical year in Cambridge with Berridge and it was decided to see if the hypothesis of Michell[11] applied to the question of how flies spit. The blowfly salivary gland turned out to be an ideal system in which to investigate the role of phosphatidylinositol hydrolysis in Ca^{2+} gating.[20,171;40,172] The major features which facilitate research in this area are: 1. The intracellular pool of inositol is very small and inositol taken up by the gland appears to rapidly equilibrate with medium inositol. 2. The synthesis of phosphatidylinositol occurs in a compartment in equilibrium with the cytosol Ca^{2+} pool since it is readily inhibited by A-23187 or 5-HT which both elevate cytosol Ca^{2+}. 3. The phosphatidylinositol synthesized in the presence of added $(^{32}P)P_i$ or (^{3}H) inositol rapidly equilibrates with the small pool of phosphatidylinositol which is hydrolyzed by the hormone 5-HT. 4. A large fraction (80–90 per cent) of the newly synthesized phosphatidylinositol can be hydrolysed during subsequent exposure of the gland to 5-HT, but less than 5 per cent of total phosphatidylinositol is hydrolyzed. 5. Phosphatidylinositol hydrolysis can be measured without the complication of resynthesis of phosphatidylinositol which is actually blocked by 5-HT in the fly salivary gland. 6. Cell-free systems can be obtained in which a direct effect of 5-HT on the breakdown of phosphatidylinositol in isolated membranes is readily demonstrable. The defects of the gland are primarily related to its small size and the lack of commercial source for the flies. However, it is relatively easy to raise the flies providing accurate temperature control is provided and a colony is now functioning in my laboratory.

A 1981 model for the regulation of cyclic AMP and Ca^{2+} in blowfly salivary glands is depicted in Figure 6. Probably the receptors involved in cyclase activation by 5-HT are distinct from those involved in phosphatidylinositol turnover. Evidence for at least two types of serotonin receptors has been found in membrane preparation from rat brain based on binding studies.[173] Fain and Garcia-Sainz[133] suggested that, by analogy with catecholamines, the receptors for 5-HT involved in adenylate cyclase regulation be called beta receptors and those involved in elevation of cytosol Ca^{2+} and phosphatidylinositol turnover be described as alpha.

The evidence for involvement of phosphatidylinositol hydrolysis in Ca^{2+} gating is as follows: 1. There is a breakdown of labelled phosphatidylinositol

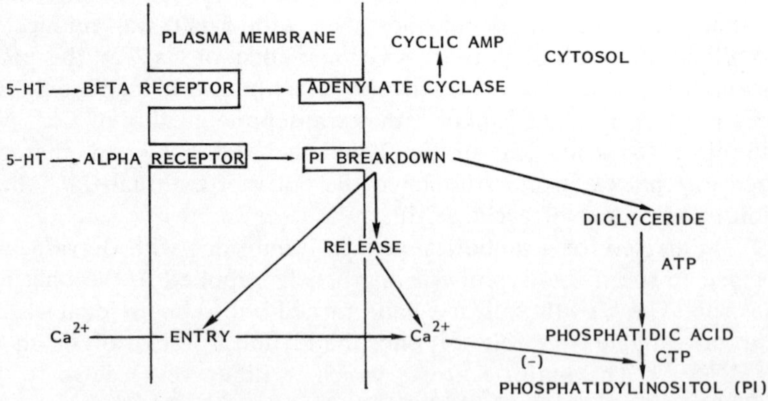

Figure 6. 5-HT activation of PI breakdown in salivary glands

which occurs at the same time that salivary secretion is stimulated. Low concentrations (10^{-9} M) of 5-HT stimulate phosphatidylinositol hydrolysis and the log dose response relationship for Ca^{2+} gating and phosphatidylinositol hydrolysis are similar. There is an increase in the accumulation of inositol 1:2 cyclic phosphate, inositol 1-phosphate and inositol in salivary glands after addition of 5-HT.[40] 2. Added cyclic AMP does not stimulate Ca^{2+} transport or accelerate phosphatidylinositol hydrolysis. Methyl xanthines potentiate the increase in salivary secretion of fluid due to 10^{-9} M 5-HT without affecting phosphatidylinositol breakdown.[40] 3. In glands incubated with 5-HT in the absence of extracellular Ca^{2+} there is an initial secretion of fluid which returns to near basal values after 30 minutes stimulation with 5-HT. However, the breakdown of phosphatidylinositol is unaffected by the absence of extracellular Ca^{2+}.[41] 4. The addition of the ionophore A-23187 accelerates transepithelial flux of Ca^{2+} and has a slight stimulatory effect on fluid secretion. The breakdown of phosphatidylinositol is unaffected by A-23187.[40] 5. The conversion of phosphatidylinositol to phosphatidic acid which is postulated to be a Ca^{2+} ionophore[174,62] does not occur in fly salivary glands. During the first two hours after addition of 5-HT the accumulation of ^{32}P in phosphatidic acid is unaffected by 5-HT.[20] 6. The synthesis of phosphatidylinositol is actually inhibited by 5-HT. This inhibition appears secondary to an elevation of cytosol Ca since it can be mimicked by the addition of the ionophore A-23187.[20] 7. In fly salivary glands there is an 80–90 per cent breakdown of pre-labelled phosphatidylinositol under conditions where there is little breakdown of total phosphaditylinositol.[172] These data indicate that the pool of phosphatidylinositol involved in Ca^{2+} gating is quite small. 8. If glands are incubated in a large volume of medium under conditions where 80 per cent or more of the prelabelled phosphatidylinositol has been hydrolysed there is no increase in Ca^{2+} gating by 5-HT. If the glands are washed free of hormone and incubated for an hour there is recovery of 5-HT responsive Ca^{2+} flux only in the presence of added inositol.[172] The incorporation of inositol into phosphatidylinositol is four-fold higher in glands previously exposed to 5-HT. In contrast, the incorporation of choline or ethanolamine at a similar concentration was unaffected by prior exposure to 5-HT.[172] 9. The addition of 5-HT to membranes obtained after centrifugation of a cell-free salivary gland homogenate at $100,000 \times g$ for 10 minutes results in a significant increase in the breakdown of the phosphatidylinositol present in the membranes.[175] 10. The phosphatidylinositol phosphohydrolase activity in salivary gland homogenates or membranes does not require Ca^{2+}.[171] In fact, there is more reproducible effect of 5-HT on homogenates incubated in Ca^{2+}-free medium containing 1 EGTA.[175]

The link between phosphatidylinositol hydrolysis and the entry of extracellular Ca^{2+} remains to be elucidated. Our lack of information about the nature of the proteins involved is a severe handicap. However, in the blowfly salivary gland it is clear that breakdown of phosphatidylinositol is not a consequence of a rise in cytosol Ca^{2+} since fortuitously the phospholipase C of this gland is Ca^{2+} insensitive.

Two mechanisms are compatible with the available data. The hormone could directly activate phospholipase C as a result of interaction with the alpha receptor for serotonin in the plasma membrane. The activated enzyme then hydrolyses phosphatidylinositol associated with the Ca^{2+} gates which results in a phasic opening of the putative gates. When the hormone content in the medium drops and it dissociates from the receptor, the Ca^{2+} gate collapses to an inactive state. Resynthesis of phosphatidylinositol and association with the inactive gate is required to return to gate to a primed or hormone-sensitive state. While there is no direct proof for this theory, it is clear that loss of phosphatidylinositol from critical sites in salivary glands results in an inability of 5-hydroxytryptamine to stimulate the entry of Ca^{2+}.

The second mechanism suggests that the complex of 5-HT bound to its alpha receptor binds to the calcium gate and induces a conformation change which permits entry of Ca^{2+}. In the open state the phosphatidylinositol bound to plasma membrane Ca^{2+} gate is more accessible to the hydrolytic action of phospholipase C. Whether the phospholipase C is bound to the membrane or in the cytosol is not known but isolated membrane fragments can hydrolyse their phosphatidylinositol in the absence of cytosol except for what might be inside membrane vesicles. In this mechanism the hydrolysis of phosphatidylinositol changes the conformation of the gate at Ca^{2+} no longer enters. The gate remains in the inactive state until it rebinds phosphatidylinositol and then reverts to a hormone-sensitive state. The crucial differences which remain to be elucidated are whether the hormone activates the phospholipase C or the substrate and whether the hydrolysis of phosphatidylinositol opens or shuts the Ca^{2+} gate.

11 SUMMARY

This chapter has considered selected aspects of the relationship between phosphatidylinositol breakdown and Ca^{2+} gating in which particular progress has been made in the past few years. Michell[11,41] has reviewed the many systems in which effects of stimuli on phosphatidylinositol turnover can be demonstrated. Hawthorne and Picard[176] have reviewed phosphatidylinositol turnover in synaptosomes which have not been mentioned in this chapter since the observed effects appear to be secondary to changes in Ca^{2+}. Some of the other effects of stimuli on phosphatidylinositol are clearly secondary to Ca^{2+} and in systems such as platelets an initial breakdown of phosphatidylinositol may elevate trigger Ca^{2+} which also influences phosphatidyl-inositol hydrolysis. Another possible explanation for the phospholipid effect is that phosphatidylinositol is hydrolysed to diglyceride and then converted to phosphatidic acid which regulates Ca^{2+} permeability. Salmon and Honeyman[171] based on studies of cholinergic action on amphibian smooth muscle and Putney *et al.*[62] from studies on rat parotids suggested that phosphatidic acid mediates the inward movement of Ca^{2+}. Irvine and Dawson[34] found that phosphatidate accelerated phosphatidylinositol degradation by brain cytosol and suggested that increased formation of phosphatidic acid might further accelerate phosphatidylinositol hydrolysis. Possibly, in systems where large amounts of phosphatidylinositol disappear and are replaced by equivalent increases in phosphatidic acid these changes act in a cumulative fashion to regulate membrane function.

In the blowfly salivary gland the breakdown of phosphatidylinositol seen in the presence of hormones is restricted to a pool which is less than 5 per cent of the total phosphatidylinositol and is probably involved in the entry of extracellular Ca^{2+}. The breakdown of phosphatidylinositol is not secondary to elevation of Ca^{2+} since in this tissue the phospholipase C is a Ca^{2+}-independent enzyme. It has been possible to demonstrate direct activation of phosphatidylinositol breakdown by addition of 5-HT to a cell-free homogenate of salivary glands incubated in Ca^{2+}-free buffer.

In the pancreas and the parotid there is a breakdown of almost half of the total phosphatidylinositol with saturating concentrations of hormones which is accompanied by a corresponding increase in phosphatidic acid. This large breakdown could be secondary to breakdown of trigger phosphatidylinositol in the plasma membrane which results in release of arachidonic acid. Possibly prostaglandins, phosphatidic acid or Ca^{2+} all activate phospholipase C which further accelerates breakdown of phosphatidylinositol.

In platelets the release of arachidonic acid initially seen in response to thrombin or other stimuli appears to be derived from breakdown of phosphatidylinositol. In platelets the resulting rise in cytosol Ca^{2+} also further activates phospholipase C.

In the adrenal as in platelets too much emphasis has probably been placed on the hypothesis that arachidonic acid release is due to activation of a phospholipase A_2. One explanation for the increase in arachidonic acid turnover of phosphatidylinositol which is seen in cells exposed to ACTH is that the hormone activates a phosphatidylinositol specific phospholipase C. The resulting elevation of cytosol Ca^{2+} and accumulation of arachidonic acid increases the formation of polyphosphoinositides which may accelerate mitochondrial pregnenolone formation.

In hepatocytes and adipocytes the maximal increase in phosphatidylinositol breakdown which can be seen in the presence of alpha catecholamines is less than 5 per cent. In these cells the breakdown of phosphatidylinositol may be limited to the plasma membrane pool which is involved in the release of trigger Ca^{2+} into the cytosol from the endoplasmic reticulum and the mitochondria. The role of arachidonic acid in elevation of Ca^{2+} remains to be elucidated. However, it is possible that the increase in synthesis of phosphatidylinositol is secondary to a drop in Ca^{2+} content of the endoplasmic reticulum since the synthesis of this phospholipid is inhibited by Ca^{2+}. This explanation for the large increase in ^{32}P uptake is outlined in Figure 5. A summary of the effects of hormones on hepatic glycogen phosphorylase and phospholipid turnover is outlined in Table 1 above.

The hypothesis of Michell that phosphatidylinositol turnover is involved in some causal relationship to the elevation of cytosol Ca^{2+} is still the best explanation. However, the increase in synthesis of phosphatidylinositol seen in many tissues may be secondary to a drop of Ca^{2+} in the endoplasmic reticulum because of Ca^{2+} release to the cytosol. In some cells, particularly those where appreciable breakdown of total phosphatidylinositol can be observed, the resulting elevation in cytosol Ca^{2+} and the accumulation of arachidonic acid may activate phosphatidylinositol specific phospholipase C. This enzyme also appears to be far more important in arachidonic acid release than has previously been recognized.

REFERENCES

1. Kuksis, A., and Mookerjea, S. (1978). Inositol. *Nutrition Reviews*, **36**, 233–238.
2. Dawson, R. M. C., and Freinkel, N. (1961). The distribution of free meso inositol in mammalian tissues including some observations on the lactating rat. *Biochem. J.*, **78**, 606–610.
3. Sherman, W. R., Stewart, M. A., Kurien, M. M., and Goodwin, S. L. (1968). Measurement of myoinositol, myo-inosose-2 and scyllo-inositol in mammalian tissues. *Biochem. Biophys. Acta*, **158**, 197–205.
4. Eagle, H., Oyama, V. I., Levy, M., and Freeman, A. E. (1957). Myo-inositol as an essential growth factor for normal and malignant human cells in tissue culture. *J. Biol. Chem.*, **226**, 191–205.
5. Akino, T., and Shimojo, T. (1970). On the metabolic heterogeneity of rat liver phosphatidylinositol. *Biochim. Biophys. Acta*, **210**, 343.
6. Holub, B. J. (1978). Studies on the metabolic heterogeneity of different molecular species of phosphatidylinositols. In *Cyclitols and Phosphoinositides*, edited by W. W. Wells and F. Eisenberg, Jr., Academic Press, New York, 1978, pp. 523–534.
7. Blackwell, G. J., Duncombe, W. G., Flower, R. J., Parsons, M. F., and Vane, J. R. (1977). Distribution and metabolism of arachidonic acid in rabbit platelets during aggregation and its modification by drugs. *Brit. J. Pharmacol.*, **59**, 353–366.
8. Maude, M. B., Anderson, R. E., Armstrong, K. J., and Stouffer, J. E. (1974). Membrane phospholipids and hormone-sensitive lipolysis in fat cells. *Arch. Biochem. Biophys.*, **161**, 628–631.
9. Marshall, P. J., Dison, J. F., and Hokin, L. E. (1980). Evidence for a role in

stimulus-secretion coupling of prostaglandins derived from release of arachidonoyl residues as a result of phosphatidylinositol breakdown. *Proc. Nat. Acad. Sci. U.S.A.*, **77**, 3292–3296.

10. Bell, R. L., and Majerus, P. W. (1980). Thrombin-induced hydrolysis of phosphatidylinositol in human platelets. *J. Biol. Chem.*, **255**, 1790–1792.
11. Michell, R. H. (1975). Inositol phospholipids and cell surface receptor function. *Biochim. Biophys. Acta*, **415**, 81–147.
12. Wells, W. W., and Eisenberg, F., Jr. (1978). *Cyclitols and phosphoinositides*, Academic Press, New York, 607 pp.
13. Daleo, G. R., Piras, M. M., and Piras, R. (1976). Diglyceride kinase activity of microtubules. Characterization and comparison of the protein kinase and ATPase activities associated with vinblastine-isolated tubulin of chick embryonic muscles. *Eur. J. Biochem.*, **68**, 339–346.
14. Kanoh, H., and Akesson, B. (1978). Properties of microsomal and soluble diacylglycerol kinase. *Eur. J. Biochem.*, **85**, 225–232.
15. Agranoff, B. W., Bradley, R. M., and Brady, R. O. (1958). The enzymatic synthesis of inositol phosphatide. *J. Biol. Chem.*, **233**, 1077–1083.
16. Paulus, H., and Kennedy, E. P. (1960). The enzymatic synthesis of inositol monophosphatide. *J. Biol. Chem.*, **235**, 1303–1311.
17. Takenawa, T., Saito, M., Nagai, Y., and Egawa, K. (1977). Solubilization of the enzyme catalyzing CDP-diglyceride-independent incorporation of myo-inositol into phosphatidylinositol and its comparison to CDP-diglyceride:inositol transferase. *Arch. Biochem. Biophys.*, **182**, 244–250.
18. Jungalwala, F. B., Freinkel, N., and Dawson, R. M. C. (1971). The metabolism of phosphatidylinositol in the thyroid gland of the pig. *Biochem. J.*, **123**, 19–33.
19. Wootton, J. A., and Kinsella, J. E. (1977). Properties of cytidinediphosphodiacyl-sn-glycerol: Myoinositol transferase of bovine mammary tissue. *Int. J. Biochem.*, **8**, 449–456.
20. Berridge, M. J., and Fain, J. N. (1979). Inhibition of phosphatidylinositol synthesis and the inactivation of calcium entry after prolonged exposure of the blowfly salivary gland to 5-hydroxytryptamine. *Biochem. J.*, **178**, 59–69.
21. Wallace, M., and Fain, J. N. (1981). Unpublished studies.
22. Bleasdale, J. E., Wallis, P., MacDonald, P. C., and Johnston, J. M. (1979). Characterization of the forward and reverse reactions catalyzed by CDP-diacylglycerol:inositol transferase in rabbit lung tissue. *Biochim. Biophys. Acta*, **575**, 135–147.
23. Daniels, C. J., and Palmer, F. B. St. C. (1980). Biosynthesis of phosphatidylinositol in crithidia fasciculata. *Biochim. Biophys. Acta.*, **618**, 263–272.
24. Takenawa, T., and Egawa, K. (1977). CDP-diglyceride:inositol transferase from rat liver. *J. Biol. Chem.*, **252**, 5419–5423.
25. Holub, B. J. (1974). The Mn²⁺-activated incorporation of inositol into molecular species of phosphatidylinositol in rat liver microsomes. *Biochim. Biophys. Acta*, **369**, 111–122.
26. Holub, B. J. (1976). Specific formation of arachidonoyl phosphatidylinositol from 1-acyl-sn-glycero-3-phosphorylinositol in rat liver. *Lipids*, **11**, 1–5.
27. Tolbert, M. E. M., White, A. C., Aspry, K., Cutts, J., and Fain, J. N. (1980). Stimulation by vasopressin and alpha-catecholamines of phosphatidylinositol formation in isolated rat liver parenchymal cells. *J. Biol. Chem.*, **255**, 1938–1944.
28. Yandrasitz, J. R., and Segal S. (1979). The effect of MnCl₂ on the basal and acetylcholine-stimulated turnover of phosphatidylinositol in synaptosomes. *FEBS Lett.*, **108**, 279–282.

29. Fain, J. N., and Kabnick, K. (1981). Activation of phosphatidylinositol break-down in adipocyte homogenates by deoxycholate. Unpublished.

30. Fain, J. N. (1981). Regulation of lipid metabolism. In *Handbook of Experimental Pharmacology*, Cyclic Nucleotides (Nathanson, J. A., and Kebabian, J. W., eds.), Springer Verlag, in press.

31. Gryan, G., and Czech, M. P. (1979). Hormonal control of inositol transport and its incorporation into phosphatidyl inositol in adipocyte plasma membranes. *Fed. Proceedings*, **38,** Abstract # 958, p. 410.

32. Hauser, G. (1969). Myoinositol transport in slices of rat kidney cortex. 1. Effect of incubation conditions and inhibitors. *Biochim. Biophys. Acta.*, **173,** 257–266.

33. Diringer, H, and Rott, R. (1977). Different pools of free myoinositol in chick-embryo cells as indicated by infection with Newcastle-disease virus. *Eur. J. Biochem.*, **79,** 451–457.

34. Fain, J. N., and Berridge, M. J. (1979). Relationship between hormonal activation of phosphatidylinositol hydrolysis, fluid secretion and calcium flux in the blowfly salivary gland. *Biochem. J.*, **178,** 45–58.

35. Holub, B. J., and Kuksis, A. (1971). Differential distribution of orthophosphate-^{32}P and glycerol-^{14}C among molecular species of phosphatidyl-inositols of rat liver in vivo. *J. Lipid Res.*, **12,** 699–705.

36. Holub, B. J., and Kuksis, A. (1972). Further evidence for the interconversion of monophosphoinositides in vivo. *Lipids*, **7,** 78–80.

37. Irvine, R. F., and Dawson, R. M. C. (1979). Transfer of arachidonic acid between phospholipids in rat liver microsomes. *Biochem. Biophys. Res. Commun.*, **4,** 1399–1405.

38. Bell, R. L., Kennerly, D. A., Stanford, N., and Majerus, P. W. (1979). Diglyceride lipase: a pathway for arachidonate release from human platelets. *Proc. Nat. Acad. Sci. USA*, **76,** 3238–3241.

39. Hokin, M. R., and Hokin, L. E. (1953). Enzyme secretion and the incorporation of P^{32} into phospholipids of pancreas slices. *J. Biol. Chem.*, **203,** 967–977.

40. Hokin, L. E., and Hokin, M. R. (1955). Effects of acetylcholine on the turnover of phosphoryl units in individual phospholipids of pancreas slices and brain cortex slices. *Biochim. Biophys. Acta*, **18,** 102–110.

41. Eichberg, J., Shein, H. M., Schwartz, M., and Hauser, G. (1973). Stimulation of ^{32}P$_i$ incorporation into phosphatidylinositol and phosphatidylglycerol by catecholamines and beta-adrenergic receptor blocking agents in rat pineal organ cultures. *J. Biol. Chem.*, **248,** 3615–3622.

42. Freinkel, N., El Younsi, C., and Dawson, R. M. C. (1975). Inter-relations between the phospholipids of rat pancreatic islets during glucose stimulation and their response to medium inositol and tetracaine. *Eur. J. Biochem.*, **59,** 245–252.

43. Michell, R. H. (1979). Inositol phospholipids in membrane function. *Trends in Biochem. Sci.*, **4,** 128–131.

44. Michell, R. H., Jafferji, S. S., and Jones, L. M. (1977). The possible involvement of phosphatidylinositol breakdown in the mechanism of stimulus-response coupling at receptors which control cell-surface calcium gates. *Advances in Exp. Biol. and Med.*, **83,** 447–464.

45. Hokin-Neaverson, M. (1974). Acetylcholine causes a net decrease in phosphatidylinositol and a net increase in phosphatidic acid in mouse pancreas. *Biochem. Biophys. Res. Commun.*, **58,** 763–768.

46. Hokin-Neaverson, M. (1977). Metabolism and role of phosphatidylinositol in acetylcholine-stimulated membrane function. *Advances in Exp. Biol. Med.*, **83,** 429–446.

47. Banschbach, M. W., Geison, R. L., and Hokin-Neaverson, M. (1974). Acetylcholine increases the level of diglyceride in mouse pancreas. *Biochem. Biophys. Res. Commun.*, **58**, 714–718.

48. Garcia-Sainz, J. A., and Fain, J. N. (1980). Effect of insulin, catecholamines, and calcium ions on phospholipid metabolism in isolated white fat cells. *Biochem. J.*, **186**, 781–789.

49. Marshall, P. J., Dixon, J. F., and Hokin, L. E. (1980). Evidence for a role in stimulus-secretion coupling of prostaglandins derived from release of arachidonoyl residues as a result of phosphatidylinositol breakdown. *Proc. Nat. Acad. Sci. U.S.A.*, **77**, 3292–3296.

50. Clements, R. S., Jr., Rhoten, W. B., and Starnes, W. R. (1977). Subcellular localization of the alterations in phosphatidylinositol metabolism following glucose-induced insulin release from rat pancreatic islets. *Diabetes*, **26**, 1109–1116.

51. Farese, R. V., Sabir, A. M., and Larson, R. E. (1980c). Net changes in pancreatic phospholipids during stimulation by glucose and cAMP *in vitro*. *Diabetes*, **29**, Suppl. 2, Abstract 120.

52. Malaisse, W. J., Herchuelz, A., Devis, G., Somers, G., Boschero, A. C., Hutton, J. C., Kawazu, S., and Sener, A. (1978). Regulation of calcium fluxes and their regulatory roles in pancreatic islets. *Ann. N.Y. Acad. Sci.*, **307**, 562–582.

53. Malaisse, W. J., Sener, A., Herschuelz, A., and Hutton, J. C. (1979). Insulin release: the fuel hypothesis, *Metabolism*, **28**, 373–386.

54. Jones, L. M., and Michell, R. H. (1974). Breakdown of phosphatidylinositol provoked by muscarinic cholinergic stimulation of rat parotid-gland fragments. *Biochem. J.*, **142**, 583–590.

55. Putney, J. W., Jr. (1979). Stimulus-permeability coupling: Role of calcium in the receptor regulation of membrane permeability. *Pharmacol. Rev.*, **30**, 209–245.

56. Gardner, J. D. (1979). Regulation of pancreatic exocrine function in vitro: initial steps in the actions of secretagogues. *Ann. Rev. Physiol.*, **41**, 55–66.

57. Williams, J. A. (1980). Regulation of pancreatic acinar cell function by intracellular calcium. *Am. J. Physiol.*, **238**, G269–G279.

58. Ashcroft, S. J. H. (1980). Glucoreceptor mechanisms and the control of insulin release and biosynthesis. *Diabetologia*, **18**, 5–15.

59. Kirtland, S. J., and Baum, H. (1972). Prostaglandin E₁ may act as a 'calcium ionophore'. *Nature New Biology*, **236**, 47–49.

60. Chauvelot, L., Heisler, S., Huot, J., and Gagon, D. (1979). Prostaglandins and enzyme secretion from dispersed rat pancreatic acinar cells. *Life Sci.*, **25**, 913–920.

61. Calderon, P., Furnelle, J., and Christophe, J. (1979). In vitro lipid metabolism in the rat pancreas. III. Effects of carbamylcholine and pancreozymin on the turnover of phosphatidylinositols, 1,2-diacylglycerols and phosphatidylcholines. *Biochim. Biophys. Acta*, **574**, 404–413.

62. Calderon, P., Furnelle, J., and Christophe, J. (1980). Phosphatidylinositol turnover and calcium movement in the rat pancreas. *Am. J. Phys.*, **238**, G247–G254.

63. Hokin, L. E. (1966). Effects of calcium omission on acetylcholine-stimulated amylase secretion and phospholipid synthesis in pigeon pancreas slices. *Biochim. Biophys. Acta*, **115**, 219–221.

64. Putney, J. W., Jr., Weiss, S. J., Van de Walle, C., and Haddas, R. A. (1980). Is

phosphatidic acid a calcium ionophore under neurohumoral control? *Nature*, **284**, 345–347.

65. Oron, Y., Löwe, M., and Selinger, Z. (1975). Incorporation of inorganic [^{32}P]phosphate into rat parotid phosphatidylinositol. *Mol. Pharmacol.*, **11**, 79–86.

66. Jones, L. M., and Michell, R. H. (1975). The relationship of calcium to receptor-controlled stimulation of phosphatidylinositol turnover. *Biochem. J.*, **148**, 479–485.

67. Gorman, R. R. (1979). Modulation of human platelet function by prostacyclin and thromboxane A$_2$. *Federation Proc.*, **38**, 83–88.

68. Lapetina, E. G. Chandrabose, K. A., and Cuatrecasas, P. (1978). Ionophore A-23187 and thrombin-induced aggregation: independence from cycloxygenase products. *Proc. Nat. Acad. Sci. U.S.A.*, **75**, 818–822.

69. Gorman, R. R., Bunting, S., and Miller, O. V. (1977). Modulation of human platelet adenylate cyclase by prostacyclin (PGX), *Prostaglandins*, **13**, 377–388.

70. Tateson, J. E., Moncada, S., and Vane, J. R. (1977). Effects of prostacyclin (PGX) on cyclic AMP concentrations in human platelets. *Prostaglandins*, **13**, 389–397.

71. Lapetina, E. G., Schmitges, C. J., Chandrabose, K., and Cuatrecasas, P. (1977). Cyclic adenosine 3′,5′-monophosphate and prostacyclin inhibit membrane phospholipase activity in platelets. *Biochem. Biophys. Res. Commun.*, **76**, 828–835.

72. Mauco, G., Chap, H., and Douste-Blazy, L. (1979). Characterization and properties of a phosphatidylinositol phosphodiesterase (phospholipase C) from platelet cytosol. *FEBS Lett.*, **100**, 367–370.

73. Billah, M. M., Lapetina, E. G., and Cuatrecasas, P. (1979). Phosphatidylinositol-specific phospholipase C of platelets: Association with 1,2-diacylglycerol-kinase and inhibition by cyclic AMP. *Biochem. Biophys. Res. Commun.*, **90**, 92–98.

74. Rittenhouse-Simmons, S. (1979). Production of diglyceride from phosphatidylinositol in activated human platelets. *J. Clin. Invest.*, **63**, 580–587.

75. Kishimoto, A, Takai, Y., Mori, T., Kikkawa, U., and Nishizuka, Y. (1980). Activation of calcium and phospholipid-dependent protein kinase by diacylglycerol, its possible relation to phosphatidylinositol turnover. *J. Biol. Chem.*, **255**, 2273–2276.

76. Call, F. L., and Rubert, M. (1973). Diglyceride kinase in human platelets. *J. Lipid Res.*, **14**, 466–474.

77. Vanderhoek, J. Y., and Feinstein, M. B. (1979). Local anesthetics, chlorpromazine and propanolol inhibit stimulus-activation of phospholipase A$_2$ in human platelets. *Mol. Pharmacol.*, **16**, 171–180.

78. Schoene, N. W. (1978). Properties of platelet phospholipase A$_2'$. *Adv. Prostaglandin Thromb. Res.*, **3**, 121–125.

79. Vallee, E., Gougat, J., Navarro, J., and Delahayes, J. F. (1979). Antiinflammatory and platelet anti-aggregant activity of phospholipase-A$_2$ inhibitors. *J. Pharm. Pharmacol.*, **31**, 588–592.

80. Pointer, R. H. and Fain, J. N. (1973). Response of isolated white fat cells to anti-malarial drugs. *Biochem. Pharmacol.*, **22**, 3025–3032.

81. Reijngoud, D.-J., and Tager, J. M. (1976). Chloroquine accumulation in isolated rat liver lysosomes. *FEBS Letters*, **64**, 231–235.

82. Apitz-Castro, R. J., Mas, M. A., Cruz, M. R., and Jain, M. K. (1979). Isolation

of homogeneous phospholipase A_2 from human platelets. *Biochem. Biophys. Res. Commun.*, **91**, 63–71.

83. Jesse, R. L., and Franson, R. C. (1979). Modulation of purified phospholipase A_2 activity from human platelets by calcium and indomethacin. *Biochim. Biophys. Acta*, **575**, 467–470.

84. Wong, P. Y.-K., and Cheung, W. Y. (1979). Calmodulin stimulates human platelet phospholipase A_2. *Biochem. Biophys. Res. Commun.*, **90**, 473–480.

85. Freinkel, N. (1957). Pathways of thyroidal phosphorus metabolism: The effect of pituitary thyrotropin upon the phospholipids of the sheep thyroid gland. *Endocrinology*, **61**, 448–460.

86. Altman, M., Oka, H., and Field, J. B. (1966). Effect of TSH acetylcholine, epinephrine, serotonin and synkavite on ^{32}P incorporation into phospholipids in dog-thyroid slices. *Biochim. Biophys Acta*, **116**, 586–588.

87. Haye, B., Marcy, G., and Jacquemin, C. (1979). Relationship between the 'phospholipid effect' and calcium in the thyroid. *Biochimie*, **61**, 905–912.

88. Boeynaems, J. M., Waelbroeck, M., and Dumont, J. E. (1979). Cholinergic and alpha-adrenergic stimulation of prostaglandins release by dog thyroid in vitro. *Endocrinology*, **105**, 988–995.

89. Haye, B., Champion, S., and Jacquemin, C. (1973). Control by TSH of a phospholipase A_2 activity, a limiting factor in the biosynthesis of prostaglandins in the thyroid. *FEBS Lett.*, **30**, 253–260.

90. Haye, B., and Jacquemin, C. (1974). Stimulation par la thyreostimuline de la production de l'inositol 1-2 phosphate cyclique. *Biochimie*, **56**, 1283–1285.

91. Boeynaems, J. M., Van Sande, J., and Dumont, J. E. (1975). Blocking of dog thyroid secretion in vitro by inhibitors of prostaglandin synthesis. *Biochem. Pharmacol.*, **24**, 1333–1337.

92. Thompson, M. E., Orcziyk, G. P., and Hedge, G. A. (1977). In vivo inhibition of thyroid secretion by indomethacin. *Endocrinology*, **100**, 1060–1067.

93. Schrey, M. P., and Rubin, R. P. (1979). Characterization of a calcium-mediated activation of arachidonic acid turnover in adrenal phospholipids by corticotropin. *J. Biol. Chem.*, **254**, 11234–11241.

94. Rubin, R. P., and Laychock, S. G. (1978). Prostaglandins and calcium-membrane interactions in secretory glands. *Ann. N.Y. Acad. Sci.*, **307**, 377–390.

95. Farese, R. V., Sabir, A. M., and Larson, R. E. (1980b). ACTH and cAMP rapidly increase adrenal levels of phosphatidic acid, phosphatidylinositol, and polyphosphoinositides by a cycloheximide-sensitive process. *Endocrine Society*, Abstract # 812, p. 277.

96. Farese, R. V., Sabir, A. M., and Vandor, S. L. (1979). Adrenocorticotropin acutely increases adrenal polyphosphoinositides. *J. Biol. Chem.*, **254**, 6842–6844.

97. Laychock, S. G., Shen, J. C., Carmines, E. L., and Rubin, R. P. (1978). The effect of corticotropin on phospholipid metabolism in isolated adrenocortical cells. *Biochim. Biophys. Acta*, **528**, 355–363.

98. Farese, R. V., Sabir, A. M., Vandor, S. L., and Larson, R. E. (1980). Are polyphosphoinositides the cycloheximide-sensitive mediator in the steroidogenic actions of adrenocorticotropin and adenosine-3',5'-monophosphate? *J. Biol. Chem.*, **255**, 5728–5734.

99. De Torrontegui, G., and Berthet, J. (1966). The action of adrenalin and

glucagon on the metabolism of phospholipids in rat liver. *Biochim. Biophys. Acta*, **116,** 467–476.

100. Kirk, C. J., Verrinder, T. R., and Hems, D. A. (1977). Rapid stimulation by vasopressin and adrenaline, of inorganic phosphate incorporation into phosphatidylinositol in isolated hepatocytes. *FEBS Letters*, **83,** 267–271.

101. Sutherland, E. W., and Rall, T. W. (1958). Fractionation and characterization of cyclic adenine ribonucleotide formed by tissue particles. *J. Biol. Chem.*, **232,** 1077–1091.

102. Rall, T. W., Sutherland, E. W. W., and Berthet, J. (1957). The relationship of epinephrine and glucagon to liver phosphorylase. *J. Biol. Chem.*, **224,** 463–475.

103. Sherline, P., Lynch, A., and Glinsmann, W. H. (1972). Cyclic AMP and adrenergic receptor control of rat liver glycogen metabolism. *Endocrinology*, **91,** 680–690.

104. Tolbert, M. E. M., Butcher, F. R., and Fain, J. N. (1973). Lack of correlation between catecholamine effects on cyclic adenosine 3′,5′-monophosphate and gluconeogenesis in isolated rat liver cells. *J. Biol. Chem.*, **248,** 5686–5692.

105. Birnbaum, M. J., and Fain, J. N. (1977). Activation of protein kinase and glycogen phosphorylase in isolated rat liver cells by glucagon and catecholamines. *J. Biol. Chem.*, **252,** 528–535.

106. Malbon, C. C., Gilman, H. R., and Fain, J. N. (1980). Hormonal stimulation of cyclic AMP accumulation and glycogen phosphorylase activity in calcium-depleted hepatocytes from hypothyroid rats. *Biochem. J.*, **188,** 593–599.

107. Pointer, R. H., Butcher, F. R., and Fain, J. N. (1976). Studies on the role of cyclic GMP and extracellular Ca^{2+} in the regulation of glycogenolysis in rat liver cells. *J. Biol. Chem.*, **251,** 2987–2992.

108. Hutson, N. J., Brumley, F. T., Assimacopoulos, F. D., Harper, S. C., and Exton, J. H. (1976). Studies on the alpha-adrenergic activation of hepatic glucose output. *J. Biol. Chem.*, **251,** 5200–5208.

109. Cherrington, A. D., Assimacopoulos, F. D., Harper, S. C., Corbin, J. D., Park, C. R., and Exton, J. H. (1976). Studies on the alpha-adrenergic activation of hepatic glucose output. *J. Biol. Chem.*, **251,** 5209–5218.

110. Keppens, S., Vandenheede, J. R., and De Wulf, H. (1977). On the role of calcium as second messenger in liver for the hormonally induced activation of glycogen phosphorylase. *Biochim. Biophys. Acta*, **496,** 448–457.

111. Fain, J. N. (1978). *Hormones, Membranes and Cyclic Nucleotides*, Vol. 6A of the series *Receptors and Recognition*. Edited by P. Cuatrecasas and M. Greaves, Chapman and Hall, London, pp. 1–61.

112. Hems, D. A., and Whitton, P. D. (1980). Control of hepatic glycogenolysis. *Physiol. Rev.*, **60,** 1–50.

113. Exton, J. H. (1979). Mechanisms involved in effects of catecholamines on liver carbohydrate metabolism. *Biochem. Pharmacol.*, **28,** 2237–2240.

114. Exton, J. H. (1979). Mechanisms involved in alpha-adrenergic effects of catecholamines on liver metabolism. *J. Cyclic Nucl. Res.*, **5,** 277–287.

115. Assimacopoulos-Jeannet, F. D., Blackmore, P. F., and Exton, J. H. (1977). Studies on alpha-adrenergic activation of hepatic glucose output. *J. Biol. Chem.*, **252,** 2662–2669.

116. Chen, J.-L. J., Babcock, D. F., and Lardy, H. A. (1978). Norepinephrine, vasopressin, glucagon, and A-23187 induce efflux of calcium from an exchangeable pool in isolated rat hepatocytes. *Proc. Nat. Acad. Sci. U.S.A.*, **75,** 2234–2238.

117. Babcock, D. F., Chen, J.-L. J., Yip, B. P., and Lardy, H. A. (1979). Evidence

for mitochondrial localization of the hormone-responsive pool of Ca^{2+} in isolated hepatocytes. *J. Biol. Chem.*, **254**, 8117–8120.

118. Blackmore, P. F., Dehaye, J. P., and Exton, J. H. (1979). Studies on alpha-adrenergic activation of hepatic glucose output. *J. Biol. Chem.*, **254**, 6945–6950.

119. Murphy, E., Coll, K., Rich, T. L., and Williamson, J. R. (1980). Hormonal effects on calcium homeostasis in isolated hepatocytes. *J. Biol. Chem.*, **255**, 6600–6608.

120. Foden, S., and Randle, P. J. (1978). Calcium metabolism in rat hepatocytes. *Biochem. J.*, **170**, 615–625.

121. Althaus-Salzmann, M., Carafoli, E., and Jakob, A. (1980). Ca^{2+}, K^+ redistributions and alpha-adrenergic activation of glycogenolysis in perfused rat livers. *Eur. J. Biochem.*, **106**, 241–248.

122. Barritt, G. J., Parker, J. C., and Wadsworth, J. C. (1981). A kinetic analysis of the effects of adrenaline on calcium distribution in isolated rat liver parenchymal cells. *J. Physiol.*, **312**, 29–55.

123. Stubbs, M., Kirk, C. J., and Hems, D. A. (1976). Role of extracellular calcium in the action of vasopressin on hepatic glycogenolysis. *FEBS Letters*, **69**, 199–202.

124. Blackmore, P. F., Brumley, F. T., Marks, J. L., and Exton, J. H. (1978). Studies on alpha-adrenergic activation of hepatic glucose output. *J. Biol. Chem.*, **253**, 4851–4858.

124[A]. Chan, T. M., and Exton, J. H. (1977). Alpha-adrenergic-mediated accumulation of adenosine $3':5'$-monophosphate in calcium-depleted hepatocytes. *J. Biol. Chem.*, **252**, 8645–8651.

125. Billah, M. M., and Michell, R. H. (1979). Phosphatidylinositol metabolism in rat hepatocytes stimulated by glycogenolytic hormones. *Biochem. J.*, **182**, 661–668.

126. Lin, S., and Fain, J. N. (1981). Vasopressin and epinephrine stimulation of phosphatidylinositol breakdown in the plasma membrane of rat hepatocytes. *Life Sci.*, **18**, 1905–1912.

127. Kirk, C. J., Verrinder, T. R., and Hems, D. A. (1978). The influence of extracellular calcium concentration on the vasopressin-stimulated incorporation of inorganic phosphate into phosphatidylinositol in hepatocyte suspensions. *Biochem. Soc. Trans.*, **6**, 1031–1034.

128. Miach, P. J., Dausse, J. P., and Meyer P. (1978). Direct biochemical demonstration of two types of alpha-adrenoceptor in rat brain. *Nature*, **274**, 492–494.

129. U'Prichard, D. C., and Snyder, S. H. (1979). Distinct alpha-noradrenergic receptors differentiated by binding and physiological relationships. *Life Sci.*, **24**, 79–88.

130. Hoffman, B. B., De Lean, A., Wood, C. L., Schoken, D. D., and Lefkowitz, R. J. (1979). Alpha-adrenergic receptor subtypes: quantitative assessement by ligand binding. *Life Sci.*, **24**, 1739–1746.

131. Berthelsen, S., and Pettinger, W. A. (1977). A functional basis for classification of alpha-adrenergic receptors. *Life Sci.*, **21**, 595–606.

132. Wikberg, J. E. S. (1979). The pharmacological classification of adrenergic $alpha_1$ and $alpha_2$ receptors and their mechanisms of action. *Acta Physiol. Scandinavia Supplementum*, **468**, 35–47.

133. Fain, J. N., and Garcia-Sainz, J. A. (1980). Role of phosphatidylinositol turnover in $alpha_1$ and of adenylate cyclase inhibition in $alpha_2$ effects of catecholamines. *Life Sci.*, **26**, 1183–1194.

134. Jones, L. M., and Michell, R. H. (1978). Stimulus-response coupling at alpha-adrenergic receptors. *Biochem. Soc. Trans.*, **6,** 672–688.
135. Sabol, S. L., and Nirenberg, M. (1979). Regulation of adenylate cyclase of neuroblastoma × glioma hybrid cells by alpha-adrenergic receptors. *J. Biol. Chem.*, **254,** 1913–1920.
136. Kneer, N. M., Wagner, M. J., and Lardy, H. A. (1979). Regulation by calcium of hormonal effects on gluconeogenesis. *J. Biol. Chem.*, **254,** 12160–12168.
137. Aggerbeck, M., Guellaen, G., and Hanoune, J. (1980). Adrenergic receptor of the $alpha_1$-subtype mediates the activation of the glycogen phosphorylase in normal rat liver. *Biochem. Pharmacol.*, **29,** 643–645.
138. Hoffman, B. B., Michell, T., Kilpatrick, M. D., Lefkowitz, R. J., Tolbert, M. E. M., Gilman, H., and Fain, J. N. (1980). Agonist versus antagonist binding to alpha-adrenergic receptors. *Proc. Nat. Acad. Sci. U.S.A.*, **77,** 4569–4573.
139. Hirata, F., Strittmatter, W. J., and Axelrod, J. (1979). Beta-adrenergic receptor agonists increase phospholipid methylation, membrane fluidity, and beta-adrenergic receptor-adenylate cyclase coupling. *Proc. Nat. Acad. Sci. U.S.A.*, **76,** 368–372.
140. Schneider, W. J., and Vance, D. E. (1979). Conversion of phosphatidylethanolamine to phosphatidylcholine in rat liver. *J. Biol. Chem.*, **254,** 3886–3891.
141. Bremer, J., and Greenberg, D. M. (1961). Methyl transferring enzyme system of microsomes in the biosynthesis of lecithin (phosphatidylcholine). *Biochim. Biophys. Acta*, **46,** 205–216.
142. Klevan, J., and Fain, J. N. (1981). Inhibition of phospholipid methylation in rat hepatocytes by catecholamines. Unpublished.
143. Strittmatter, W. J., Hirata, F., Axelrod, J., Mallorga, P., Tallman, J. F., and Henneberry, R. C. (1979). Benzodiazepine and beta-adrenergic receptor ligands independently stimulate phospholipid methylation. *Nature*, **282,** 857–859.
144. Stein, J. M., and Hales, C. N. (1972). Effect of adrenaline on ^{32}P incorporation into rat fat cell phospholipids. *Biochem. J.*, **128,** 531–541.
145. Perry, M. C., and Hales, C. N. (1970). Factors affecting the permeability of isolated fat cells from the rat to [^{42}K]potassium and [^{36}Cl] chloride ions. *Biochem. J.*, **117,** 615–621.
146. Fain, J. N., and Butcher, F. R. (1976). Cyclic guanosine 3′:5′-monophosphate and the regulation of lipolysis in rat fat cells. *J. Cycl. Nucl. Res.*, **2,** 71–78.
147. Hales, C. N., Campbell, A. K., Luzio, J. P., Siddle, K. (1977). Calcium as mediator of hormone action. *Biochem. Soc. Trans.*, **5,** 866–872.
148. Garcia-Sainz, J. A., Hoffman, B. B., Li, S.-Y., Lefkowitz, R. J., and Fain, J. N. (1980). Role of $alpha_1$ adrenoceptors in the turnover of phosphatidylinositol and of $alpha_2$ adrenoceptors in the regulation of cyclic AMP accumulator in hamster adipocytes. *Life Sci.*, **27,** 953–961.
149. Burns, T. W., Langley, P. E., Terry, B. E., Bylund, D. B., Hoffman, B. B., Tharp, M. D., Lefkowitz, R. J., Garcia-Sainz, J. A., and Fain, J. N. (1981). Pharmacological characterization of adrenergic receptors in human adipocytes. *J. Clin. Investig.*, **67,** 467–475.
150. Lawrence, J. C., Jr., and Larner, J. (1977). Evidence for alpha-adrenergic activation of phosphorylase and inactivation of glycogen synthase in rat adipocytes. *Mol. Pharmac.*, **13,** 1060–1075.
151. Lawrence, J. C., Jr., and Larner, J. (1978). Effects of insulin, methoxamine, and calcium on glycogen synthase in rat adipocytes. *Mol. Pharmacol.*, **14,** 1079–1091.

152. Garcia-Sainz, J. A., and Fain, J. N. (1980). Effect of adrenergic amines on phosphatidylinositol labelling and glycogen synthase activity in fat cells from euthyroid and hypothyroid rats. *Mol. Pharmacol.*, **18,** 72–77.

153. McDonald, J. M., Bruns, D. E., and Jarett, L. (1976). Characterization of calcium binding to adipocyte plasma membranes. *J. Biol. Chem.*, **251,** 5345–5351.

154. McDonald, J. M., Bruns, D. E., and Jarett, L. (1976). Ability of insulin to increase calcium binding to adipocyte plasma membranes. *Proc. Nat. Acad. Sci. U.S.A.*, **73,** 1542–1546.

155. McDonald, J. M., Bruns, D. E., and Jarett, L. (1978). Ability of insulin to increase calcium uptake by adipocyte endoplasmic reticulum. *J. Biol. Chem.*, **253,** 3504–3508.

156. Lewis, G. P., Piper, P. J., and Vigo, C. (1979). The effects of glucocorticoids on the distribution and mobilisation of arachidonic acid in fat cell ghosts. *Brit. J. Pharmacol.*, **67,** 393–400.

157. Christ, E. J., and Nugteren, D. H. (1970). The biosynthesis and possible function of prostaglandins in adipose tissue. *Biochim. Biophys. Acta*, **218,** 296–307.

158. Hales, C. N., Luzio, J. P., Chandler, J. A., and Herman, L. (1974). Localization of calcium in the smooth endoplasmic reticulum of rat isolated fat cells. *J. Cell Sci.*, **15,** 1–15.

159. Lapetina, E. G., and Michell, R. H. (1973). A membrane-bound activity catalysing phosphatidylinositol breakdown to 1,2-diacylglycerol, D-myoinositol 1:2-cyclic phosphate and D-myoinositol 1-phosphate. *Biochem. J.*, **131,** 433–442.

160. Irvine, R. F., and Dawson, R. M. C. (1978). The distribution of calcium-dependent phosphatidylinositol-specific phosphodiesterase in rat brain. *J. Neurochem.*, **31,** 1427–1434.

161. Majumder, A. L., and Eisenberg, F., Jr. (1974). The formation of cyclic inositol, 1,2-monophosphate, inositol 1-phosphate, and glucose 6-phosphate by brain preparations stimulated with deoxycholate and calcium: a gas chromatographic study. *Biochem. Biophys. Res. Commun.*, **60,** 133–139.

162. Lapetina, E. G., Grosman, M., and Canessa de Scarnati, O. (1976). Phosphatidylinositol-cleaving activity in smooth muscle from rat vas deferens. *Int. J. Biochem.*, **7,** 507–513.

163. Mollay, C., Kreil, G., and Berger, H. (1976). Action of phospholipases on the cytoplasmic membrane of *escherichia coli. Biochim. Biophys. Acta*, **426,** 317–324.

164. Yunes, R., Goldhammer, A. R., Garner, W. K., and Cordes, E. H. (1977). Phospholipases: Melittin facilitation of bee venom phospholipase A₂-catalysed hydrolysis of unsonicated lecithin liposomes. *Arch. Biochem. Biophys.*, **183,** 105–112.

165. Berridge, M. J., and Patel, N. G. (1968). Insect salivary glands: Stimulation of fluid secretion by 5-hydroxytryptamine and adenosine-3′,5;-monophosphate. *Science*, **162,** 462–463.

166. Berridge, M. J. (1970). The role of 5-hydroxytryptamine and cyclic AMP in the control of fluid secretion by isolated salivary glands. *J. Exp. Biol.*, **53,** 171–186.

167. Rasmussen, H. (1970). Cell communication, calcium ion and cyclic adenosine monophosphate. *Science*, **170,** 404–412.

168. Prince, W. T., Berridge, M. J., and Rasmussen, H. (1972). Role of calcium and

adenosine-3′:5′-cyclic monophosphate in controlling fly salivary gland secretion. *Proc. Nat. Acad. Sci. U.S.A.*, **69**, 553–557.

169. Hansen-Bay, C. M. (1978). The control of enzyme secretion from fly salivary glands. *J. Physiol.*, **274**, 421–435.

170. Berridge, M. J., and Lipke, H. (1979). Changes in calcium transport across calliphora salivary glands induced by 5-hydroxytryptamine and cyclic nucleotides. *J. Exp. Biol.*, **78**, 137–148.

171. Berridge, M. J., and Fain, J. N. (1979). Phosphatidylinositol metabolism and calcium gating, Proceedings of the VIth International Symposium on Medicinal Chemistry, edited by A. Simkin, Cotswold Press, Oxford, p. 117–127.

172. Fain, J. N., and Berridge, M. J. (1979). Relationship between phosphatidylinositol synthesis and recovery of 5-hydroxytryptamine-responsive Ca^{2+} flux in blowfly salivary glands. *Biochem. J.*, **180**, 655–661.

173. Peroutka, S. J., and Snyder, S. H. (1979). Multiple serotonin receptors: Differential binding of [³H]5-hydroxytryptamine, [³H]lysergic acid diethylamide and [³H]spiroperiodol. *Mol. Pharmacol.*, **16**, 687–699.

174. Salmon, D. M., and Honeyman, T. W. (1980). Proposed mechanism of cholinergic action in smooth muscle. *Nature*, **284**, 344–345.

175. Litosch, I., and Fain, J. N. (1981). Acceleration of phosphatidylinositol breakdown in a cell-free system from blowfly salivary glands. *Fed. Proceed.*, **40**, 366.

176. Hawthorne, J. N., and Pickard, M. R. (1979). Phospholipids in synaptic function. *J. Neurochem.*, **32**, 5–14.

12 Receptor modulation of calcium and cyclic AMP: differences in coupling to specific hormone-receptor complexes in cells responsive to several hormones

J. D. Gardner and **R. T. Jensen**

1 INTRODUCTION

Of the many secretory processes that occur in mammalian cells, the process that results in the secretion of digestive enzymes from pancreatic acinar cells is among the best characterized.[1,2] On the other hand, although the morphologic and biochemical events that characterize the synthesis, packaging and secretion of digestive enzymes are reasonably well established, the mechanisms by which various secretagogues *stimulate* the secretory process have only begun to be elucidated during the past 5–10 years. These studies of the biochemical basis of action of pancreatic secretagogues have shown that acinar cells possess two functionally distinct mechanisms by which secretagogues can increase enzyme secretion (Figure 1 and for review see References 3, 4, 5, 6, 7). One mechanism involves binding of the secretagogue to its receptors, mobilization of cellular calcium and, after a series of presently undefined steps, stimulation of enzyme secretion. Those secretagogues that cause release of cellular calcium also increase cellular cyclic GMP; however, cyclic GMP does not mediate the actions of these secretagogues on enzyme secretion.[6] The other mechanism involves binding of the secretagogue to its receptors, activation of adenylate cyclase, increased cellular cyclic AMP, activation of cyclic AMP-dependent protein kinase and, after a series of undefined steps, stimulation of enzyme secretion. The initial steps in these two pathways are functionally distinct.

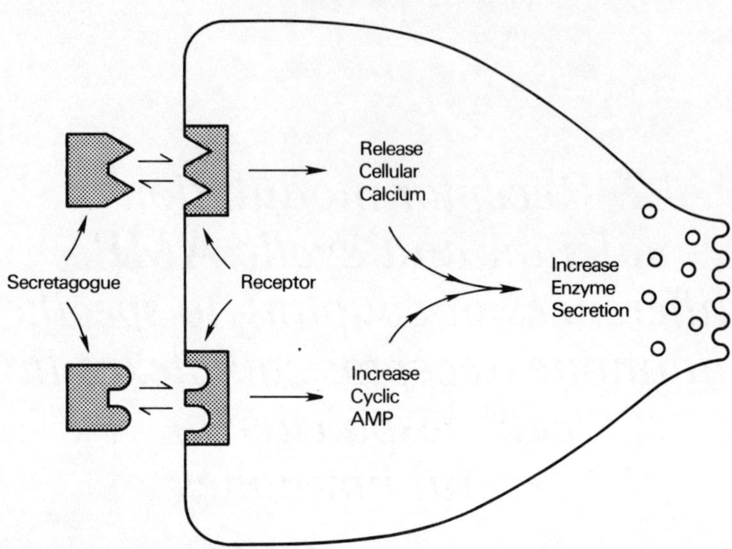

Figure 1. Mechanisms of action of secretagogues on acinar cells from guinea pig pancreas. Secretagogues by interacting with their membrane receptors on pancreatic acinar cells can cause one of two functionally distinct sequences of changes that produce stimulation of enzyme secretion. One sequence is characterized by mobilization and release of cellular calcium; the other involves activation of adenylate cyclase and increased cellular cyclic AMP. Although the two sequences are initially separate, they converge at some presently unidentified step that is distal to the mobilization of calcium and to the generation of cyclic AMP. The nature of this convergence is such that potentiation of enzyme secretion occurs when acinar cells are incubated with a secretagogue that causes release of cellular calcium plus a secretagogue that increases cellular cyclic AMP

Secretagogues that cause release of cellular calcium do not increase cyclic AMP and do not alter the increase in cyclic AMP caused by other secretagogues. Similarly, secretagogues that increase cellular cyclic AMP do not alter calcium transport and do not alter the increase in calcium outflux caused by other secretagogues.

2 RECEPTORS FOR SECRETAGOGUES ON PANCREATIC ACINAR CELLS

The initial step in the mechanism of action of secretagogues on pancreatic enzyme secretion is reversible binding of the secretagogue to receptors located on the outer surface of the plasma membrane of the pancreatic acinar cells. Binding of the secretagogue to its receptor is both necessary and sufficient to initiate the sequence of events that ultimately causes stimulation of pancreatic enzyme secretion.

With the development of techniques for preparing radiolabelled secretagogues of high specific activity it has become possible to measure directly the interaction of secretagogues with their receptors and to explore the quantitative relation between the number of receptors occupied and the corresponding changes in cellular function. In general, binding of the radiolabelled secretagogue is reversible, temperature-dependent and to a finite number of sites located on the plasma membrane. Because there are a finite number of receptors, one can monitor the interaction of a particular agent with the receptors by measuring that agent's ability to compete with the radiolabelled secretagogue for occupation of the binding sites and by so doing inhibit binding of radioactivity.[8,4,5] The apparent affinity of a secretagogue for its receptors is reflected in the range of concentrations over which it inhibits binding of the radiolabelled ligand—the higher the affinity of the secretagogue for the receptor, the lower the secretagogue concentration required to inhibit binding of radioactivity. From the standpoint of classifying different pancreatic secretagogues in terms of the receptors with which they interact, those secretagogues that inhibit binding of a given radiolabelled ligand interact with the same class of receptors, whereas those secretagogues that do not inhibit binding of the ligand do not interact with the same receptors as the ligand. The availability of a radiolabelled ligand, however, is not essential for distinguishing different classes of receptors (for detailed discussion see Reference 4). For example, competitive antagonists can be used to distinguish different classes of receptors in that those secretagogues whose actions can be inhibited competitively by a particular antagonist, interact with the same class of receptors, whereas those secretagogues whose actions are not inhibited by the antagonist, interact with other receptors. The classification of receptors that mediate the actions of adrenergic agents (α and β), cholinergic agents (muscarinic and nicotinic) and histamine (H_1 and H_2) developed primarily from studies using selective, competitive antagonists.

By comparing the ability of a secretagogue to inhibit binding of a radiolabelled ligand with the accompanying secretagogue-induced changes in acinar cell function one can investigate the relation between receptor occupation and the secretagogue-induced response. In some instances, there is a one-to-one correlation between receptor occupation and secretagogue-induced changes in cell function. In other instances, however, there are 'spare receptors' in that occupation of only a fraction of the receptors is sufficient to produce a maximal secretagogue-induced change in function and occupation of the remaining receptors by the secretagogue is not accompanied by a secretagogue-induced change in cell function.[8,4]

2.1 Receptors for secretagogues that cause mobilization of cellular calcium

2.1.1 *Receptors for cholecystokinin (CCK) and structurally related peptides*

CCK is a peptide that was originally isolated and purified from hog upper small intestine and found to contain 33 amino acid residues.[7,9] In the course of investigating the structure of CCK, Mutt and Jorpes[7] also detected a peptide that differed from CCK in that although it was retained by a CM-cellulose column, it was eluted by 0.2 M but not by 0.02 M ammomium bicarbonate. This peptide, which has been termed 'CCK-variant', possesses 39 amino acids. The C-terminal 33 amino acid residues are identical to those of CCK and the N-terminal hexapeptide sequence is Try–Ile–Gln–Gln–Ala–Arg.[10] Subsequent studies of the structure of CCK showed that the C-terminal heptapeptide–amide possessed all of the biological activity of the native peptide and that the C-terminal octapeptide had the same efficacy as CCK but was approximately 10-times more potent than CCK (for example, see References 11 and 12). In CCK, as in all of the peptides that can increase pancreatic enzyme secretion, the C-terminal amino acid residue is amidated (Figures 2–5). There are two other naturally occurring peptides that are structurally similar to CCK and that can increase pancreatic enzyme secretion by causing release of cellular calcium: gastrin and caerulein (Figure 2). Gastrin occurs naturally in several different chemical forms[13] and shares a common C-terminal pentapeptide–amide sequencewith CCK. Gastrin, like CCK, also possesses a sulfated tyrosine residue; however, in CCK the sulfated tyrosine is the seventh residue from the C-terminus, whereas in gastrin the sulfated tyrosine is the sixth residue from the C-terminus (Figure 2). Caerulein is a decapeptide originally isolated from the skin of the

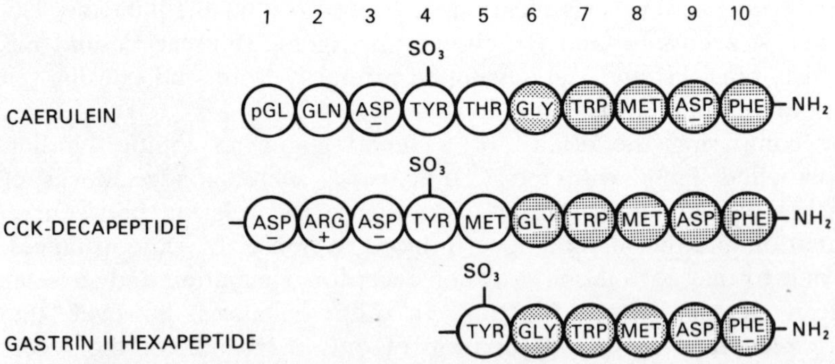

Figure 2. Amino acid sequences for caerulein, the C-terminal decapeptide of CCK and the C-terminal hexapeptide of gastrin II. Shaded residues indicate identical corresponding amino acids in each of the three peptides

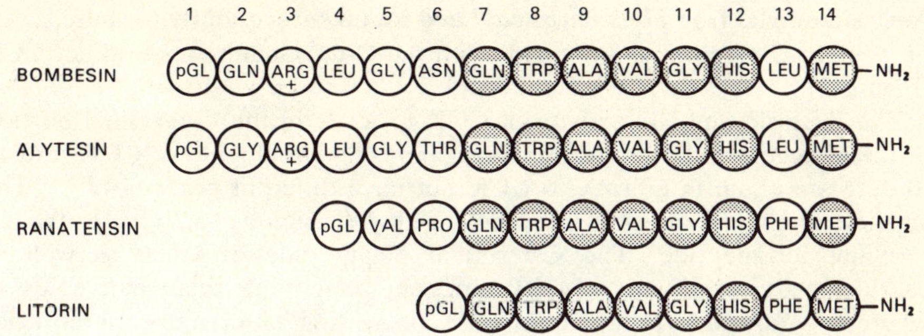

Figure 3. Amino acid sequences for bombesin, alytesin, ranatensin and litorin. Shaded residues indicate identical corresponding amino acids in each of the four peptides

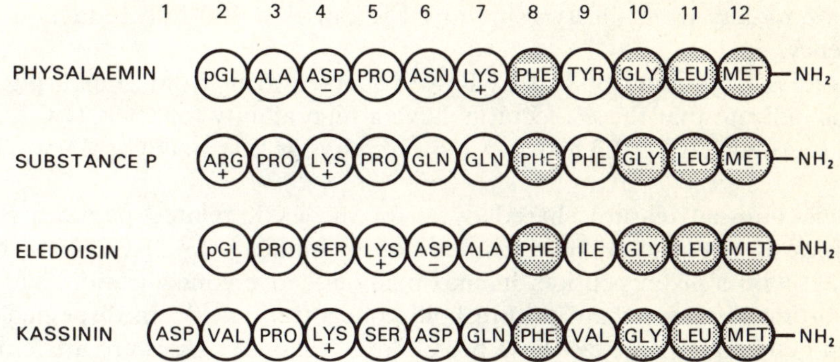

Figure 4. Amino acid sequences for physalaemin, substance P, eledoisin and kassinin. Shaded residues indicate identical corresponding amino acids in each of the four peptides

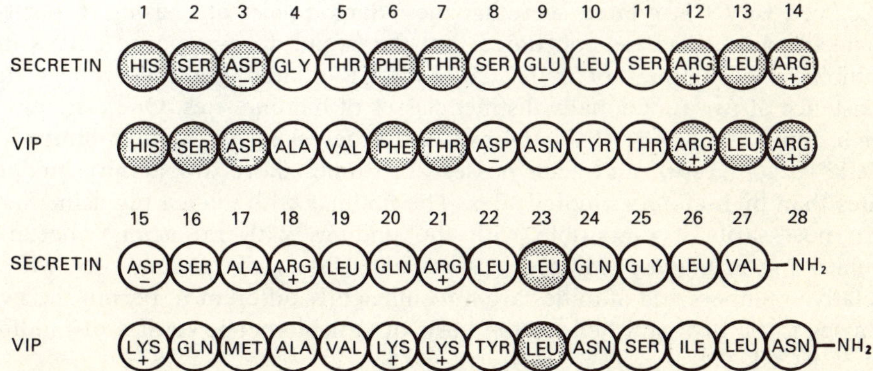

Figure 5. Amino acid sequences for secretin and VIP. Shaded residues indicate identical corresponding amino acids in both peptides

Australian hylid frog *Hyla caerulea*[14] and found subsequently to share seven of its eight C-terminal amino acids with the C-terminal octapeptide of CCK (Figure 2).

The intrinsic biologic activity of CCK residues in the C-terminal portion of the molecule and all C-terminal fragments and analogues of CCK studied to date are equal in efficacy to CCK but have different potencies.[4,5,11] The smallest fragment of CCK that possesses full biologic activity is the C-terminal tetrapeptide. The C-terminal octapeptide of CCK as well as caerulein are approximately 10-times more potent than native CCK, whereas C-terminal fragments of CCK possessing 12 or more amino acids (including CCK variant) have the same potencies as does CCK. Gastrin and the N-terminal hexapeptide of CCK are approximately 1000-times less potent than CCK. Although removing the sulfate moiety from the tyrosine in gastrin causes little or no change in its potency or efficacy, removing the sulfate moiety from the tyrosine in CCK causes a 500-fold reduction in its potency.

Studies of binding of ^{125}I-CCK to dispersed acini from guinea pig pancreas indicate that these receptors have a high affinity for CCK (EC_{50} 2 nM) and caerulein (EC_{50} 0.3 nM) and a low affinity for gastrin (EC_{50} 2 μM) (Table 1).

One unusual feature shared by all of the CCK-related peptides is that their dose-response curves for inhibition of binding of ^{125}I-CCK are broad in that with a given peptide, its maximally effective concentration is at least 1000-times greater than its threshold concentration.[11,15] In dispersed acini from guinea pig pancreas these broad dose-response curves are attributable, at least in part, to the abilities of CCK-related peptides to accelerate the rate of dissociation of receptor-bound ^{125}I-CCK.[11] That is, acini from guinea pig pancreas possess at least one class of interacting receptors whose affinities appear to be influenced by the extent to which these receptors are occupied by CCK-related secretagogues. Sankaran *et al.*[15] using dispersed acini from rat pancreas concluded that the broad dose-response curves for inhibition of binding of ^{125}I-CCK by CCK-related peptides reflect the existence of two functionally distinct classes of binding sites. One class has a high affinity for CCK (EC_{50} 64 pM), the other class has a low affinity for CCK (EC_{50} 21 nM), and acini possess 160-times more low-affinity binding sites than high-affinity binding sites. The findings with guinea pig acini[11] are not necessarily incompatible with the findings with rat acini,[15] because guinea pig acini may, in fact, possess two classes of binding sites whose relative numbers and affinities are not sufficiently different to permit the two classes to be distinguished on the basis of stoichiometric studies of binding of ^{125}I-CCK.

Studies comparing the abilities of various C-terminal fragments and analogues of CCK to inhibit binding of ^{125}I-CCK with their abilities to cause

Table 1. Secretagogue receptors on acinar cells from guinea pig pancreas: agents that mobilize cellular calcium

Class of receptors	Ligand used for binding studies	Number of receptors per acinar cell	Naturally occurring agonist (EC_{50})	Competitive antagonists (EC_{50})
CCK	^{125}I-CCK	9000	Caerulein (0.2 nM) CCK (2 nM) Gastrin (2 μM)	Bt$_2$cGMP (0.1 mM)
Bombesin	^{125}I-[Tyr^4]Bombesin	5000	Bombesin (4 nM) Litorin (40 nM) Ranatensin (12 nM) Alytesin (12 nM)	None known
Physalaemin	^{125}I-Physalaemin	500	Physalaemin (2 nM) Substance P (5 nM) Eledoisin (100 nM) Kassinin (250 nM)	None known
Cholinergic	[^{3}H]QNB?	Not known	Muscarinic cholinergic agonists	Muscarinic cholinergic antagonists

EC_{50} refers to the concentration required to occupy fifty-per cent of the receptors.
Bt$_2$cGMP, dibutyryl cyclic GMP.
QNB, quinuclidinyl benzilate.
CCK, cholecystokinin.

changes in acinar cell function have shown that the region of CCK molecule that possesses intrinsic biologic activity also determines the affinity of the peptide for the CCK receptors.[4,5,11] To date, no peptide has been found that will occupy the CCK-receptors and cause anything other than a full biological response; therefore, the differences in the potencies with which various analogues and fragments of CCK alter acinar cell function are attributable entirely to differences in their affinities for the CCK receptor.[11]

In general, the structural requirements for occupation of a peptide hormone receptor are quite stringent in that only other peptides will bind to the hormone receptor and the peptides that do bind are usually fragments of the hormone or analogues with a similar chemical structure. One striking exception to this general pattern occurs with the opiate receptors.[16,17] These receptors can interact not only with peptides (the enkephalins and other opioid peptides) but also with morphine and structurally related alkaloid compounds. A second apparent exception to the general principle that only peptides will bind to peptide hormone receptors is the finding that butyryl derivatives of cyclic GMP can competitively inhibit the interaction of CCK and structurally related peptides with their receptors on pancreatic acinar cells.[11,18] With dibutyryl cyclic GMP as well as with each of two different monobutyryl derivatives, the dose–response curves for inhibition of CCK-stimulated enzyme secretion are the same as the corresponding dose–response curves for inhibition of binding of ^{125}I-CCK. Because native cyclic GMP does not inhibit binding of ^{125}I-CCK[11,18] or the effects of CCK on acinar cell function, the presence of at least one butyryl moiety appears to be essential for cyclic GMP to inhibit interaction of CCK with its receptors.[11,18] In inhibiting the actions of CCK, N^2-monobutyryl cyclic GMP is less potent than the $O^{2'}$-monobutyryl derivative which, in term, is less potent than the dibutyryl derivative.[11,18]

2.1.2 *Receptors for bombesin and structurally related peptides*

Another group of naturally occurring peptides that can stimulate pancreatic enzyme secretion by causing release of cellular calcium comprises bombesin, alytesin, ranatensin and litorin (Table 1, Figure 3). These peptides, like caerulein, were originally isolated from the skins of various frogs and were named after the particular class of frog from which they were isolated.[19] For example, bombesin was isolated from the skin of *Bombina bombina*.[20] Like CCK and structurally related peptides, the bombesin-related peptides possess a C-terminal amide; however, unlike CCK, which possesses a C-terminal phenylalanine-amide, each of the bombesin-related peptides has a C-terminal methionine-amide (Figure 3). In bombesin, alytesin, ranatesin and litorin, seven of their eight C-terminal amino acids are identical. In bombesin and alytesin the penultimate C-terminal residue is leucine, whereas in ranatesin and litorin it is phenylalanine (Figure 3). These peptides probably have structural counterparts in mammals because bombesin-like immunoreactivity has been detected in mammalian tissues[18,21] and a heptacosapeptide with gastrin-releasing action has recently been isolated from procine intestinal mucosa and found to be structurally similar to bombesin.[22]

A radiolabelled analogue of bombesin, ^{125}I-[Tyr^4]bombesin, can be used to identify binding sites that interact with bombesin and structurally related peptides but not with other pancreatic secretagogues.[23] There is a good correlation between the concentrations at which bombesin and related peptides inhibit binding of ^{125}I-[Tyr^4]bombesin and the concentrations at which these same secretagogues produce changes in acinar cell function.[19] Each pancreatic acinar cell has approximately 5000 receptors for bombesin and structurally related peptides, and these receptors have a higher affinity for bombesin (EC$_{50}$ 4 nM) than for ranatensin (EC$_{50}$ 12 nM), alytesin (EC$_{50}$ 12 nM) or litorin (EC$_{50}$ 40 nM) (Table 1).

As is the case with CCK, the intrinsic biologic activity of bombesin and structurally related peptides is a property of the C-terminal portion of the molecule, and the C-terminal non-apeptide of bombesin has the same potency and efficacy as does the native tetradecapeptide.[18] Shorter C-terminal fragments of bombesin still possess full intrinsic biologic activity, but their potencies are less than that of native bombesin. The region of the bombesin molecule that possesses biologic activity also determines the affinity of the peptide for its receptors. That is, the variations in potencies among various fragments and analogues of the bombesin reflect variations in their receptor affinities (Table 1). Of the fragments and analogues of bombesin that have been tested, none has been found to occupy the receptor and not cause a full biologic response. Thus, as is also true of CCK, there are no known peptides that will competitively antagonize the actions of bombesin and structurally related secretagogues.

2.1.3 *Receptors for physalaemin and structurally related peptides*

A third group of naturally occurring peptides that can stimulate pancreatic enzyme secretion by causing release of cellular calcium includes physalaemin, substance P, eledoisin and kassinin. Physalaemin and kassinin, like caerulein and the bombesin-related peptides, were originally isolated from frog skin.[24,25] Eledoisin was originally isolated from the posterior salivary gland of a Mediterranean octopod, *Eledone moschata*.[18] Substance P was originally isolated from mammalian gastrointestinal mucosa and brain.[26] Thus, physalaemin, eledoisin and kassinin probably represent phylogenetic precursors of substance P. The physalaemin-related peptides, like the bombesin-related peptides, have a C-terminal methionine-amide

(Figure 4). In the physalaemin-related peptides, four of the five C-terminal amino acids are identical and each peptide has a hydrophobic residue in the fourth position from the C-terminus (Figure 4). Radioiodinated physalaemin, ^{125}I-physalaemin, can be used to identify a class of binding sites that interact with physalaemin and structurally related peptides but not with other pancreatic secretagogues.[27] There is a good correlation between the concentrations at which physalaemin and structurally related peptides inhibit binding of ^{125}I-physalaemin and the concentrations at which these same secretagogues produce changes in acinar cell function. Each pancreatic acinar cell has approximately 500 receptors for physalaemin and these receptors have a relatively high affinity for physalaemin (EC_{50} 2 nM) and for substance P (EC_{50} 5 nM) and a relatively low affinity for eledoisin (EC_{50} 100 nM) and kassinin (EC_{50} 250 nM) (Table 1).

As occurs with CCK-related peptides and with bombesin-related peptides, the intrinsic biologic activity of physalaemin and structurally related peptides is a property of the C-terminal portion of the molecule. The smallest C-terminal fragment of physalaemin that has been examined and found to retain biologic activity is the pentapeptide. Eledoisin is approximately 50 per cent more effective than physalaemin or substance P in stimulating pancreatic enzyme secretion.[27,28] Physalaemin and eledoisin have also been found to have different efficacies for producing hypotension *in vivo* and contraction of large intestine, small intestine and rat uterus *in vitro*.[29,30,31] These findings suggest that with physalaemin and related peptides, the portion of the molecule that possesses intrinsic biologic activity is not congruent with the region of the molecule that influences the affinity of the peptide for its receptor. The findings also raise the possibility that a peptide exists or can be synthesized which will be able to occupy the physalaemin receptor but which will be devoid of agonist activity; and therefore, can function as a competitive antagonist of the action of physalaemin-related peptides.

2.1.4 *Receptors for muscarinic cholinergic agents*

Pancreatic acinar cells possess receptors that interact with muscarinic cholinergic agents and cause mobilization of cellular calcium as well as stimulation of enzyme secretion (Table 1). To date, however, this class of receptors has been identified only by the abilities of atropine and other muscarinic antagonists to competitively inhibit the actions of acetylcholine and other muscarinic agonists. Tritiated quinuclidinyl benzilate, [³H]QNB, as well as the iodinated hydroxy derivative of quinuclidinyl benzilate, ^{125}I-OH-QNB, have been used with broken cell preparations from several different tissues,[32,33,34] including pancreas,[35,36] to detect binding sites that have the properties of muscarinic cholinergic receptors. Because no corresponding agonist-induced functional change has been measured in these broken cell preparations, the conclusion that these binding sites are, in fact, muscarinic cholinergic receptors has been based on functional changes in intact cell preparations incubated under different conditions.[35,36] Attempts to measure binding of radiolabelled QNB to muscarinic cholinergic receptors in dispersed intact cells or in tissue fragments have been unsuccessful primarily because the relatively high lipid solubility of QNB causes the radiolabelled compound to partition into the plasma membrane and by so doing masks detection of its binding to muscarinic cholinergic receptors.

2.2 Receptors for secretagogues that increase cyclic AMP (Table 2)

2.2.1 *Receptors for cholera toxin*

In acinar cells from guinea pig pancreas cholera toxin increases enzyme secretion by virtue of its ability to activate adenylate cyclase and increase celgular cyclic AMP.[37] Cholera toxin is a protein with a molecular weight of about 83,000 and is composed of 3 subunits. (For a comprehensive review of the structure and mode of action of cholera toxin see Reference 38.) The 'B subunit' of cholera toxin, also referred to as 'choleragenoid' or 'aggregated B subunit', has a molecular weight of approximately 54,000 and is thought to be composed of 4–6 identical monomers. The 'A_1 subunit' has a molecular weight of 23,000 and the 'A_2 subunit' has a molecular weight of 6000. A simplified view of the functions of the different portions of the cholera toxin molecule is that the B subunit, which has no intrinsic biologic activity, binds to monosialogangliosides on the outer surface of the plasma membrane. The A_2 subunit, in some unknown way, facilitates the passage of the A_1 subunit across the membrane to its inner surface where the A_1 subunit causes activation of adenylate cyclase.

[125]I-labelled cholera toxin can be used to identify binding sites on dispersed acini from guinea pig pancreas that interact with cholera toxin but not with other secretagogues.[37] There is a good correlation between the concentrations of cholera toxin required to inhibit binding of ^{125}I-toxin and the concentrations at which cholera toxin produces changes in acinar cell function.[37] As given in Table 2, each pancreatic acinar cell has approximately 21,000 receptors for cholera toxin and 1 nM toxin causes occupation of 50 per cent of these receptors. Choleragenoid, which unlike the native toxin is devoid of intrinsic efficacy, can also bind to the cholera toxin receptors on pancreatic acinar cells and by so doing function as a competitive antagonist of the action of cholera toxin (Table 2).

2.2.2 *Receptors for secretin and vasoactive intestinal peptide (VIP)*

Secretin is a heptacosapeptide[39] originally isolated from the upper small intestine and is the substance for which the term 'hormone' was coined[40,41]

Table 2. Secretagogue receptors on acinar cells from guinea pig pancreas: agents that increase cyclic AMP

Class of receptors	Ligand used for binding studies	Number of receptors per acinar cell	Naturally occurring agonists (EC_{50})	Competitive antagonist (EC_{50})
VIP-preferring	^{125}I-VIP	9,000	VIP (0.7 nM) Secretin (7 μM)	Secretin$_{5-27}$ (6 μM)
Secretin-preferring	^{125}I-VIP	135,000	Secretin (0.3 nM) VIP (80 nM)	Secretin$_{5-27}$ (0.2 μM)
Cholera toxin	^{125}I-Cholera toxin	21,000	Cholera toxin (3 nM)	Choleragenoid (10 nM)

EC_{50} refers to the concentration required to occupy fifty per cent of the receptors.
VIP, vasoactive intestinal peptide.

VIP is an octacosapeptide, also isolated from upper small intestine, which was named for its ability to produce a decrease in blood pressure following intravenous injection in dogs.[42] VIP and secretin have a similar spectrum of biologic activities that reflect the similarities in their amino acid sequences (Figure 5 above). In both secretin and VIP, as in CCK, bombesin and physalaemin, the C-terminal amino acid is amidated. VIP and secretin share 9 amino acid identities. Eight of these identities are in the N-terminal region; one is in the C-terminal region. VIP and secretin also share 10 amino acids which, although they are not identical, confer a similar chemical atmosphere on the molecule. Of these 10 amino acid similarities, three are in the N-terminal region and seven are in the C-terminal region.

In contrast to those peptides that increase calcium release from acinar cells and have their intrinsic biologic activities in the C-terminal portion of the molecule, in VIP and secretin the intrinsic biologic activity resides in the N-terminal portion of the molecule.[43,44] For example, secretin$_{1-14}$ has an efficacy equal to that of native secretin, whereas secretin$_{5-27}$ and secretin$_{14-27}$ have efficacies that are less than 2 per cent of that of secretin.[43,44,45,46] Although C-terminal partial sequences of VIP and secretin have little or no intrinsic biologic activity, these peptides are able to interact with the receptors for VIP and secretin; and therefore, function as competitive antagonists.[43,44,45,46] Both the N-terminal and the C-terminal regions of VIP and secretin influence the affinity of the peptide for its receptors.[43,44,45,46] For example, the affinities of secretin$_{1-14}$ and secretin$_{14-27}$ for the secretin receptors are approximately 1000-fold less than the affinity of secretin for these same receptors. Finally, one feature of VIP and secretin that distinguishes them from the other peptides that stimulate pancreatic enzyme secretion is that, to date, no fragment of either peptide has been found to be as potent as the native molecule.

^{125}I-VIP can be used to detect binding sites that interact with VIP and secretin but not with other pancreatic secretagogues.[43] Acinar cells possess two classes of receptors each of which interact with VIP and secretin and each of which cause an increase in cyclic AMP.[43,44,45,46] As is summarized in Table 2, each acinar cell possesses approximately 9000 receptors that have a high affinity for VIP (EC$_{50}$ 0.7 nM) and a low affinity for secretin (EC$_{50}$ 7 μM), and approximately 135,000 receptors that have a high affinity for secretin (EC$_{50}$ 0.3 nM) and a low affinity for VIP (EC$_{50}$ 80 nM).

3 ROLES OF CALCIUM IN THE REGULATION OF PANCREATIC ENZYME SECRETION

3.1 Effects of pancreatic secretagogues on calcium transport

The actions of secretagogues such as CCK or cholinergic agents on pancreatic enzyme secretion were initially thought to result from their abilities to increase the influx of extracellular calcium into pancreatic acinar cells. (For review see References 3 and 5.) This conclusion was based primarily on the observation that stimulation of pancreatic enzyme secretion is reduced or abolished in calcium-free incubation solutions, and on the findings that in tissues such as adrenal medulla, posterior pituitary or nerve terminals, agents that cause secretion also increase calcium influx. However, measurements of ^{45}Ca transport in pancreatic fragments, dispersed acini or dispersed acinar cells showed that secretagogues such as CCK or cholinergic agents cause a significant increase in outflux of ^{45}Ca and that this increased outflux did not require extracellular calcium. To date, stimulation of ^{45}Ca outflux has been observed with muscarinic cholinergic agents as well as with CCK and structurally related peptides, bombesin and structurally related peptides, and physalaemin and structurally related peptides. Secretin, VIP and cholera toxin do not alter ^{45}Ca transport or the changes in ^{45}Ca transport caused by other secretagogues.[3,5]

Although most investigators have found that pancreatic secretagogues do not alter cellular uptake of ^{45}Ca, some[47,48,49] have found that CCK or carbachol can increase uptake of ^{45}Ca by pancreatic acinar cells. In contrast to the effects of CCK or carbachol on ^{45}Ca outflux, the stimulation of ^{45}Ca uptake caused by these secretagogues depends on the extracellular calcium concentration and can be detected only with calcium concentrations greater than 0.5 mM.[49] It remains to be determined whether the secretagogue-induced increase in ^{45}Ca uptake reflects an increase in the acinar cell membrane permeability to calcium[47,48] or an increase in the amount of cellular calcium that is available to exchange with the tracer.[49] Regardless of the basis for the secretagogue-induced increase in ^{45}Ca uptake or its significance to acinar cell function, most investigators believe that the initial stimulus to pancreatic enzyme secretion results from mobilization of cellular calcium.

3.2 Intracellular sites from which calcium is released by secretagogues

Although the cellular sites from which calcium is released by pancreatic secretagogues is not known, most studies do indicate that secretagogues cause release of cellular calcium that exists in a bound or sequestered state. Three types of experimental approaches have been used to explore the cellular sites from which calcium is released by various pancreatic secretagogues.

One approach has been to load pancreatic acinar cells with ^{45}Ca and then use subcellular fractionation and ultrastructural techniques to analyse the cellular distribution of radioactivity.[50,51] Although this approach has the major technical shortcoming that there is substantial redistribution of ^{45}Ca during the time required for subcellular fractionation, it has indicated that after loading acinar cells with ^{45}Ca, most of the cellular radioactivity is bound to smooth microsomes, plasma membranes, zymogen granule membranes and mitochondria.[50,51] Incubating the tissue with caerulein decreases the amount of ^{45}Ca associated with mitochondria but does not alter the amount of radioactivity associated with the other subcellular fractions.[50,51]

A second approach has been to load acinar cells with ^{45}Ca and then measure total cellular radioactivity as well as the ^{45}Ca that remains associated with cellular particulate material after the cells are lysed with a hypotonic solution and the particulate material collected rapidly by vacuum filtration.[52] Although this approach does not allow one to determine the subcellular location of the ^{45}Ca, the rapid processing minimizes the redistribution of bound radioactivity that can occur following disruption of the cells. In pancreatic acinar cells that have been loaded with ^{45}Ca, approximately 70 per cent of the total cellular radioactivity exists in a bound or sequestered state. Adding CCK causes a 40–50 per cent decrease in bound ^{45}Ca during the first 10 minutes of incubation after which bound ^{45}Ca remains decreased for the duration of the incubation. Thus, the CCK-induced decrease in total cellular ^{45}Ca results from loss of bound ^{45}Ca.[52]

A third experimental approach to defining the source from which cellular calcium is released by pancreatic secretagogues involves the use of chlorotetracycline.[53] This fluorescent compound forms complexes with membrane-associated calcium and magnesium, and thereby enables one to monitor changes in these ions by fluorometry and fluorescent microscopy. This technique has the advantage of allowing one to analyse the subcellular location of membrane-associated calcium without disrupting the cell and the disadvantage of not providing information about total cellular calcium. In pancreatic acinar cells that have been preloaded with chlorotetracycline, caerulein as well as bethanechol, a muscarinic cholinergic agonist, cause a rapid decrease in fluorescence within the first five minutes after adding the secretagogue. Analysis of the fluorescence spectrum before and after adding the secretagogue indicates that the decrease in fluorescence reflects a reduction in membrane-associated calcium. This secretagogue-induced decrease in fluorescence is not altered by removing extracellular calcium and appears to reflect loss of calcium from mitochondria.

3.3 Relation between release of cellular calcium and stimulation of enzyme secretion

The secretagogue-induced release of cellular calcium is thought to be one of, if not the initial step in the sequence of events that serves to couple the stimulus to secretion. Several types of experimental observations support this conclusion. As illustrated in Figure 6, there is a close correlation between the time-course for secretagogue-stimulated calcium outflux and

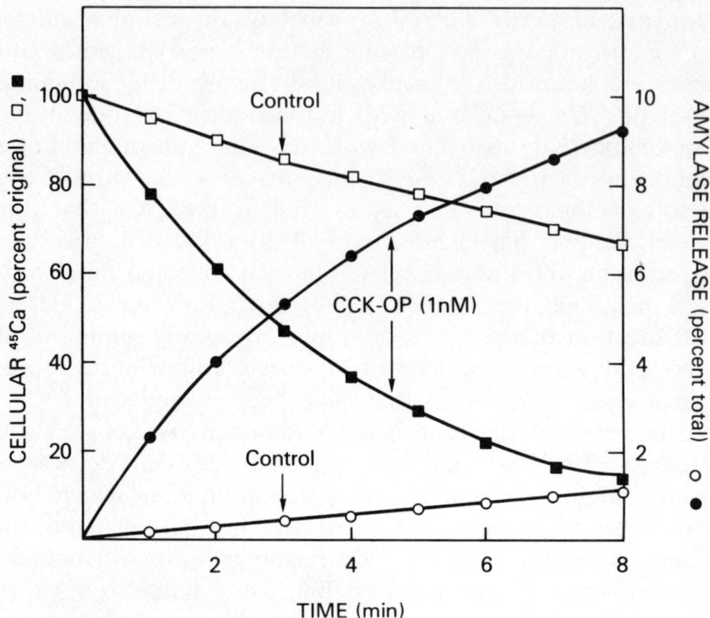

Figure 6. Effect of cholecystokinin on amylase release and calcium outflux in dispersed acini from guinea pig pancreas. Acini were preincubated with or without ^{45}Ca for 90 minutes at 37 °C, washed, resuspended in fresh incubation solution containing 0.5 mM calcium and reincubated for up to 8 minutes at 37 °C with or without 1 nM CCK-OP. Cellular ^{45}Ca, expressed as the per cent of the radioactivity originally present in the cells, and amylase secretion, expressed as per cent release, were determined at 1-minute intervals as described in Reference 23 and 53. CCK-OP, C-terminal octapeptide of cholescystokinin.

that for secretagogue-stimulated amylase release in that significant stimulation of both functions can be detected at the earliest times it is technically convenient to take samples. In the absence of extracellular calcium, the magnitude of secretagogue-stimulated enzyme secretion is not altered during the first 10 minutes of incubation. Thus, under conditions that preclude inward movement of extracellular calcium but still allow outflux of cellular calcium, the magnitude of secretagogue-stimulated enzyme secretion is the same as it is in the presence of extracellular calcium. There is a good correlation between the dose–response curves for secretagogue-stimulated calcium outflux and those for stimulation of enzyme secretion.[5] Furthermore, there is a good correlation between a secretagogue's efficacy for increasing calcium outflux and its efficacy for stimulating enzyme secretion. For example, physalaemin and structurally related peptides are less efficacious than cholinergic agents, CCK-related peptides and bombesin-related peptides in stimulating enzyme secretion[4,5] as well as in stimulating calcium outflux.[5,26]

The antibiotic A23187 promotes transfer of divalent cations across biological as well as artificial membranes[54,55] and stimulates pancreatic enzyme secretion (for review see Reference 5). This divalent cation ionophore, like cholinergic agents, CCK, bombesin and physalaemin, increases calcium outflux.[56,57] A23187 also increases influx of extracellular calcium into pancreatic acinar cells.[56,57] In the presence of extracellular calcium, the A23187-induced increase in calcium influx is greater than the increase in calcium outflux and the net effect is that the ionophore increases cellular uptake of calcium. As occurs with other pancreatic secretagogues, A23187 can stimulate enzyme secretion during the initial minutes of incubation of acinar cells in a calcium-free medium.[57]

Although for a given secretagogue the dose–response curve for stimulation of calcium outflux occurs over the same range of secretagogue concentrations as does the dose–response curve for enzyme secretion, with cholinergic agents, CCK or bombesin, the secretagogue concentration that produces half-maximal stimulation of amylase secretion is significantly lower than that required to produce half-maximal stimulation of calcium outflux.[19,25,11,21] These findings that maximal stimulation of enzyme secretion occurs with a secretagogue concentration that causes submaximal stimulation of calcium outflux suggest that these secretagogues may be able to cause more calcium to be released than is necessary for maximal stimulation of enzyme secretion. With physalaemin, which is less efficacious than carbachol, CCK or bombesin in stimulating enzyme secretion and in stimulating calcium outflux[5,25] the dose–response curve for stimulation of enzyme release is superimposed on that for stimulation of calcium outflux. This relation presumably occurs because physalaemin cannot cause sufficient release of cellular calcium to achieve maximal stimulation of the enzyme secretory process.

3.4 Other effects of secretagogue-induced calcium release

3.4.1 *Electrical changes in pancreatic acinar cells*

Those pancreatic secretagogues that cause calcium release also cause depolarization of pancreatic acinar cells.[58] This secretagogue-induced depolarization of acinar cells is accompanied by a reduction in the cell surface membrane resistance and by changes in the membrane conductance of sodium, potassium and chloride ions. One laboratory has reported that in rat acinar cells, acetylcholine and CCK cause hyperpolarization and increase surface membrane resistance (for discussion see Reference 58). The basis for this discrepancy is not clear; however, it is not due to a species difference because others using rat acinar cells have observed depolarization with acetylcholine and CCK.

The secretagogue-induced electrical changes in pancreatic acinar cells appear to result from the abilities of the secretagogues to cause release of cellular calcium. Muscarinic cholinergic agents, CCK, bombesin or A23187 cause depolarization and a decrease in the surface membrane resistance of pancreatic acinar cells, whereas secretin, a secretagogue that does not alter calcium transport in acinar cells, does not cause depolarization. The electrical changes caused by various secretagogues do not require extracellular calcium. Because intracellular injection of calcium also causes depolarization and decreased surface membrane resistance, the secretagogue-induced electrical changes probably result from an increased concentration of cytoplasmic calcium caused by secretagogue-induced mobilization of intracellular calcium stores. Thus, the secretagogue-induced electrical changes do not result from the release of cellular calcium *per se*, but instead, from the consequent increase in the cytoplasmic calcium concentration.

The secretagogue-induced depolarization of pancreatic acinar cells does not of itself stimulate enzyme secretion. High concentrations of extracellular potassium also cause depolarization of pancreatic acinar cells but do not cause stimulation of pancreatic enzyme secretion provided that atropine is added to block the action of acetylcholine released from nerve terminals.[1] In the presence of depolarizing concentrations of extracellular potassium, the increase in enzyme secretion caused by CCK is the same as it is with normal concentrations of extracellular potassium.[1] It should be noted that although depolarization of pancreatic acinar cells can be produced by high concentrations of extracellular potassium as well as by those pancreatic secretagogues that cause release of cellular calcium, only the secretagogue-induced depolarization is accompanied by a decrease in the surface membrane resistance.

Pancreatic acinar cells are electrically coupled because the junctional cell membranes have a specific resistance that is substantially lower than that of the surface cell membranes.[58] Furthermore, each acinar cell is electrically coupled to other cells within the same acinus but not to acinar cells in adjacent acini; therefore, from the electrical standpoint each acinus behaves as a single functional unit. At relatively high concentrations, those pancreatic secretagogues that cause depolarization can also cause electrical uncoupling of acinar cells.[58] Repeated injections of intracellular calcium, like high concentrations of secretagogues, also cause electrical uncoupling of pancreatic acinar cells.[58] These findings suggest that at high concentrations, secretagogues may be able to increase the concentration of cytoplasmic calcium to an extent sufficient to cause electrical uncoupling. The functional significance of electrical coupling of pancreatic acinar cells is not known; however, this coupling does provide a potential amplification mechanism whereby signals initiated by interaction of a secretagogue with a few cells could be transmitted to all of the cells within the same acinus. Just as electrical coupling may amplify the response of the system to low secretagogue concentrations, electrical uncoupling could conceivably serve as an attenuation mechanism to buffer the response of the system to high secretagogue concentrations.

3.4.2 *Cyclic GMP in pancreatic acinar cells*

Those pancreatic secretagogues that cause release of cellular calcium also increase cellular cyclic GMP and this increase can occur in a calcium-free incubation solution.[56,4] There is a close correlation between the time-course for secretagogue-induced mobilization of cellular calcium and that for the effect of the secretagogue on cyclic GMP in that significant increases in both functions can be detected at the earliest time it is technically possible to make measurements. In addition, with muscarinic cholinergic agents, CCK-related peptides, bombesin-related peptides or physalaemin-related peptides, the dose–response curve for the secretagogue-induced increase in cyclic GMP is the same as that for secretagogue-stimulated calcium outflux. Several findings indicate that secretagogues increase cellular cyclic GMP as a result of their abilities to cause mobilization of cellular calcium. (1) Calcium can influence guanylate cyclase activity in broken cell preparations from various tissues. (2) A23187 increases cellular cyclic GMP in pancreatic acinar cells and this action can occur in a calcium-free medium. (3) 8-Br cyclic GMP does not increase calcium outflux, but does increase enzyme secretion.

Because of the abilities of those secretagogues that cause calcium outflux to increase cellular cyclic GMP, and because of the ability of 8-Br cyclic GMP to increase pancreatic enzyme secretion, cyclic GMP has been proposed to be an intracellular mediator of the actions of secretagogues on pancreatic enzyme secretion. Several types of experimental findings, however, indicate that although some pancreatic secretagogues can increase cyclic GMP, this cyclic nucleotide does not mediate the actions of these secretagogues on enzyme secretion. To date, none of the secretagogues that increases cyclic GMP has been found to increase guanylate cyclase activity in a broken cell preparation. Pancreatic acinar cells have a cyclic GMP-dependent protein kinase activity;[59] however, the secretagogues that increase cellular cyclic GMP do not cause endogenous activation of the cyclic GMP-dependent protein kinase.[59] When pancreatic acinar cells are incubated with secretagogues such as CCK, all the cyclic GMP is in the extracellular medium after 60 minutes of incubation.[6] Although 8-Br cyclic GMP can increase pancreatic enzyme secretion, this action reflects the ability of this cyclic nucleotide derivative to mimic the action of endogenous cyclic AMP.[5,6,18] The most compelling argument against cyclic GMP mediating the actions of secretagogues on pancreatic enzyme secretion, has been obtained with agents such as hydroxylamine and sodium nitroprusside, which are able to activate guanylate cyclase in broken cell preparations and to increase cyclic GMP in intact tissues.[60] In contrast to CCK, which increases cyclic GMP, calcium outflux and enzyme secretion, nitroprusside causes a 70-fold increase in cyclic GMP, but does not alter calcium outflux or enzyme secretion.[6] In addition, nitroprusside does not alter the increase in calcium outflux or the increase in enzyme secretion caused by CCK or other secretagogues.[6]

3.5 Role of extracellular calcium in the actions of secretagogues on pancreatic enzyme secretion

Although mobilization of intracellular calcium appears to be one of the initial steps in the process by which some secretagogues increase pancreatic enzyme secretion, extracellular calcium is also clearly involved in the actions of secretagogues on enzyme secretion. Removing extracellular calcium reduces the increase in enzyme secretion caused by all classes of secretagogues i.e. those secretagogues that increase calcium outflux as well as those that increase cyclic AMP.[61] During the first 10 minutes of incubation, removing extracellular calcium does not alter the time-course for the increase in enzyme secretion caused by CCK (a secretagogue that mobilizes cellular calcium) or by VIP (a secretagogue that increases cellular cyclic AMP). After 10 minutes of incubation, however, secretagogue-stimulated amylase release is reduced significantly in a calcium-free incubation medium.[61,62] Thus, the effect of extracellular calcium on secretagogue-stimulated enzyme secretion does not appear to depend on the mode of action of the secretagogue. In the absence of extracellular calcium there are two phases of secretagogue-stimulated pancreatic enzyme secretion:[61,62] (1) an initial phase lasting approximately 10 minutes during which secretagogue-stimulated enzyme secretion does not depend on extracellular calcium and (2) a later phase during which extracellular calcium is required for optimal stimulation of enzyme secretion. During this later phase, the role of extracellular calcium appears to be to support the stimulation of enzyme secretion caused by all pancreatic secretagogues. What is not known is whether during the later phase of stimulation of enzyme secretion, the dependence on extracellular calcium reflects a direct effect of extracellular calcium or whether extracellular is simply required to maintain the calcium content of some cellular compartment that functions to support secretagogue stimulation of the secretory process.

4   ROLE OF CYCLIC AMP IN THE REGULATION OF PANCREATIC ENZYME SECRETION

VIP, secretin and cholera toxin can increase pancreatic enzyme secretion in some but not all species.[63] Several types of experimental findings indicate that in those species in which these peptides increase enzyme secretion, cyclic AMP functions to couple the stimulus to the secretory response. In guinea pig pancreas, VIP, secretin and cholera toxin increase cyclic AMP and enzyme secretion, and secretin and VIP can activate adenylate cyclase activity in partially purified plasma membranes from pancreatic acinar cells (for review see Reference 5). Theophylline, which inhibits cyclic nucleotide phosphodiesterase, augments the increase in cyclic AMP and in enzyme secretion caused by VIP, secretin or cholera toxin. Figure 7 summarizes the salient features of the time course of the action of VIP on amylase secretion and on cyclic AMP in dispersed acini from guinea pig pancreas as well as the

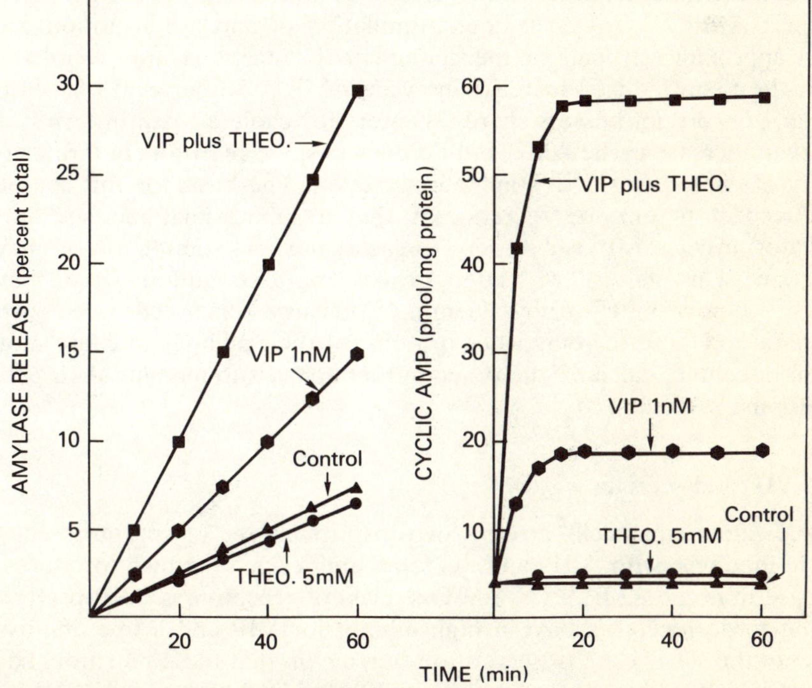

Figure 7. Effect of VIP and theophylline on amylase secretion and cyclic AMP in dispersed acini from guinea pig pancreas. Acini were incubated for up to 60 minutes at 37 °C with the agents indicated. Amylase secretion expressed as per cent release, and cyclic AMP were determined at the times indicated as described in references 29 and 53

ability of theophylline to augment these actions of VIP. Additional evidence that cyclic AMP mediates the actions of VIP, secretin and cholera toxin on pancreatic enzyme secretion is that exogenous derivatives of cyclic AMP reproduce the actions of these secretagogues.[5,18] Finally, cyclic AMP exerts its effects on cellular functions by activating cyclic AMP-dependent protein kinase to cause phosphorylation of cellular proteins by ATP.[64] Pancreatic acinar cell possess a cyclic AMP-dependent protein kinase, and incubating acinar cells with VIP or secretin causes endogenous activation of this enzyme.[59]

4.1 Cholera toxin

The dose–response curve for binding of cholera toxin to its membrane receptors is the same as that for the toxin-induced increase in cyclic AMP. On the other hand, the dose–response curve for toxin-stimulated amylase secretion is to the left of those for cholera toxin binding and for the increase in cyclic AMP.[5,35] Thus, maximal stimulation of amylase secretion occurs when approximately half of the cholera toxin receptors are occupied and when the toxin induced increase in cyclic AMP is 50 per cent of maximal. Acinar cells do not possess spare receptors for cholera toxin in terms of its ability to increase cyclic AMP, but do possess spare receptors in terms of the ability of cholera toxin to stimulate secretion. The basis for this spareness with respect to enzyme secretion is that a submaximal increase in the mediator, cyclic AMP, is sufficient to cause maximal stimulation of enzyme secretion. This pattern of action of cholera toxin differs from that of CCK,[5,11] bombesin[5,21] or physalaemin,[5,25] because acinar cells possess spare receptors for these secretagogues in terms of their abilities to cause mobilization of cellular calcium, the function that appears to mediate their actions on enzyme secretion.

4.2 VIP and secretin

Pancreatic acinar cells possess two distinct classes of receptors each of which interacts with VIP and secretin and each of which produces an increase in cyclic AMP.[49,42,5,43,41] One class of receptors is 'VIP-preferring' in that these receptors have a high affinity for VIP and a low affinity for secretin; the other class is 'secretin-preferring' in that these receptors have a high affinity for secretin and a low affinity for VIP. The magnitude of the increase in cyclic AMP caused by the secretin-preferring receptor is approximately 10-times greater than that caused by the VIP-preferring receptors.[43] Although both classes of receptors mediate an increase in cellular cyclic AMP, the secretagogue-induced increase in enzyme secretion correlates with secretagogue occupation of the VIP-preferring receptors.[5,43]

The dose–response curve for binding of VIP to the VIP-preferring receptors is the same as that for the VIP-induced increase in cyclic AMP, whereas the dose–response curve for VIP-stimulated enzyme secretion is to the left of those for binding and for cyclic AMP.[5,43] Thus, as also occurs with cholera toxin,[36] maximal stimulation of enzyme secretion occurs when approximately half of the VIP-preferring receptors are occupied by VIP because the resulting increase in cyclic AMP is sufficient to cause maximal stimulation of enzyme secretion.

The finding that although both VIP-preferring receptors and secretin-preferring receptors mediate an increase in cyclic AMP, secretagogue-stimulated enzyme secretion correlates only with occupation of the VIP-preferring receptors might explain the marked species variation in terms of the abilities of secretin and VIP to stimulate pancreatic enzyme secretion. That is, secretin and VIP can increase cyclic AMP in fragments of pancreas from dog, cat, rat, guinea pig and mouse, but increase enzyme secretion only with pancreatic fragments from rat and guinea pig.[63] Thus, it may be that acinar cells from dog, cat and mouse pancreas possess secretin-preferring receptors but not VIP-preferring receptors and, as occurs in guinea pig pancreas, the increase in cyclic AMP caused by occupation of the secretin-preferring receptors is not accompanied by stimulation of enzyme secretion.

5 PATTERNS OF ACTIONS OF SECRETAGOGUES ON PANCREATIC ENZYME SECRETION

5.1 Secretagogues that mobilize cellular calcium

For those secretagogues that act by causing mobilization of cellular calcium, the pattern of action of a secretagogue on enzyme secretion differs depending on the particular class of receptors that is activated. As illustrated in Figure 8, with increasing concentrations of CCK, enzyme secretion increases, becomes maximal and then decreases as the secretagogue concentration is increased further. Although other secretagogues that are structurally related to CCK (e.g. caerulein, gastrin and various C-terminal fragments or analogues of CCK) have different potencies for stimulating secretion, they are equal in efficacy to CCK, and the configuration of their dose–response curves are the same as that for CCK illustrated in Figure 8 (for example, see Reference 11). The dose–response curves for the action of muscarinic cholinergic agents on enzyme secretion, like those for CCK and structurally related peptides, also show submaximal stimulation with supra-maximal secretagogue concentrations; however, with cholinergic agents the magnitude of the decrement in the dose–response curve is usually not as great as it is with CCK-related peptides (Figure 8). Bombesin and structurally related peptides also show a small, but statistically significant decrease

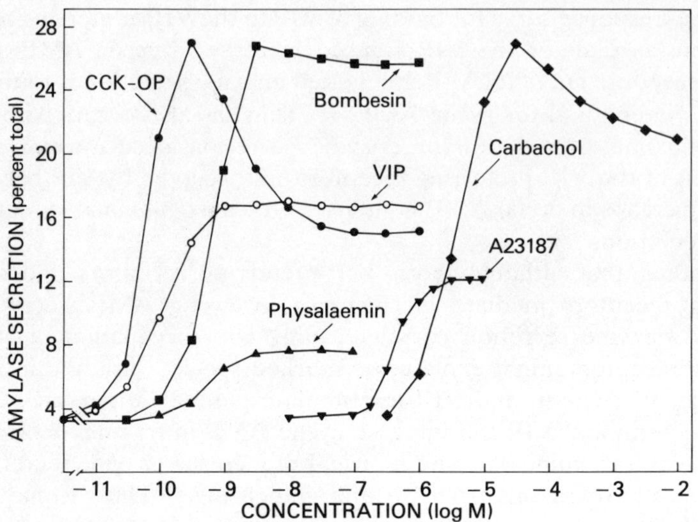

Figure 8. Dose-response curves for the secretagogue-induced increases in amylase secretion by dispersed acini from guinea pig pancreas. Acini were incubated with the secretagogue specified for 30 minutes at 37 °C and amylase secretion is expressed as per cent release. Results are from References 27, 29, 30, 35, 36, 37, 52, 53 and 65. CCK-OP, C-terminal octapeptide of cholecystolinin; VIP, vasoactive intestinal peptide

in enzyme secretion with supramaximal secretagogue concentrations, but the magnitude of this decrease is substantially less than that which occurs with cholinergic agents or with CCK-related peptides (Figure 8). As occurs with CCK-related peptides as well as with cholinergic agents, the bombesin-related peptides have different potencies for stimulating enzyme secretion, but are equal in efficacy and the configuration of their dose–response curves is the same as that for bombesin illustrated in Figure 8. Furthermore, with maximally effective secretagogue concentrations, the increase in enzyme secretion caused by bombesin or one of its structurally related peptides is the same as that caused by a maximally effective concentration of a cholinergic agent or a CCK-related peptide (Figure 8). The pattern of action of the physalaemin-related peptides on secretion differs in several respects from those for other classes of secretagogues that cause release of cellular calcium. Neither physalaemin nor its structurally related peptides show submaximal stimulation of enzyme secretion with supramaximal concentrations of the secretagogue (Figure 8). The efficacies of physalaemin and substance P are equal and are approximately 30% less than that of eledoisin.[26,27] In terms of the maximal increase in amylase secretion caused by CCK, bombesin or cholinergic agents, the physalaemin-related peptides are substantially less efficacious (Figure 8). The reduced efficacy with which physalaemin and structurally related peptides stimulate enzyme secretion can be accounted for by the lower efficacies of these peptides for stimulating calcium release.[28,27,65] On the other hand, the basis for the differences in the configurations of the dose–response curves for CCK-related peptides, bombesin-related peptides and muscarinic cholinergic agents is not clear. Each of these classes of peptides is equally efficacious in causing mobilization of cellular calcium[5] and the configurations of their dose–response curves for stimulating calcium mobilization are the same.[5]

The stoichiometric pattern of the changes in enzyme secretion with two secretagogues each of which cause calcium release depends on the particular secretagogues that are combined and the concentrations at which they are tested. With maximally effective concentrations of two secretagogues each of which causes calcium release, the increase in enzyme secretion is the same as that caused by the more effective secretagogue acting alone. With supramaximal concentrations of a secretagogue that shows submaximal stimulation of enzyme secretion at supramaximal concentrations, adding another secretagogue that causes calcium release will not alter enzyme secretion.

5.2 Secretagogues that increase cyclic AMP

In contrast to those secretagogues that cause release of cellular calcium, the two classes of secretagogues that act by increasing cyclic AMP show the same pattern of action on enzyme secretion. That is, although VIP, secretin and cholera toxin have different potencies for increasing enzyme secretion, these three secretagogues are equal in efficacy and their dose–response curves have the same configuration as that illustrated for VIP in Figure 8. The increase in amylase secretion caused by maximally effective concentrations of two secretagogues each of which increases cyclic AMP is the same as that caused by either secretagogue acting alone.[5,17] Thus, with secretagogues that increase cyclic AMP, the changes in enzyme secretion are what one would expect from a system that has a common limited effector mechanism that can be activated by multiple agents.

5.3 Potentiation of enzyme secretion

Although the two different mechanisms for stimulating secretion have initial steps that are functionally distinct, these two mechanisms do interact at some presently unknown step and one consequence of this interaction is that there is potentiation of enzyme secretion.[66,5,36,67,28] That is, the increase secretion caused by a secretagogue that increases cyclic AMP plus a secretagogue that increases calcium release is substantially greater than the sum of the effect of each secretagogue acting alone. Figure 9 illustrates two characteristic features of the pattern of amylase secretion that occurs with

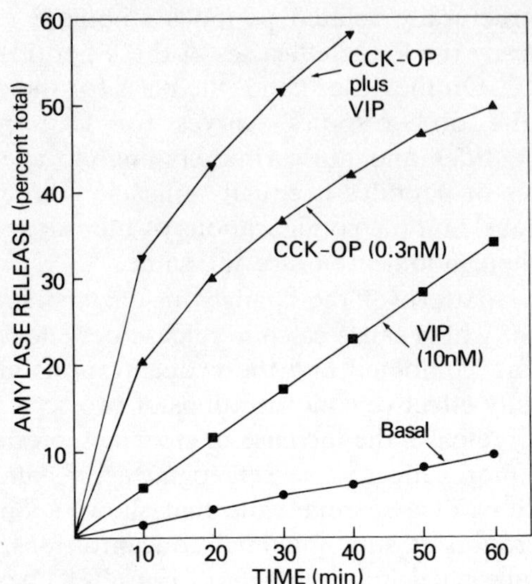

Figure 9. Time-course for amylase secretion from dispersed acini from guinea pig pancreas. Acini were incubated at 37 °C with the secretagogues indicated and amylase secretion, expressed as per cent release, was determined at the times indicated. Results are from experiments performed as in Reference 53. VIP, vasoactive intestinal peptide; CCK-OP, C-terminal octapeptide of cholecystokinin

the combination of a secretagogue that increases cyclic AMP plus a secretagogue that causes release of cellular calcium. First, the kinetic pattern of enzyme secretion with the two secretagogues in combination is the same as that for the secretagogue which causes calcium release, alone. That is, the rate of amylase secretion with CCK plus VIP, like that with CCK alone, is greater during the initial minutes of incubation than it is at later times. Second, the increase in enzyme secretion caused by the combination of a secretagogue that increases cyclic AMP plus a secretagogue that causes calcium release is significantly greater than the sum of the increase caused by each secretagogue acting alone. Although the biochemical basis of potentiation is not known, this phenomenon serves to amplify the signal generated in response to occupation of a particular class of receptors by a secretagogue.

5.4   Desensitization of enzyme secretion

5.4.1   CCK

First incubating pancreatic acini with CCK causes desensitization of pancreatic enzyme secretion measured during a subsequent incubation.[68]

CCK-induced desensitization does not depend on protein synthesis, is temperature-dependent and reversible, and the rate of onset as well as the rate of reversal depend on the concentration of CCK.[68] CCK-induced desensitization is not restricted to those secretagogues that interact with the CCK receptor but also reduces the subsequent response to all secretagogues that increase pancreatic enzyme secretion by causing mobilization of cellular calcium (i.e. CCK, carbachol, bombesin, physalaemin and A23187).[68] The desensitization induced by CCK is selective, however, because first incubating pancreatic acini with CCK does not reduce the subsequent stimulation of enzyme secretion caused by secretagogues such as VIP, secretin or 8-Br cyclic AMP, whose actions are mediated by cyclic AMP.[68]

As illustrated in Figure 10 below, the dose–response curve for CCK-induced desensitization occurs over a range of concentrations of CCK that are supramaximal for stimulating enzyme secretion and that actually cause submaximal stimulation. That is, the lowest concentration of the secretagogue that causes detectable desensitization of amylase secretion, 0.3 nM, is the concentration that causes maximal stimulation of amylase secretion. Increasing the concentration of CCK above 0.3 nM causes submaximal stimulation of enzyme secretion but causes greater desensitization. Although, the mechanism by which CCK causes desensitization of pancreatic enzyme secretion is not known, this action appears to occur at some step after receptor occupation, because first incubating acini with CCK reduces the subsequent stimulation of enzyme secretion caused by all secretagogues that cause mobilization of cellular calcium.

5.4.2 *Bombesin*

As occurs with CCK and cholinergic agents, first incubating pancreatic acini with bombesin causes desensitization of pancreatic enzyme secretion measured during a subsequent incubation. Bombesin-induced desensitization, like that induced by CCK, does not depend on protein synthesis and is temperature-dependent as well as reversible.[48] Bombesin-induced desensitization appears to develop somewhat more slowly than does that induced by CCK in that the former requires 2 hours of incubation for maximal induction[48] whereas the latter is maximal after 40–60 minutes of incubation.[68] Another difference between bombesin-induced desensitization and CCK-induced desensitization is that first incubating pancreatic acini with bombesin reduces the subsequent response to bombesin but not that to CCK. In fact, bombesin induces desensitization only to those peptides that can interact with the bombesin receptor (i.e. bombesin, litorin and ranatensin[48]), but does not cause desensitization to other secretagogues that cause mobilization of cellular calcium (i.e. carbachol, CCK, physalaemin or A23187). Bombesin, like CCK and carbachol, does not

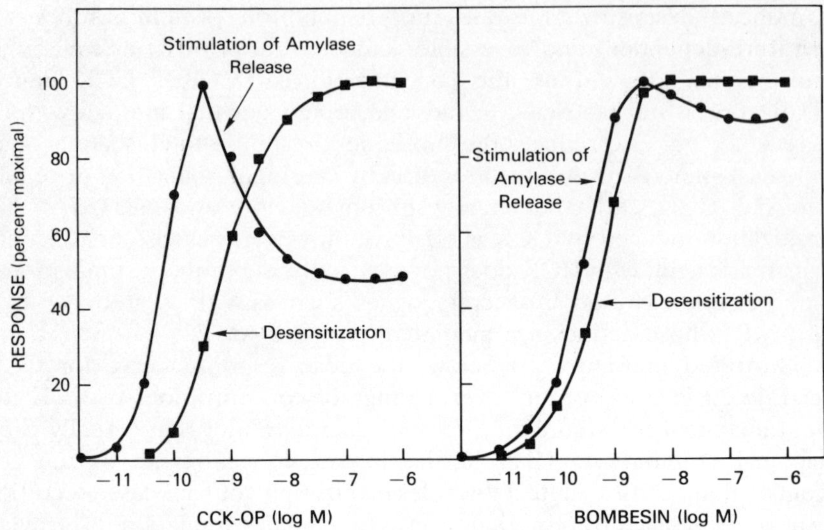

Figure 10. Comparison of the dose-response curve for secretagogue-stimulated anylase secretion with that for secretagogue-induced desensitization of amylase secretion in dispersed acini from guinea pig pancreas. Amylase secretion was determined at the end of a 30-minute incubation at 37 °C. Desensitization was measured as described in References 1 and 42. Values are expressed as the per cent of the maximal effect caused by the secretagogue indicated

cause desensitization to pancreatic secretagogues that act by increasing cyclic AMP (i.e. VIP, secretin or 8-Br cyclic AMP).

As illustrated in Figure 10, the relation between the dose–response curve for bombesin-stimulated amylase secretion and that for bombesin-induced desensitization is different from the relation between the dose–response curves for the effect of CCK on these two functions. That is, the dose–response curve for bombesin-induced desensitization is nearly congruent with that for bombesin-stimulated enzyme secretion.

Although the mechanism by which bombesin causes desensitization is not known, this phenomenon appears to reflect changes that occur at or close to the bombesin receptor.[22,69] Effects of bombesin that occur at steps following receptor occupation cannot account for the selectivity of bombesin-induced desensitization, and the close correlation between the dose–response curve for bombesin-stimulated enzyme secretion and that for bombesin-induced desensitization favour an action at the receptor for bombesin and related peptides.

REFERENCES

1. Case, R. M. (1978). Synthesis, intracellular transport and discharge of exportable proteins in the pancreatic acinar cell and other cells. *Biol. Rev.*, **53,** 211–354.

2. Jamieson, J. D. (1973). The secretory process in the pancreatic exocrine cell: morphologic and biochemical aspects. In *Secretin, Cholecystokinin, Pancreozymin and Gastrin*, edited by J. E. Jorpes and V. Mutt, pp. 195–217. Springer-Verlag, Inc., New York.

3. Gardner, J. D. (1979). Regulation of pancreatic exocrine function *in vitro*. Initial steps in the action of secretagogues. *Ann. Rev. Physiol.*, **41,** 55–66.

4. Gardner, J. D., and Jensen, R. T. (1980). Receptors for secretagogues on pancreatic acinar cells. *Am. J. Physiol.*, **238,** G63–G66.

5. Gardner, J. D., and Jensen, R. T. (1981). Regulation of pancreatic enzyme secretion *in vitro*. In *Physiology of the Digestive Tract*, [Vol. 2, 831–871, edited by L. Johnson, J. Christensen, M. Grossman, E. D. Jacobson and S. G. Schultz, Raven Press, New York.

6. Gardner, J. D., and Rottman, A. J. (1980). Evidence against cyclic GMP as a mediator of the actions of secretagogues on amylase release from guinea pig pancreas. *Biochim. Biophys. Acta*, **627,** 230–243.

7. Mutt, V., and Jorpes, E. (1971). Hormonal polypeptides of the upper intestine. *Biochem. J.*, **125,** 57–58.

8. Gardner, J. D. (1979). Receptors for gastrointestinal hormones. *Gastroenterology*, **76,** 202–214.

9. Mutt, V., and Jorpes, J. E. (1968). Structure of porcine cholecystokinin-pancreozymin 1. Cleavage with thrombin and trypsin. *Eur. J. Biochem.*, **6,** 156–162.

10. Mutt, V. (1976). Further investigations on intestinal hormonal polypeptides. *Clin. Endocrinol.*, **5,** 175s–183s.

11. Jensen, R. T., Lemp, G. F., and Gardner, J. D. (1980). Interaction of cholecystokinin with specific membrane receptors on pancreatic acinar cells. *Proc. Nat. Acad. Sci. U.S.A.*, **77,** 2079–2083.

12. Sjodin, L., and Gardner, J. D. (1977). Effect of cholecystokinin variant (CCK_{39}) on dispersed acinar cells from guinea pig pancreas. *Gastroenterology*, **73,** 1015–1018.

13. Walsh, J. H. (1978). Gastrointestinal peptide hormones and other biologically active peptides. In *Gastrointestinal Disease*, edited by M. H. Sleisenger and J. S. Fordtran, pp. 107–157. W. B. Saunders and Co., Philadelphia.

14. Anastasi, A., Erspamer, V., and Endean, R. (1968). Isolation and amino acid sequence of caerulein, the active decapeptide of the skin of *hyla caerulea*. *Archives. Biochem. Biophys.*, **125,** 57–68.

15. Sankaran, H., Goldfine, I. D., Deveney, C. W., Wong, K.-Y., and Williams, J. A. (1980). Binding of cholecystokinin to high affinity receptors on isolated rat pancreatic acini. *J. Biol. Chem.*, **255,** 1849–1853.

16. Goldstein, A. (1976). Opioid peptides (endorphins) in pituitary and brain. *Science.*, **193,** 1081–1086.

17. Snyder, S. H. (1977). Opiate receptors in the brain. *New Eng. J. Med.*, **296,** 266–271.

18. Peikin, S. R., Costenbader, C. L., and Gardner, J. D. (1979). Actions of derivatives of cyclic nucleotides on dispersed acini from guinea pig pancreas: Discovery of a competitive antagonist of the action of cholecystokinin. *J. Biol. Chem.*, **254,** 5321–5327.

19. Erspamer, V., and Melchiorri, P. (1973). Active polypeptides of the amphibian skin and their synthetic analogs. *Pure and Appl. Chem.*, **35,** 463–494.

20. Erspamer, V., Falconieri Erspamer, G., Inselvini, M., and Negri, L. (1972). Occurrence of bombesin and alytesin in extracts of the skin of three European

discoglossid frogs and pharmacological actions of bombesin on extravascular smooth muscle. *Brit. J. Pharmacol.*, **45**, 333–348.

21. Erspamer, V., and Melchiorri, P. (1975). Actions of bombesin on secretions and motility of the gastrointestinal tract. In *Gastrointestine Hormones*, edited by J. C. Thompson, pp. 575–589. University of Texas Press, Austin, Texas.
22. McDonald, T. J., Jörnvall, H., Nilsson, G., Vagne, M., Ghatei, M., Bloom, S. R., and Mutt, V. (1979). Characterization of a gastrin releasing peptide from porcine non-antral gastric tissue. *Biochem. Biophys. Res. Commun.*, **90**, 227–233.
23. Jensen, R. T., Moody, T., Pert, C., Rivier, J. E., and Gardner, J. D. (1978). Interaction of bombesin and litorin with specific membrane receptors on pancreatic acinar cells. *Proc. Nat. Acad. Sci. U.S.A.*, **75**, 6139–6143.
24. Anastasi, A., Erspamer, V., and Cei, J. M. (1964). Isolation and amino acid sequence of physalaemin, the main active polypeptide of the skin of *physalaemus fuscumaculatus*. *Arch. Biochem. and Biophys.*, **108**, 341–348.
25. Erspamer, V., Erspamer, G. F., and Linari, G. (1977). Occurrence of tachykinins (physalaemin- or substance P-like peptides) in the amphibian skin and their actions on smooth muscle preparations. In *Substance P*, edited by U. S. v Euler and B. Pernow, pp. 67–74. Raven Press, New York.
26. v Euler, U. S., and Gaddum, J. H. (1931). An unidentified depressor substance in certain tissue extracts. *J. Physiol.*, **72**, 74–87.
27. Jensen, R. R., and Gardner, J. D. (1979). Interaction of physalaemin, substance P and eledoisin with specific membrane receptors on pancreatic acinar cells. *Proc. Natl. Acad. Sci. USA*, **76**, 5679–5683.
28. Uhlemann, E. R., Rottman, A. J., and Gardner, J. D. (1979). Actions of peptides isolated from amphibian skin on amylase release from dispersed pancreatic acini. *Am. J. Physiol.*, **236**, E571–E576.
29. Bernardi, L., Bosisio, G., Chillemi, F., de Caro, G., de Castiglione, R., Erspamer, V., Glaesser, A., and Goffredo, O. (1946). Synthetic peptides related to eledoisin. *Experientia*, **20**, 306–309.
30. Bernardi, L., Bosisio, G., Chillemi, F., de Caro, G., de Castiglione, R., Erspamer, V., Glaesser, A., and Goffredo, O. (1965). Synthetic peptides related to eledoisin. *Experimentia*, **21**, 695–696.
31. Bernardi, L., Bosisio, G., Chillemi, F., de Caro, G., de Castiglione, R., Erspamer, V., and Goffredo, O. (1966). Synthetic peptides related to eledoisin. Physalaemin-like peptides. *Experimentia*, **22**, 29–32.
32. Flanagan, S. D., and Storni, A. (1979). An ^{125}I-labeled binding probe for the muscarinic cholinergic receptor. *Brain Res.*, **168**, 261–274.
33. Yamamura, H. I., and Snyder, S. H. (1974). Muscarinic cholinergic binding in rat brain. *Proc. Nat. Acad. Sci. U.S.A.*, **71**, 1725–1729.
34. Yamamura, H. I., and Snyder, S. H. (1974). Muscarinic cholinergic receptor binding in the longitudinal muscle of the guinea pig ileum with [^{3}H]quinuclidinyl benzilate. *Mol. Pharmacol.*, **10**, 861–867.
35. Larose, L., Lanoe, J., Morrisset, J., Geoffrion, L., Dumont, Y., Lord, A., and Poirier, G. G. (1979). Rat pancreatic muscarinic cholinergic receptors. In *Hormone Receptors in Digestion and Nutrition*, edited by G. Rosselin, P. Fromageot and S. Bonfils, pp. 229–238. Elsevier/North Holland Press, Amsterdam.
36. Ng, K. H., Morisset, J., and Poirier, G. G. (1979). Muscarinic receptors of the pancreas: a correlation between displacement of [^{3}H]-quinuclidinyl benzilate binding and amylase secretion. *Pharmacol.*, **18**, 263–270.

37. Gardner, J. D., and Rottman, A. J. (1979). Action of cholera toxin on dispersed acini from guinea pig pancreas. *Biochim. Biophys. Acta*, **585**, 250–265.
38. Gill, D. M. (1977). Mechanism of action of cholera toxin. In *Advances in Cyclic Nucleotide Research*, edited by P. Greengard and G. A. Robison, Vol. 8, pp. 85–118. Raven Press, New York.
39. Mutt, V., Jorpes, J. E., and Magnusson, S. (1970). Structure of Porcine Secretin. The Amino Acid Sequence. *European Journal of Biochemistry*, **15**, 513–519.
40. Bayless, W. M., and Starling, E. H. (1902). The mechanism of pancreatic secretion. *J. Physiol.*, **28**, 325–353.
41. Bayless, W. M., and Starling, E. H. (1904). The chemical regulation of the secretory process. *Proc. Royal Soc.*, **73**, 310–322.
42. Mutt, V., and Said, S. I. (1974). Structure of Porcine Vasoactive Intestinal Octacosapeptide, *European Journal of Biochemistry*, **42**, 581–589.
43. Christophe, J. P., Conlon, T. P., and Gardner, J. D. (1976). Interaction of porcine vasoactive intestinal peptide with dispersed pancreatic acinar cells from the guinea pig. Binding of radioiodinated peptide. *J. Biol. Chem.*, **251**, 4629–4634.
44. Robberecht, P., Conlon, T. P., and Gardner, J. D. (1976). Interaction of porcine vasoactive intestinal peptide with dispersed acinar cells from the guinea pig: structural requirements for effects of VIP and secretin on cellular cyclic AMP. *J. Biol. Chem.*, **251**, 4635–4639.
45. Gardner, J. D., Conlon, T. P., Fink, M. L., and Bodanszky, M. (1976). Interactions of peptides related to secretin with hormone receptors on pancreatic acinar cells. *Gastroenterology*, **71**, 965–970.
46. Gardner, J. D., Rottman, A. J., Natarajan, S., and Bodanszky, M. (1979). Interaction of secretin$_{5-27}$ and its analogues with hormone receptors on pancreatic acini. *Biochim. Biophys. Acta*, **583**, 491–503.
47. Kondo, S., and Schulz, I. (1976). Calcium ion uptake in isolated pancreas cells induced by secretagogues. *Biochim. Biophys. Acta*, **419**, 76–92.
48. Kondo, S., and Schulz, I. (1976). Ca^{++} fluxes in isolated cells of rat pancreas: effect of secretagogues and different Ca^{++} concentrations. *J. Memb. Biol.*, **29**, 185–203.
49. Renckens, B. A. M., Schrijen, J. J., Swarts, H. G. P., de Pont, J. J. H. H. M., and Bonting, S. L. (1978). Role of calcium in exocrine pancreatic secretion. IV. Calcium movements in isolated acinar cells of rabbit pancreas. *Biochim. Biophys. Acta*, **544**, 338–350.
50. Clemente, F., and Meldolesi, J. (1975). Calcium and pancreatic secretion. I. Subcellular distribution of calcium and magnesium in the exocrine pancreas of the guinea pig. *J. Cell Biol.*, **65**, 88–102.
51. Clemente, F., and Meldolesi, J. (1975). Calcium and pancreatic secretion. Dynamics of subcellular calcium pools in resting and stimulated acinar cells. *Brit. J. Pharmacol.*, **55**, 369–379.
52. Shelby, H. T., Gross, L. P., Lichty, P., and Gardner, J. D. (1976). Action of cholecystokinin and cholinergic agents on membrane-bound calcium in dispersed pancreatic acinar cells. *J. Clin. Invest.*, **58**, 1482–1493.
53. Chandler, D. E., and Williams, J. A. (1977). Fluorescent probe detects redistribution of cell calcium during stimulus-secretion coupling. *Nature*, **268**, 659–660.
54. Reed, P. W., and Lardy, H. A. (1972). A23187: A divalent cation ionophore. *J. Biol. Chem.*, **247**, 6970–6977.
55. Reed, P. W., and Lardy, H. A. (1972). Antibiotic A23187 as a probe for the

study of calcium and magnesium in biological systems. In *The Role of Membrane in Metabolic Regulation*, edited by A. M. Mehlman and R. W. Hanson, pp. 111–131. Academic Press, New York.

56. Christophe, J. P., Frandsen, E. K., Conlon, T. P., Krishna, G., and Gardner, J. D. (1976). Action of cholecystokinin, cholinergic agents and A23187 on accumulation of guanosine 3′,5′-monophosphate in dispersed guinea pig pancreatic acinar cells. *J. Biol. Chem.*, **251**, 4640–4645.

57. Gardner, J. D., Walker, M. D., and Rottman, A. J. (1980). Effect of A23187 on amylase release from dispersed acini prepared from guinea pig pancreas. *Am. J. Physiol.*, **238**, G458–G466.

58. Petersen, O. H. (1976). Electrophysiology of mammalian gland cells. *Physiol. Rev.*, **56**, 535–577.

59. Jensen, R. T., and Gardner, J. D. (1978). Cyclic nucleotide-dependent protein kinase activity in acinar cells from guinea pig pancreas. *Gastroenterology*, **75**, 806–816.

60. Mittal, C. K., and Murad, F. (1977). Properties and oxidative regulation of guanylate cyclase. *J. Cycl. Nucl. Res.*, **3**, 381–391.

61. Gardner, J. D., Costenbader, C. L., and Uhlemann, E. R. (1979). Effect of extracellular calcium on amylase release from dispersed pancreatic acini. *Am. J. Physiol.*, **236**, E754–E762.

62. Scheele, G., and Haymovits, A. (1979). Cholinergic and peptide-stimulated discharge of secretory protein in guinea pig pancreatic lobules. *J. Biol. Chem.*, **254**, 10346–10353.

63. Robberecht, P., Deschodt-Lanckman, M., Lammens, M., de Neef, P., and Christophe, J. (1977). *In vitro* effects of secretin and vasoactive intestinal polypeptide on hydrolase secretion and cyclic AMP levels in the pancreas of five animal species. A comparison with caerulein. *Gastroenterol. Clin. Biol.*, **1**, 519–525.

64. Langan, T. A. (1973). Protein kinase and protein kinase substrates. In *Advances in Cyclic Nucleotide Research*, edited by P. Greengard and G. A. Robison, Vol. 3, pp. 99–153. Raven Press, New York.

65. May, R. J., Conlon, T. P., Erspamer, V., and Gardner, J. D. (1978). Actions of peptides isolated from amphibian skin on pancreatic acinar cells. *Am. J. Physiol.*, **235**, E112–E118.

66. Gardner, J. D., and Jackson, M. J. (1977). Regulation of amylase release from dispersed pancreatic acinar cells. *J. Physiol.*, **270**, 439–454.

67. Peikin, S. R., Rottman, A. J., Batzri, S., and Gardner, J. D. (1978). Kinetics of amylase release by dispersed acini prepared from guinea pig pancreas. *Am. J. Physiol.*, **235**, E743–E749.

68. Abdelmoumene, S., and Gardner, J. D. (1981). Cholecystokinin-induced desensitization of enzyme secretion in dispersed acini from guinea pig pancreas. *Am. J. Physiol.*, in press.

69. Lee, P. C., Jensen, R. T., and Gardner, J. D. (1980). Bombesin-induced desensitization of enzyme secretion in dispersed acini from guinea pig pancreas. *Am. J. Physiol.*, **238**, G213–G218.

13 GTP-dependent stimulation and inhibition of adenylate cyclase

D. M. F. Cooper and C. Londos

1 INTRODUCTION

It is now generally accepted that three major components constitute adenylate cyclase; a hormone receptor, R; a guanine nucleotide regulatory unit, N; and a catalytic unit, C.[1,2] In the presence of GTP the N unit senses occupancy of the receptor by hormones or neurotransmitters and transmits this information to the catalytic unit—a process which generally results in an increased conversion of MgATP to cyclic AMP. The pivotal role of the N unit in both transmitting and terminating (by dissociation) the hormone signal is widely acknowledged, although the chemical nature of the various components, their spatial relationships and associations in plasma membranes, and the kinetic mechanism whereby the N unit (with GTP) promotes stimulation of enzyme activity are largely unknown.

Whereas in many systems GTP either is without effect or elicits a slight stimulation of adenylate cyclase activity (in the absence of hormone), early observations on the fat cell indicated that relatively low concentrations of this nucleotide inhibited the enzyme.[3,4,5] It has now become apparent that this inhibitory action of GTP underlies the presence of a regulatory circuit mediated by the nucleotide which is utilized by hormones and neuro-transmitters to inhibit the enzyme's action.

In this chapter we set out to establish that in many systems the classical scheme of R, N and C is inadequate and must be superseded by a more elaborate formulation consisting of stimulatory and inhibitory receptors (R_s and R_i), stimulatory and inhibitory guanine nucleotide units (N_s and N_i), in addition to the catalytic unit (Figure 1). We shall review and elaborate the evidence that such dual circuitry exists widely and summarize what can be said of the biochemical and organizational characteristics of such systems as well as their functional properties and significance. The approach taken will centre on the properties of the N components since it has been by attention to the regulatory properties of GTP that our present level of understanding

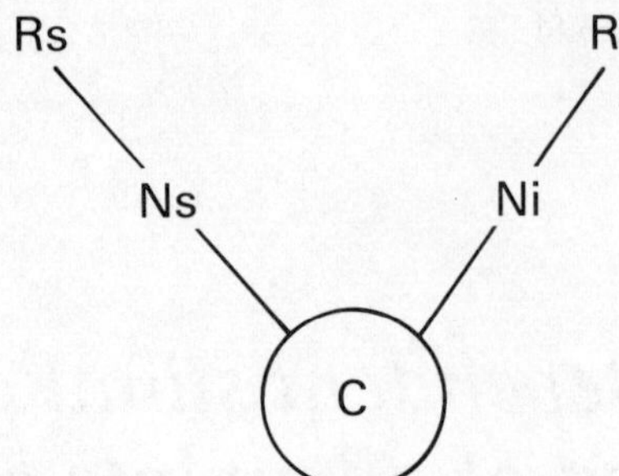

Figure 1. Functional components of adenylate cyclase. Distinct receptors, either stimulatory (Rs), or inhibitory (Ri) are linked to the catalytic unit, C, by two classes of GTP regulatory proteins, Ns and Ni respectively

of adenylate cyclase systems has been gained. The functional consequences and thus the role of inhibitory receptors derives immediately from the regulatory properties of the two N units.

2   METHODS

Previously published methods were used in the preparation of membranes from fat cells,[6] brain cortex,[7] platelet[8] and caudate nucleus.[9] In the case of the caudate nucleus, the crude ('P1') fraction was used. ATP was found not to be necessary in any of the stages of membrane purification where it had been used previously by Harwood *et al.*[10] and consequently it was excluded in all of the studies reported here. The sources of materials were as previously reported.[6,11,12] D-Ala-met enkephalin was from Calbiochem.

Adenylate cyclase was assayed using either ATP or 2′deoxy ATP as substrate and the appropriate cyclic nucleotide isolation procedure.[12,13]

3   TWO GTP-MEDIATED PROCESSES IN THE FAT CELL

For many years research on adenylate cyclase centred on the stimulation of activity by hormones. However, it is probably reasonable to suggest that much of the rapid progress in elucidating the mechanisms whereby hormones affect adenylate cyclase is due to a shift in emphasis away from the hormones and towards the actions of guanine nucleotides. Later, we review the evidence that several hormonal agents can inhibit the enzyme and, as with stimulation, that this inhibition is mediated by GTP. Thus we will defer discussion of inhibitory hormones and first take up the matter of inhibition of adenylate cyclase by GTP, since it is this phenomenon that mediates the actions of inhibitory hormones.

3.1　Characteristics

Inhibition of adenylate cyclase by GTP in the fat cell predates the demonstration of the familiar stimulatory effects of the nucleotide in other systems. However, until a few years ago it was not clear what were to be considered as the characteristic actions of GTP on the fat cell enzyme. With hindsight it is possible to say that the enzyme's reputation as unstable or susceptible to subtle variations in assay conditions largely stem from a combination of factors discussed under Methodological Considerations (Section 7). However by using highly purified plasma membranes, a substrate free of GTP, and in the absence of redox agents, a characteristic and by now standard response to GTP was encountered (Figure 2). A relatively symmetrical sharply biphasic relationship is apparent between the concentration of GTP and adenylate cyclase activity.[6] The rising phase up to the peak is referred to hereafter as the stimulatory phase; the steady decline in activity with increasing GTP concentration after the peak is referred to as the inhibitory phase. This behaviour is shared by the guanine nucleotide analogues ITP and 2'-deoxy GTP (Figure 2), which are also active in systems where GTP appears to display only stimulatory effects. By contrast, if the effect of GTP on the liver adenylate cyclase in combination with glucagon is examined, a relatively simple hyperbolic relationship between GTP concentration and activity is observed.

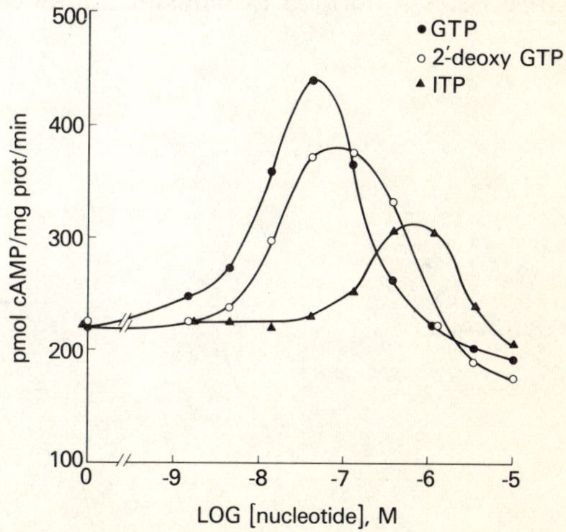

Figure 2. Effects of nucleotide triphosphates on isoproterenol-stimulated fat cell adenylate cyclase activity. Adenylate cyclase activity was assayed at 24°, for 30 min using 0.1 mM ATP as substrate as described in Reference 6, in the presence of 0.1 μM isoproterenol

Under the relatively uncomplicated conditions which we used, the biphasic response to GTP was a constant feature, withstanding mild variations in assay conditions such as temperature and divalent metal ion concentration and not altered by any of the hormones which stimulate the fat cell enzyme.[6]

3.2 Separation of the two processes

Although the biphasic behaviour was stable to simple variation in assay conditions, situations were devised in which it was possible to view only one of the actions of GTP. A schematic summary of the results of various perturbations is presented in Figure 3. A brief description of these situations follows.

3.2.1 *Divalent metal ions*

Adenylate cyclases are activated by hormones which bind to discrete receptors and by Mg^{2+} and Mn^{2+} which bind to a putative divalent cation site.[14] When stimulated by hormones the enzymes retain the biphasic response to varying GTP concentrations. By contrast, stimulation with either high Mg^{2+} (20 mM) or moderate Mn^{2+} (2–4 mM) concentrations results in a loss of GTP inhibition but retention of stimulation by the nucleotide.[6] This result has since been confirmed in hamster adipocytes.[15]

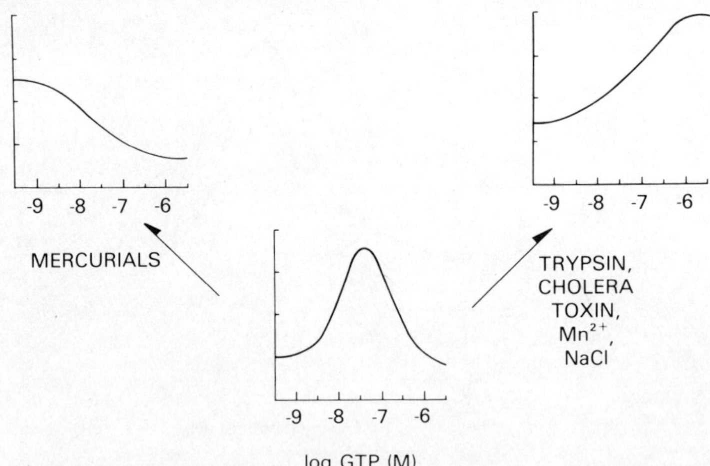

Figure 3. Schematic summary of treatments or conditions which selectively express one phase of the action of GTP in the fat cell. The central GTP titration represents the normal conditions encountered in the fat cell, assayed as described in Figure 2. The other titrations represent the same experiment performed after the treatments or under the assay conditions indicated (see text, Section 3.2)

3.2.2   *Gpp(NH)p*

The hydrolysis-resistant GTP analogue, Gpp(NH)p, activates adenylate cyclases in the presence and, in many cases, in the absence of stimulatory hormones. Although there seems to be little question on the ability of this analogue to mimic the stimulatory effects of GTP, the question of its ability to mimic the inhibitory effects of GTP is not so clear-cut. Low concentrations of Gpp(NH)p do inhibit cyclases, but such inhibition is transient, lasting at most a few minutes. Once steady state activity is achieved, the analogue evokes stimulation at all concentrations tested (cf. references 16, 4; Figure 4).

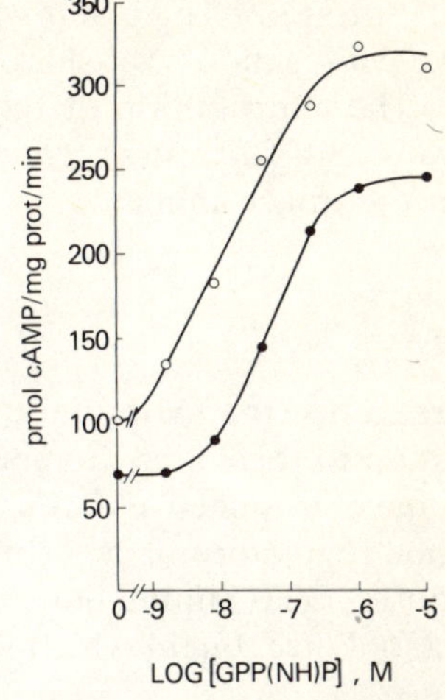

Figure 4. Fat cell adenylate cyclase activity as a function of Gpp(NH)p concentration. Adenylate cyclase was assayed under the conditions described in Figure 2 for 60 min in the presence of the indicated concentrations of Gpp(NH)p with (○) and without (●) 1 μM isoproterenol

3.2.3   *Trypsin*

Yamamura *et al.*[17] had shown that pretreatment of fat cell membranes with trypsin selectively abolished inhibition by GTP. We have extended this result and shown that mild trypsin treatment of *intact* fat cells prior to the preparation of membranes results in the selective loss of the inhibitory process with the full retention of stimulation by GTP and hormones. Anderson *et al.* have also shown that in fibroblast cells trypsin treatment will abolish GTP inhibition.[18,19]

3.2.4   *Cholera toxin*

Cholera toxin, which in combination with NAD, activates most adenylate cyclase systems, evokes an apparently unique effect on the fat cell. As in other systems, it activates the enzyme and enhances the response to stimulatory hormones. However, the unique effect is that it abolishes the expression of GTP inhibition. This observation coupled with the lack of evocation of the inhibitory process by Gpp(NH)p (see 3.2.2) invites, speculation on the role of GTP hydrolysis in GTP inhibition, (see later, Section 5).

3.2.5 *Sodium ions*

For reasons discussed in the next section we examined the effects of sodium chloride on the GTP titration. The result was a selective abolition of GTP inhibition (Figure 3). This action was shared by K^+, Li^+, NH_4^+ and choline to a lesser extent. The contribution of the anion was insignificant. NaCl is optimally effective at 80–100 mM—a concentration similar to that encountered by the cell in the intact animal.

3.2.6 *Sulphydryl reagents*

A group of sulphydryl reagents, the mercurials, have an effect on the fat cell opposite to those listed up to this point. Compounds such as *p*-hydroxy-mercuriphenylsulphonate have a selective irreversible effect on a wide variety of guanine nucleotide regulatory units.[20] In the fat cell their effect is to eliminate the expression of GTP stimulation while fully retaining GTP inhibition. So far this is the only treatment which we have discovered which exerts this action.

In view of the foregoing group of observations showing that one of the actions of GTP can be abolished while the other is fully retained, it seems reasonable to postulate that GTP mediates two distinct processes in the fat cell. The functions associated with these two processes are discussed below.

4 FUNCTION OF THE INHIBITORY PROCESS MEDIATED BY GTP

4.1 Adenosine in the fat cell

For at least ten years it has been known that adenosine acting at a cell surface receptor was one of a number of compounds which inhibit lipolysis, presumably by reducing cyclic AMP production, and it was suspected that this effect occurred *via* the cyclase. However breakage of the cells and examination of the particulate adenylate cyclase activity did not reveal the expected inhibition by nanomolar adenosine concentrations or its antilipolytic analogues.[21] This finding led to the suggestion that the adenosine receptor was uncoupled or inactivated on preparation of the membranes or else that the effect on the cyclase was indirect.[22]

The conceptual climate for observing direct inhibitory effects of hormones and neurotransmitters on adenylate cyclase had been set by two earlier observations. First, Klee and Nirenberg[23] showed that opiates directly inhibited adenylate cyclase in crude homogenates of neuroblastoma × glioma hybrid cells. Second, Jakobs *et al.*[24] found that epinephrine, acting at an alpha adrenergic receptor, inhibited the cyclase in platelet membranes, and that such inhibition was dependent on the addition of GTP to the assay medium.

Upon elimination of background adenosine from the adenylate cyclase assay medium (see Section 7) we were able to observe adenosine-receptor mediated inhibition of the cyclase in purified fat cell membranes. More importantly, such studies served to define a role for the GTP-inhibitory process associated with the enzyme. Whereas the role of the GTP-stimulatory process was to enhance the activation of cyclase by stimulatory hormones (catecholamine, ACTH, glucagon, TSH, secretin, etc.), adenosine receptor agonists inhibited activity only over the GTP concentration range where the nucleotide alone was inhibitory. That is, the adenosine receptor was intimately linked to the GTP inhibitory process discussed above (Figure 5). Further studies with a variety of adenosine analogues and known

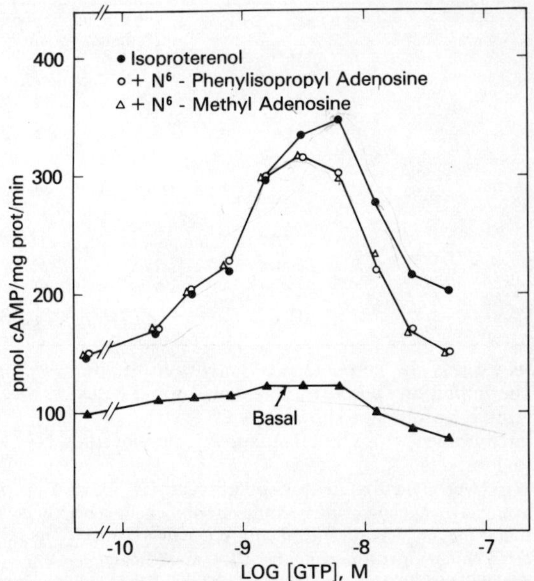

Figure 5. Effect of N^6-phenylisopropyladenosine and N^6-methyladenosine on isoproterenol-stimulated fat cell adenylate cyclase. Assays were performed as described in Figure 2 with the addition of 0.1 units/ml adenosine deaminase; basal activity, ▲; 1 μM isoproterenol in the absence, ●, and presence of either 50 μM N^6-phenylisopropyladenosine, ○, or 50 μM N^6-methyladenosine, △

adenosine receptor antagonists (methyl xanthines) revealed that the GTP-dependent inhibition by adenosine was indeed mediated by the adenosine receptor that mediates the antilipolytic actions of the nucleoside on adipocytes.[11]

To test our contention that inhibition mediated by the adenosine receptor was initimately linked to the GTP inhibitory process, the action of adenosine was examined after the treatments described previously (Section 3), which selectively eliminate the expression of either the stimulatory or the inhibitory process mediated by GTP (see Table 1). With one exception, treatments which result in the suppression of the expression of GTP inhibition also

Table 1. Modification of N^6-phenylisopropyladenosine inhibition of fat cell adenylate cyclase activity by various procedures

The adenylate cyclase activity of 0.5 to 1.5 μg aliquots of variously treated membranes was determined under standard assay conditions in the presence of 2.5 units/ml of adenosine deaminase to reduce endogenous adenosine levels. Where indicated, GTP and N^6-phenylisopropyladenosine (PIA) were present at final concentrations of 9 μM and 0.4 μM, respectively

Pretreatment or condition of assay	Isoproterenol	$-$GTP $-$PIA	$+$GTP $-$PIA	$+$GTP $+$PIA
	μM	nmol cyclic AMP/mg/30 min		
Normal	None	1.75	1.55	0.39(25)[a]
	0.4	2.67	4.60	2.52(48)
Cholera toxin pre-treated[b]	None	1.70	4.51	3.97(88)
	0.4	3.73	7.33	6.36(86)
Gpp(NH)p preacti-vated[c]	0.4	24.57	25.80	23.49(92)
Mn^{2+}(2 mM), Mg^{2+} (20 mM) in assay[d]	0.4	10.49	14.34	14.09(98)
Mercurial treatment[b]	None	2.01	0.39	0.18(48)
NaCl 100 mM in assay	None	1.8	2.5	0.6(24)
	0.4	2.8	7.4	2.9(39)
Trypsin[e]	0.4	0.77	1.33	1.26(95)

[a] Values in brackets express the percentage of activity in the presence of PIA compared to that in its absence. The inhibition caused by PIA could in all cases be reversed by 250 μM 1-methyl-3-isobutyl xanthine (results not shown).

[b] Pretreatment of membranes with cholera toxin A_1, peptide plus NAD and the mercurial MPS as described previously.[6]

[c] membranes (56 μg) were incubated for 60 min at 24 °C in the presence of 10 μM Gpp(NH)p and all adenylate cyclase assay components except α-^{32}P ATP. The activity of aliquots (1.4 μg) of this material was then determined over 10 min by its addition to a mixture containing α-^{32}P ATP, GTP, isoproterenol, and PIA as indicated.

[d] Activities were determined in the presence of 20 mM Mg^{2+} containing 2 mM Mn^{2+} under the conditions indicated in the table.

[e] During the last 5 min disaggregation of fat cells by collagenase 1 mg/ml Trypsin (247 μ/mg) from Worthington was included in the medium, then 1 mg/ml Lima Bean inhibitor (Worthington) was added prior to preparation of fat cell membranes by the usual protocol.[6]

result in the loss of the ability of adenosine to inhibit the enzyme. When the stimulatory process is suppressed by mercurial treatment, the inhibitory properties of adenosine are fully retained along with GTP inhibition. On the basis of the foregoing observations, we postulated that the ability of GTP to exhibit biphasic regulatory behaviour is a required demonstration of dual circuitry and consequently of the ability of hormones and neurotransmitters to promote inhibition. Thus we concluded that the function of the inhibitory process is to permit and promote inhibition by putative neurotransmitters and associated compounds.

The only exception to the general statement that abolition of GTP inhibition results in elimination of inhibitory effects of adenosine is incubation in the presence of NaCl, which results in a reversal of the GTP inhibition but a full retention of inhibition by adenosine. The consequence of this effect is that inhibition by the nucleoside is amplified at the expense of the inhibition by GTP. This unique effect of sodium has been observed in a number of other systems where cyclic AMP levels are reduced by putative neurotransmitters.[25–27] In such cells and membranes, sodium also affects binding of these agents—generally by decreasing the affinity of the ligand.[28–30]

In a recent reexamination of the effects of cholera toxin on the fat cell enzyme, where expression of the inhibitory process is suppressed (see Section 3), we found that in the presence of sodium, inhibition by GTP and adenosine was re-expressed. This suggests that toxin treatment may promote shifts between sodium dependent and sodium independent states of the enzyme.

4.2 Other inhibitors

Nicotinic acid and prostaglandins had long been known to be antilipolytic and to lower cyclic AMP concentrations in adipocytes, but, like adenosine, their effects in broken cell preparations remained elusive. However, applying the techniques which made possible detection of adenosine receptor-mediated inhibition of cyclase, it was possible to detect direct inhibitory effects of PGE_1 and nicotinic acid on the fat cell enzyme (Figure 6). As with adenosine, inhibition by these two antilipolytic agents was observed only over the inhibitory GTP phase and, for reasons discussed below, it was necessary to remove background adenosine from the assay medium. Further studies revealed that the three antilipolytic agents (adenosine, nicotinic acid and PGE_1) act *via* distinct receptors.

It is known that multiple stimulatory receptors exist in fat cells for ACTH, catecholamine, glucagon, secretin and TSH, and that all of these receptors converge on a common pool of catalytic activity. That is, although at submaximal doses combination of the hormones yield additive effects, at

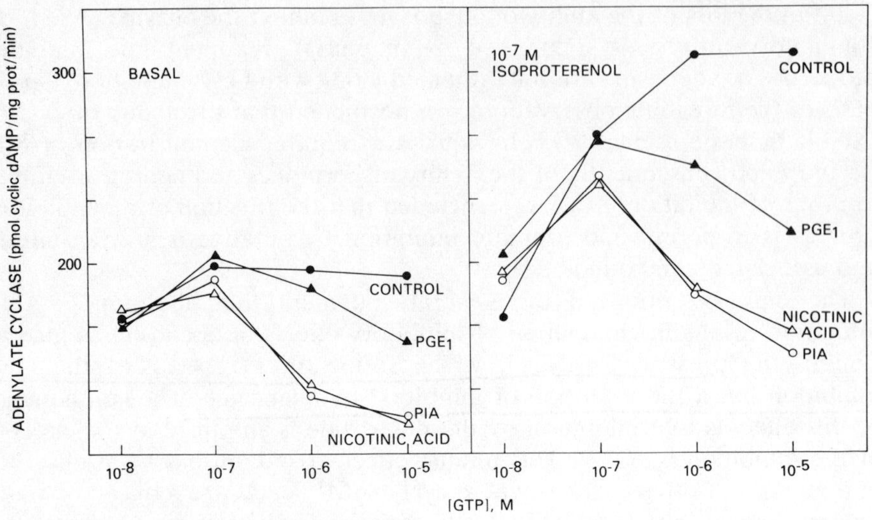

Figure 6. GTP-dependent inhibition of fat cell adenylate cyclase by PGE_1, PIA and nicotinic acid. Assays were performed with 0.1 mM ^{32}P-dATP as substrate in the presence of 100 mM NaCl at 30 °C, for 30 minutes in the absence ●, or presence of either 1 µM PGE_1, ▲, 10 µM nicotinic acid, △, or 1 µM PIA, ○. Activity was determined in the absence (left panel) or presence (right panel) of 0.1 µM isoproterenol at the GTP concentrations indicated

maximal doses their effects are non-additive, indicating that the various stimulatory receptors share a common pathway. Since each of the anti-lipolytic factors produced partial inhibition of cyclase activity, it was possible to test for additivity in inhibition. As shown in Table 2, inhibition by combinations of these agents were non-additive, since such combinations yielded no greater inhibition than was observed with the most effective agents (nicotinic acid and PIA) when tested alone. Thus, it may be concluded that, like the receptors for stimulatory hormones, the receptors for the inhibitory hormones or antilipolytic factors share a common pathway. Finally, it is this feature of non-additivity that explains the need for complete removal of adenosine from the assay medium in order to detect effects of prostaglandins and nicotinic acid.

The following picture emerges from the available evidence on the fat cell adenylate cyclase system. Stimulatory receptors and their associated guanine nucleotide regulatory components comprise one circuit for regulation of activity, inhibitory receptors and their regulatory proteins comprise a second circuit and both circuits converge on a common cyclase catalytic component. It is reasonable to assume that the suppression of cyclic AMP formation is as important to normal physiological function as is cyclic AMP formation, and,

Table 2. Inhibition of fat cell adenylate cyclase by antipolytic agents

Assays were conducted with $(\alpha\text{-}^{32}P)dATP$ as the substrate in the presence of 100 mM NaCl and 1 μM GTP with the indicated concentrations of isoproterenol as described in Figure 6. The concentrations of antilipolytic agents were: PIA and PGE$_1$, 1 μM, and nicotonic acid, 10 μM. The data show inhibition by each of these three agents alone, and a combination of the three inhibitors at the concentrations indicated above

	Basal	10^6 M isopro-terenol	10^{-4} M isopro-terenol
Control	1110	2760	4660
PGE$_1$	500	1850	2860
N^6-phenylisopropyladenosine	250	1450	2600
Nicotinic acid	200	1000	2540
All three agents	230	1050	2540

thus, the relationships between these two circuits must be understood. A notable feature of the interplay between the circuits is the limited and finite magnitude of inhibition produced by the antilipolytic agents. In the experiments discussed thus far, inhibition was evaluated when the cyclase was stimulated by the catecholamine receptor, leaving open the possibility that the inhibitory receptors may be loosely associated with the β-adrenergic receptor but more tightly with other stimulatory receptors. Moreover, some stimulatory receptors might be totally independent of inhibitory receptors. Such a scheme would suggest the presence of multiple pools of cyclase systems of varying composition in the fat cell membranes. To test this possibility experiments were performed with various combinations of stimulatory and inhibitory hormones. One example, shown in Figure 7, tested the inhibition by an adenosine receptor agonist (PIA) of activity stimulated by three different lipolytic hormones (isoproterenol, ACTH and glucagon). The data indicate that the extent of inhibition by the adenosine receptor is a function of the level of activity, but is independent of which hormone was used to achieve that activity. In further experiments, similar observations were obtained with prostaglandins or nicotonic acid in place of the adenosine analogue. The following conclusions are derived from such data: (a) there do not appear to be discrete pools of cyclases containing unique associations between selected pairs of stimulatory and inhibitory receptors; (b) all stimulatory and inhibitory receptors are proximal to a common cyclase catalytic component; (c) although receptors serve to initiate reactions, the resultant activity that reflects the interplay between the stimulatory and inhibitory circuits is independent of individual types of receptors. In our view the interaction between the two circuits reflects the actions of two types of guanine nucleotide regulatory components.

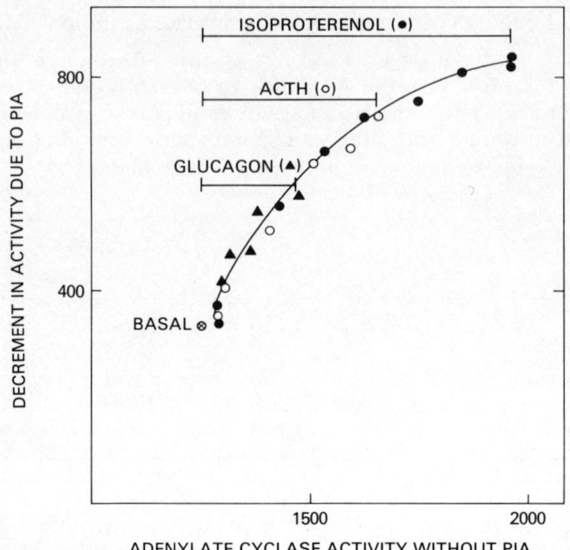

Figure 7. Extent of inhibition by PIA is independent of the hormone used to stimulate activity. Assays were performed in the absence and presence of 1 μM PIA under conditions described in Figure 6. The abscissa represents total activities produced with glucagon (▲) and ACTH$^{1-24}$ (○) both tested over the concentration range of 1 nM to 10 μM, and isoproterenol (●) over the range for each of the three lipolytic hormones, and the highest concentration of each hormone tested was sufficient to produce the maximal activity possible for that hormone. The ordinate shows the absolute decrease in activity (decrement) produced by a maximal concentration of the antilipoltyic agent, PIA (1 μM) when tested against each concentration point for each of the three lipolytic hormones

The multiple regulation to which the fat cell adenylate cyclase is subject invites speculations on the molecular architecture which governs such multiple interactions. Schemes which involve collisions of independently mobile components would seem unable to accommodate multiple non-additive and limited inhibitions converging on an activity state generated by multiple, stimulatory, non-additive, hormone receptors. Thus, currently we are inclined to envisage a large heterogenous assembly of stimulatory and inhibitory components coordinating to produce an observed activity. Evidence for such structures is available from the fat cell and is discussed in the next section (5).

4.3 Other systems

Using the insights which we had gained from our studies on the fat cell adenylate cyclase, we investigated a number of other tissues in search of further evidence of dual regulation in order to determine what general statements could be made on these systems. In the rat caudate nucleus we encountered biphasic GTP kinetics and found that opiates inhibited the enzyme (Figure 8) with similar characteristics to the inhibition of fat cell adenylate cyclase by adenosine; the inhibition was dependent on GTP concentrations in the inhibitory range and sodium reversed the inhibition promoted by GTP alone and thus also enhanced the inhibition by the opiates.[46] Similarly, in brain cortex, biphasic GTP regulation was encountered, and in the GTP inhibitory phase, adenosine, acting through a receptor very similar in properties to that in the fat cell, inhibited the enzyme in a fashion dependent on the GTP inhibitory process (Figure 9). Sodium chloride again reversed the inhibition evoked by GTP alone, causing an enhanced inhibition by adenosine and its analogues.[7] In the platelet, which is inhibited by epinephrine in a GTP dependent fashion, we found no signs of GTP inhibition in the absence of the catecholamine and sodium chloride was completely without effect.

The above data permit certain generalizations regarding the effects of Na^+ and serve as an appropriate point to address the matter of a 'sodium ion requirement' for the actions of inhibitory hormones and neurotransmitters. Inhibitory hormones act in the absence of sodium ion as long as sufficient GTP is present. However, in the absence of inhibitory hormones, different adenylate cyclases are inhibited to varying degrees by GTP *alone*, and it is this effect of the nucleotide that is modified by sodium ions. Thus, with the

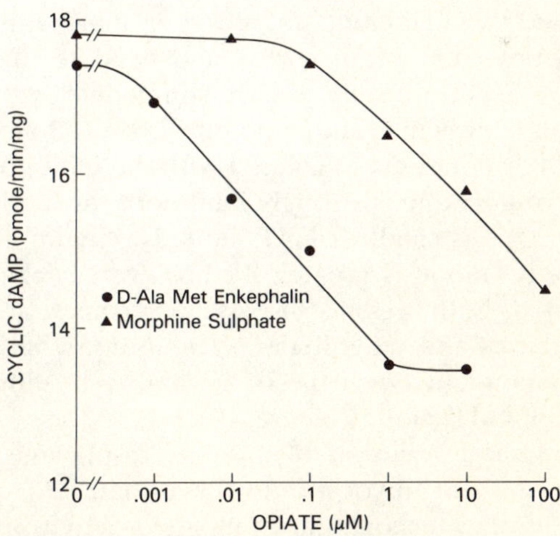

Figure 8. Inhibition of caudate nucleus adenylate cyclase by opiates. Caudate nucleus adenylate cyclase was assayed as described in Figure 2 using 0.1 mM ^{32}P-dATP as substrate in the presence of 4 µM GTP, 80 mM NaCl and the indicated concentrations of D-Ala-met enkephalin (●) or morphine sulphate (▲)

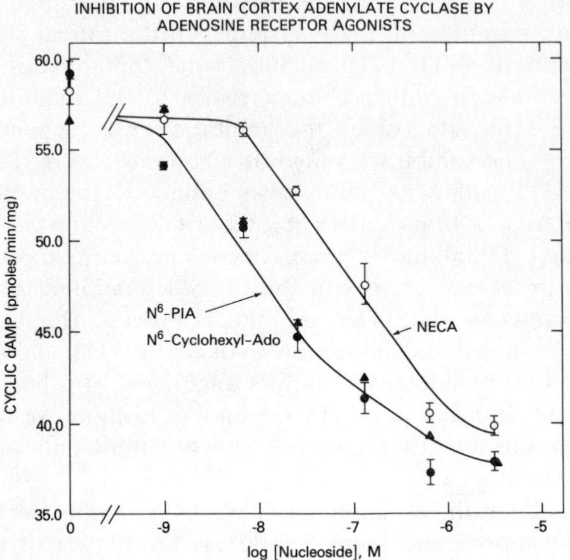

Figure 9. Inhibition of brain cortex adenylate cyclase by adenosine analogues. Brain cortex adenylate cyclase was assayed as described in Figure 8 with the indicated concentrations of N^6-phenylisopropyladenosine (●), N^6-cyclohexyladenosine (▲) and N-ethylcarboxamide adenosine (○). Standard error bars of triplicate determinations are shown

enzymes from the fat cell and the neuroblastoma × glioma hybrid cell, which are strongly inhibited by GTP alone, there is a dramatic stimulatory effect of the ion. The enzymes from brain cortex and caudate nucleus, which are weakly inhibited by GTP, are only weakly stimulated by Na^+, whereas the enzyme from the platelet, which shows no effect of GTP alone, is unaffected by Na^+. Inhibitory hormones act in concert with the inhibitory phase of GTP action. If the nucleotide alone is strongly inhibitory, the additional inhibition elicited by the hormone is small. If GTP alone is only mildly inhibitory, the effect of inhibitory hormone is greater. If, however, background inhibition by GTP is eliminated, both systems display approximately equal inhibition. Thus, sodium increases the magnitude of inhibition that is hormone- or neurotransmitter-dependent by virtue of its ability to eliminate the background inhibition by GTP alone (Figure 10).

To continue with a description of systems displaying dual regulation, recent studies in the rat myocardium have detected phenylisopropyladenosine inhibition of catecholamine-stimulated activity in a GTP dependent manner (Scibilia and Londos, in preparation). In a number of other studies, most notably those of Jakobs et al.,[25] Sabol and Nirenberg[31] and Blume et al.,[27] very similar GTP-dependent sodium-amplified inhibition has been demonstrated. In Table 3 data are summarized which represent the

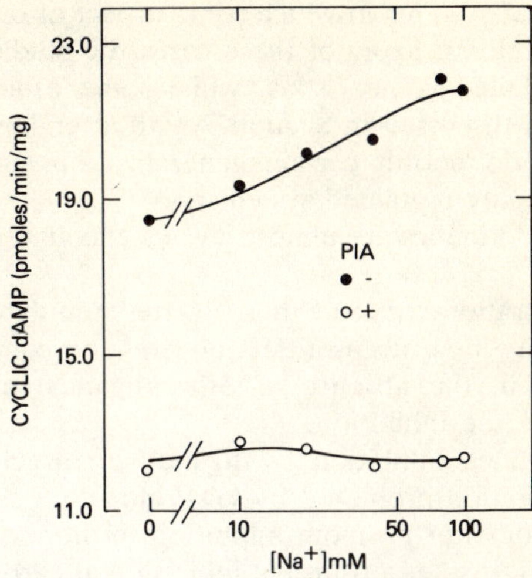

Figure 10. Cerebral cortical adenylate cyclase activity in relation to NaCl concentration. Adenylate cyclase was assayed as described in Figure 8 with 40 μM GTP in the absence (●) or presence (○) of 10 μM PIA and the indicated concentrations of NaCl (cf. Reference 7)

Table 3. Summary of dually regulated adenylate cyclase systems and some of their properties

Tissue	Inhibitory ligand	GTP requirement	Sodium amplification	Gpp(NH)p	Reference
Adipose	Adenosine, PGE$_1$, nicotinate	Yes	Yes	No	11, 34
Platelet	α-adrenergics	Yes	No	No	24
Neuroblastomax Glioma hybrids	Opiates, alpha-adrenergics muscarinic choleringic	Yes	Yes	No	31, 26, 32
Cardiac	Muscarinic choleringic adenosine	Yes	Yes	No	25 Scibilia and Londos, in preparation
Parotid	Muscarinic cholergic	Yes	—	—	
Brain cortex	Adenosine	Yes	Yes	No	7
Pituitary adenoma	Dopamine	?	?	?	35
Caudate nucleus	Opiates	Yes	Yes	No	46
Renal cortex	α-Adrenergic	Yes	?	?	
Thyroid	Muscarinic cholinergic	Yes	Yes	?	

increasing number of systems in which some aspect of dual GTP regulation has been encountered. In many of these cases the studies are too preliminary to permit detailed comparisons, while some others suffer from the technical difficulties discussed in Section 7 with attendant ambiguity. However these studies do permit certain generalizations to be made on the characteristics of dually regulated systems:

(a) Biphasic GTP kinetics are almost always encountered (in the absence of sodium ions).

(b) GTP concentrations beyond those in the stimulatory range promote inhibition by putative neurotransmitters and related compounds.

(c) Where GTP in the absence of other ligands promotes inhibition, sodium ions reverse the inhibition.

(d) Sodium amplifies inhibition by ligands in direct proportion to its reversal of the inhibition promoted by GTP alone.

(e) Gpp(NH)p does not promote inhibition by hormones or neurotransmitters, even when it evokes a transient inhibition at early incubation times.

(f) Inhibition by neurotransmitters is generally less than 60 per cent, except when directed against basal activities when it may reach 90 per cent.

(g) Where multiple inhibitory effectors operate, their effects are non-additive.[31,32]

(h) As with stimulatory ligands, the binding of inhibitory ligands is modulated by GTP.

(i) Sodium always appears to affect binding of inhibitory ligands, even when no effect of sodium is demonstrable on activity, as in the platelet.

5 THE RELATIONSHIP BETWEEN Ns AND Ni

Given the foregoing evidence on the existence of dual circuitry and functionally distinct regulatory components, one of the most important questions from a biochemical and regulatory viewpoint is what, if any, are the relationships between these components. In terms of our introductory contention that R, N and C inadequately described the components of adenylate cyclases, the question arises as to whether the subdivision suggested by the preceding pages, Rs, Ns, Ri, Ni (where Rs and Ri represent distinct receptors mediating stimulation and inhibition, respectively) and C, is warranted. That is to say, are these components the same, very similar, interconvertible forms or unrelated? The options which we feel are worth presenting are listed in Table 4.

It is not possible to unequivocally discriminate between the various models presented in Table 4. However, to permit some evaluation of the relative merits of the various possibilities, selected pertinent arguments are presented below under each heading.

Table 4. Possible relationships between Ns and Ni and conditions for their interconversion

	Relationships	Interconverted by
I	RsN = 'RsNs'; RiN = 'RiNi'	Rs and Ri
II	Ni = Ns[a]	GTP
III	Ns ≠ Ni	Not interconverted
IV	Ni = f(Ns)	Perturbations[b] such as mercurial treatment, Mn^{2+} etc.

[a] No scheme is considered in which the regulatory components do not converge on a common pool of activity.

[b] It is assumed that not all perturbations interconvert the two forms; some treatments may simply suppress or 'uncouple' one circuit, permitting the expression of the other.

5.1 A common pool of N exists which has the potential to mediate both stimulation and inhibition—the form assumed by N is dependent on its association with either Rs or Ri

In many cases, pharmacologically distinct receptor subgroups are identifiable which correlate with either stimulation or inhibition of adenylate cyclase; for instance, alpha adrenergic receptors have a different potency series for catecholamines than beta adrenergic receptors; stimulatory adenosine receptors have a different potency series for a selected group of adenosine analogues than inhibitory adenosine receptors.[36] Thus it is reasonable to consider that these receptors are distinct proteins which might elicit different behaviour from N proteins. Some observations which would not be anticipated in a scheme involving a receptor controlled expression of either Ns or Ni follow. In the absence of receptor agonists a biphasic response is evoked by GTP; in the presence of either a stimulatory or an inhibitory ligand, biphasic kinetics continue to be observed in many systems. If receptors dictated what form of N was expressed, it would be expected that the GTP titration could be shifted to the right or the left by varying concentrations of stimulatory and inhibitory ligands—this does not occur. This suggests that GTP, *via* pre-existing forms of N, dictates the function which is expressed, independent of receptor occupancy.

The converse of this question arises with receptors such as those for adenosine, catecholamine or prostaglandins, which have been shown to be capable of mediating either stimulation or inhibition depending on the tissue; i.e. is the function expressed by the receptor merely a function of the N unit to which it is coupled? Support for this contention comes from recent studies by Kather.[37] These showed biphasic GTP effects in human fat cell and interestingly showed stimulation by a catecholamine receptor in the stimulatory GTP phase and epinephrine amplified inhibition in the inhibitory phase. This raises the possibility that Rs and Ri are forms of receptors which are dictated by Ns or Ni.

5.2 Are the stimulatory and inhibitory functions of N merely different activity states of one protein promoted by a different range of GTP concentrations?

The biphasic GTP titration in the fat cell (Figure 2) may be the most suggestive evidence that GTP promotes a transition from one activity state which amplifies hormonal stimulation to, at higher concentrations, another state which amplifies inhibition. A possible kinetic basis for this suggestion derives from the observation that Gpp(NH)p preferentially expresses Ns activity with only a transient expression of inhibition; in addition, cholera toxin, which is often assumed to act by the inhibition of a GTP-ase activity (although this has only been demonstrated in turkey erythrocyte adenylate cyclase[38]) suppresses Ni expression. Both lines of evidence have been invoked to support the hypothesis that GTP-mediated stimulation (Ns) is terminated by hydrolysis to GDP at the nucleotide binding site. Since neither substitution of Gpp(NH)p for GTP nor activation by cholera toxin permits expression of the inhibitory process, it could be argued that hydrolysis of GTP to GDP is essential for the Ni function. Following from this suggestion it might be predicted that GDP or GDP-S might preferentially express the inhibitory process. However, these nucleotides are without such a selective effect on the fat cell adenylate cyclase in a variety of experimental designs, involving either preincubation or direct addition of either GDP or GDP-S. GDP-S did however appear to shift the entire GTP titration of the fat cell enzyme to the right without selectively affecting either phase (Cooper and Schlegel, unpublished) which would seem not to indicate a selective role for GTP hydrolysis in the inhibitory phase.

If the biphasic GTP titration represented two activity states and the second state required GTP hydrolysis for its expression, then mercurial treatment, which abolished stimulation while retaining GTP inhibition (Section 3), might be considered to have increased the rate of GTP hydrolysis, such that the residence time of the nucleotide triphosphate is insufficient to sustain activation. In such circumstances it would be expected that Gpp(NH)p would continue to activate the enzyme. Indeed, when a Gpp(NH)p titration is performed after mercurial treatment, full stimulation is achieved by the nucleotide (Cooper, unpublished). These observations would support a two state model for Ns/Ni.

5.3 and 5.4 Are Ns and Ni distinct proteins which may or may not be interconverted by traumatic treatments?

The possibility that these regulatory units are distinct and either unique and unrelated or related but interconverted only by extreme perturbation appears likely. It is probably not possible to discern the precise degree of the relationship between the regulatory units with our current knowledge.

However a presentation of pertinent data may help to formulate a working scheme.

The various treatments outlined in Section 4 'which result in the suppression of either one or other phase of the action of GTP and also result in a transition from a sharply biphasic to a simple hyperbolic GTP titration' is certainly compatible with the notion that distinct though possibly related proteins are involved; the various effects could result from either interconversions or uncoupling of a component from the regulatory complex.

In addition, irradiation inactivation studies of fat cell adenylate cyclase indicate that larger structures mediate inhibition than those which mediate stimulation of activity.[39] This conclusion derives from observations which showed that at low irradiation doses inhibition by GTP was abolished while stimulation was retained.

From this study and others of Schlegel *et al.*[40] it is concluded that a very large molecular aggregate ($>1.2 \times 10^6$ daltons) is required for the expression of inhibition. It is also conceivable that Ni may be a simple association of Ns with an unidentified factor, which confers unique properties on the guanine nucleotide regulatory protein. In such a situation the inhibitory process could be suppressed by any agent acting on the factor or dissociating the factor from Ns. Support for this type of notion comes from preliminary studies on hippocampal adenylate cyclase, which displays biphasic responses to GTP and an adenosine-mediated GTP-dependent inhibition of activity. EGTA treatment suppresses expression of inhibition which can be partially restored by the addition of calmodulin (Girardot and Cooper, in preparation). Thus, it appears that calmodulin may be selectively associated with the inhibitory process in this tissue.

Cholera toxin (with ^{32}P-NAD) has been profitably used to locate and identify N proteins in a wide range of purification studies of adenylate cyclase components.[2] Since cholera toxin also exerts differential effects on GTP stimulation and inhibition (see Section 4), it was hoped that this would permit the labelling of a unique protein. Comparison of the labelling patterns of tissues containing both Ni and Ns with those apparently containing only Ns displayed additional bands on SDS/polyacrylamide gel electrophoresis (Jagus and Cooper, unpublished). However tissues containing only Ns displayed considerable variation in the number and location of labelled proteins, so that detection of subtle (qualitative or quantitative) changes would be difficult. Additionally the recent report by Gilman (41) showing that the purified N (Ns) protein from liver contains three bands on SDS electrophoresis, two of which are labelled by toxin and NAD, increases the possibility that different labelling patterns will arise among tissues. Such a scenario makes it likely that quantitative or qualitative redistribution of N components, reflecting the presence or absence of Ni, will not be detected readily by toxin labelling.

The foregoing has summarized most of the arguments on the biochemical distinctions between the various formulations considered for Ns and Ni at the beginning of this section. A clear answer is not available. However it is possible to present a tentative scheme around which most of the available data can be gathered and which provides a framework from which further experimentation may be devised. Our tentative scheme can take two formulations; a mosaic model and a multimeric aggregation/disaggregation model (Figure 11(a) and (b) respectively).

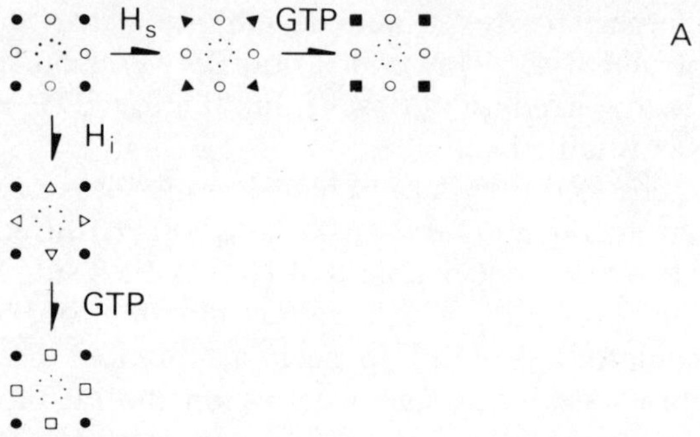

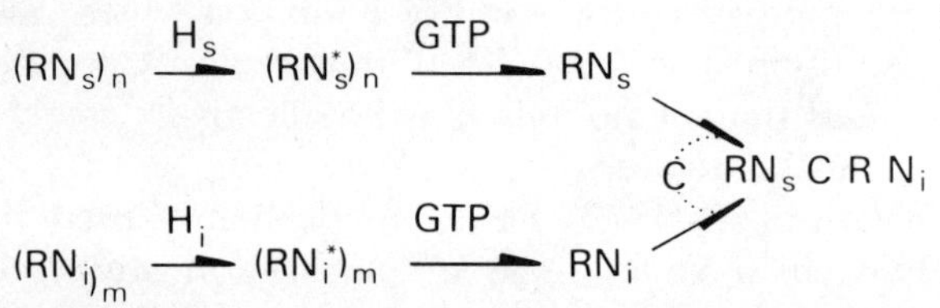

$$H_s \quad GTP \qquad A$$

$$H_i$$

$$GTP$$

$$(RN_s)_n \xrightarrow{H_s} (RN_s^*)_n \xrightarrow{GTP} RN_s \qquad B$$

$$\underset{C}{RN_s\, C\, R\, N_i}$$

$$(RN_i)_m \xrightarrow{H_i} (RN_i^*)_m \xrightarrow{GTP} RN_i$$

Figure 11. Tentative scheme for dually regulated adenylate cyclase systems. Both models envisage a large precursor; in formulation (a) the precursor is composed of stimulatory and inhibitory receptors, Rs and Ri in association with their respective guanine nucleotide regulatory proteins Ns and Ni (●, ▲, ■ and ○, △, □ respectively) which impinge on a common pool of catalytic activity. Occupancy of the stimulatory or inhibitory receptors by appropriate ligands induces a conformational change in the receptor-nucleotide regulatory complex (● → ▲ or ○ → △) to a form on which GTP can act, which results in a further conformational change (▲ → ■ or △ → □) which causes either increased or decreased activity; in formulation (b) stimulatory and inhibitory hormones induce similar conformational transitions of the receptor-regulatory unit complexes (RsNs → RsNs* or RiNi → RiNi*) upon which GTP may act to dissociate the oligomer to monomers and permit its interaction with the catalytic unit C.

A combination of different stimulatory or inhibitory hormones yield non-additive effects if the receptors in the complexes of either formulation are heterologous and if partial occupancy of the receptors in a complex can result in full conformational transitions

These schemes are purely tentative formulations which accommodate much of the data presented on dually regulated adenylate cyclase systems. Both schemes envisage a large precursor; the postulation of such a structure derives from the irradiation studies on liver and fat cell adenylate cyclase[39,40] discussed above. The precise nature of the components and their interactions, is of course, still unknown (as is the case for systems where only GTP stimulation occurs). The structures may involve an oligomer of regulatory units, as presented above, or monomers may be involved in association with repeating units of remnants of the cytoskeleton from the intact cell. The large structures also provide the means whereby multiple stimulatory and inhibitory interactions can occur, as in the fat cell. An important feature of both of these schemes is that a range of activity states is available and this permits a different distribution of these activities to occur in different systems. Thus GTP alone may permit expression of inhibition in one system due to an adequate concentration of the intermediate form of Ni or another system may be totally dependent on addition of hormone to accumulate sufficient quantities of this form on which GTP could then act. In the former case a GTP titration in the absence of another ligand will be biphasic, while in the latter the curve will show no inhibitory phase. In addition, in the former case, it is easy to suggest that sodium ions act to reverse the distribution of the GTP dependent inhibitory form, while in the latter insufficient quantities of this form exist, so that sodium is without effect.

6 CONCLUSIONS AND FUTURE DIRECTIONS

In this chapter we have tried to draw together the main lines of evidence from a variety of adenylate cyclase systems on the existence of an inhibitory regulatory circuit promoted by GTP and hormonal elements in addition to the familiar stimulatory regulation of the enzyme. We have suggested that in a growing number of systems the enzyme will have to be considered to be composed of Rs, Ri, Ns, Ni and C units. An appendix follows summarizing important technical considerations which require attention for the detection and adequate study of such systems. The fat cell appears to be a model for the complexity and flexibility achieved in such systems. The pivotal role of GTP in mediating these various positive and negative signals attests to the need for a full appreciation of the nature of the proteins mediating the effects of GTP. A tabulation of the widespread occurrence and a summary of the gross characteristics of bimodally regulated adenylate cyclase systems has been presented. Evidence has been provided that these two processes are functionally distinct, however the level at which this distinction can be made has not been defined, although it seems likely that distinct or uniquely organized proteins are involved.

The full biochemical characterization of Ns and Ni promises to be a more difficult problem than that posed by the purification and full functional reconstitution of a simple Ns unit. However the conceptual advances made on the simpler problem, particularly with regard to the application of a genetic approach involving the detection of mutants lacking in some of these components, should facilitate a solution. In addition a number of abnormal physiological and disease conditions including hypothyroidism, pseudohypoparathyroidism and adrenalectomy in which hormonal response is diminished without alteration in receptor concentration, appear to involve N-unit located lesions, which may also prove useful in the characterization of these components.[42–44]

It is likely that in the next few years many more adenylate cyclase systems will be seen to be dually regulated and the role of such regulation will be widely appreciated. In neural tissues, particularly, it appears likely that awareness of the existence of such circuitry will permit the detection of actions of many neurotransmitters *in vitro*, hitherto considered only as putative agents from electrophysiological studies.

7 APPENDIX; METHODOLOGICAL CONSIDERATIONS

Many of the regulatory features mediated by GTP and the detailed characteristics of dualistic GTP regulation may be overlooked if appropriate attention is not given to a number of potential technical complications. The major difficulties arise from a lack of control of either GTP concentrations or inhibitory ligands.

7.1 Endogenous GTP

Many particulate preparations and tissue homogenates are heavily contaminated with GTP. In view of the high intracellular GTP concentration (0.1–1 mM), this is not surprising. Consequently, it is often difficult to observe a GTP requirement for the hormonal stimulation of adenylate cyclase and by inference the regulatory contributions of low concentrations of GTP are obscured. Purified particulate fractions from density gradients and repeatedly washed preparations are less likely to be subject to this problem.

A second source of endogenous GTP is the ATP substrate employed. Kimura and Nagata[45] showed that commercially available highly purified equine muscle ATP was significantly contaminated with GTP. Material synthesized from adenosine is free of this problem.

7.2 Endogenous inhibitors

A number of tissues (neural, adrenal, platelets, etc.) store granules containing neurotransmitters in high concentration. Crude membrane preparations from these sources may well be significantly contaminated with inhibitory neurotransmitters, which could consequently obscure details of inhibitory regulation. Washing, etc., as outlined above may ameliorate this problem.

A variation of this problem is the generation of inhibitory substances during assays. A consequence of the generation of endogenous inhibitors is that unstable kinetics such as non-linearity with time, are encountered. A good example of such a situation is the generation of adenosine from the adenylate cyclase substrate, ATP (or the cyclic AMP 'trap'). Even in the presence of a regenerating system, it is quite conceivable that 1 per cent of the substrate may be metabolized to adenosine. Thus, from 0.1 mM substrate 1 μM adenosine can be formed. Such quantities fully saturate inhibitory receptors and consequently obscure inhibitory regulation. Two solutions are available;

(a) The use of adenosine deaminase which deaminates the nucleoside to inosine, which is ineffective at adenosine receptors.[21] This method permitted the first identification of inhibitory regulation by adenosine analogues in the fat cell.[11] However adenosine or its analogues that are susceptible to deaminase cannot be studied using this method.

(b) The use of 2'deoxy ATP as substrate with 2'deoxy cyclic AMP as protection against phosphodiesterase. This method has been detailed.[12] Metabolism of deoxy ATP and deoxy cyclic AMP yields deoxyadenosine which is ineffective at adenosine receptors. This provides a preferable solution to the use of deaminase since deaminatable analogues and adenosine itself can be studied and competition does not arise between adenosine receptors and adenosine deaminase for endogenous adenosine and added adenosine analogues. The cyclic AMP purification methodology of Salomon *et al.*[13] is readily modified to deal with the deoxy ATP system.[12]

A group of compounds which is also likely to be generated during adenylate cyclase assays are prostaglandins and their relatives. Many tissues synthesize these compounds, which can be effective GTP-dependent inhibitors of adenylate cyclase (see Section 4). Attempts at blocking their synthesis with indomethacin etc. can be considered.

A combination of both types of problems, i.e. endogenous GTP and endogenous inhibitors would probably totally obscure the regulatory capabilities of GTP and the presence of inhibitory circuitry. Therefore adequate delineation of the regulatory features of any adenylate cyclase system requires careful attention to these possibilities.

A further, more subtle problem which is difficult to control, is encountered when the effect of redox agents are considered. Certain agents, including ascorbate, which is commonly employed as a reducing agent in the study of catecholamines, suppress the expression of GTP inhibition. Stronger redox agents such as DTT reverse this effect (Cooper and Girardot, unpublished). Although these effects proved too variable for detailed study or determination of the mechanisms involved, this variability points to the need for care in the use of such reagents.

8 ACKNOWLEDGEMENTS

We would like to thank our colleagues Drs. D. E. Wright, R. Jagus and M. C. Lin for their useful suggestions on the manuscript, James Oden for expert technical assistance, Ms. Bonnie Richards for her excellent secretarial work and Dr. Martin Rodbell for his enthusiastic support and encouragement throughout these studies.

REFERENCES

1. Rodbell, M. (1980). *Nature*, **284** 17–22.
2. Ross, E. M., and Gilman, A. G. (1980). *Annual Review of Biochemistry*, **49,** 533–564.
3. Cryer, P. E., Jarett, L., and Kipnis, D. M. (1969). *Biochim. Biophys. Acta*, **177,** 586–590.
4. Rodbell, M. (1975). *J. Biol. Chem.*, **250,** 5826–5834.
5. Pairault, J. (1976). *Eu. J. Biochem.*, **62,** 323–334.
6. Cooper, D. M. F., Schlegel, W., Lin, M. C., and Rodbell, M. (1979). *J. Biol. Chem.*, **254,** 8927–8931.
7. Cooper, D. M. F., Londos, C., and Rodbell, M. (1980). *Mol. Pharmacol.*, **18,** 598–601.
8. Cooper, D. M. F., and Rodbell, M. (1979). *Nature*, **282,** 517–518.
9. Premont, J., Tassin, J.-P., Blanc, G., and Bockaert, J. (1979). In *Physiological and Regulatory Functions of Adenosine and Adenine Nucleotides*. Eds. H. P. Baer and G. I. Drummond, pp. 259–269, Raven Press, New York.
10. Harwood, J. P., Low, H., and Rodbell, M. (1973). *J. Biol. Chem.*, **248,** 6239–6245.
11. Londos, C., Cooper, D. M. F., Schlegel, W., and Rodbell, M. (1978). *Proc. Nat. Acad. Sci. (USA)*, **75,** 5362–5366.
12. Cooper, D. M. F., and Londos, C. (1979). *J. Cyc. Nuc. Res.*, **5,** 289–302.
13. Salomon, Y., Londos, C., and Rodbell, M. (1979). *Anal. Biochem.*, **58,** 541–548.
14. Londos, C., and Preston, M. S. (1977). *J. Biol. Chem.*, **252,** 5957–5961.
15. Jakobs, K. H., Aktories, K., and Schultz, G. (1981). *Adv. Cyc. Nuc. Res.*, **14,** 173–187.
16. Ebert, R., and Schwabe, U. (1974). *Naunyn-Schmied. Arch. Pharmakol.*, **286,** 297–313.
17. Yamamura, H., Lad, P. M., and Rodbell, M. (1977). *J. Biol. Chem.*, **252,** 7964–7966.

18. Pinkett, M. O., Jaworski, C. J., Evain, D., and Anderson, W. B. (1980). *J. Biol. Chem.*, **255,** 7716–7721.
19. Evain, D., and Anderson, W. B. (1979). *J. Biol. Chem.*, **254,** 8726–8729.
20. Lin, M. C., Cooper, D. M. F., and Rodbell, M. (1980). *J. Biol. Chem.*, **255,** 7250–7254.
21. Londos, C., Wolff, J., and Cooper, D. M. F. (1979). In *Physiological and Regulatory Functions of Adenosine and Adenine Nucleotides*, Ed. H. P. Baer and G. I. Drummond, pp. 271–281. Raven Press, New York.
22. Fain, J. N., and Malbon, C. C. (1979). *Mol. Cell Biochem.*, **23,** 1–27.
23. Sharma, S. K., Nirenberg, M., and Klee, W. (1975). *Proc. Nat. Acad. Sci., USA*, **72,** 590–594.
24. Jakobs, K. H., Saur, W., and Schultz, G. (1978). *FEBS Letters*, **85,** 167–170.
25. Jakobs, K. H., Aktories, K., and Schultz, G. (1979). *Naunyn-Schmied. Arch. Pharmakol.*, **310,** 113–119.
26. Lichshtein, D., Boone, G., and Blume, A. J. (1979). *J. Cyc. Nuc. Res.*, **5,** 367–375.
27. Blume, A. J., Lichshtein, D., and Boone, G. (1979). *Proc. Nat. Acad. Sci. USA*, **76,** 5626–5630.
28. Pert, C. B., and Synder, S. H. (1974). *Mol. Pharmacol.*, **10,** 868–879.
29. Tsai, B. S., and Lefkowitz, R. J. (1978). *Mol. Pharmacol.*, **14,** 540–548.
30. U'Prichard, D. C., and Snyder, S. H. (1978). *J. Biol. Chem.*, **253,** 3444–3452.
31. Sabol, S. K., and Nirenberg, M. (1979). *J. Biol. Chem.*, **254,** 1913–1920.
32. Lichshtein, D., Boone, G., and Blume, A. J. (1979). *Life Sciences*, **25,** 985–992.
33. Oron, Y., Creacy, S., Kellogg, J., and Larner, J. (1980). *J. Cyc. Nuc. Res.*, **6,** 105–120.
34. Aktories, K., Jakobs, K. H., and Schultz, G. (1980). *FEBS Letters*, **115,** 11–14.
35. DeCamilli, P., Macconi, D., and Spada, A. (1979). *Nature*, **278,** 252–254.
36. Londos, C., Cooper, D. M. F., and Wolff, J. (1980). *Proc. Nat. Acad. Sci. USA*, **77,** 2551–2554.
37. Kather, H., and Simon, B. (1981). *Adv. Cyc. Nuc. Res.*, **14,** 664.
38. Cassel, D., and Selinger, Z. (1977). *Proc. Nat. Acad. Sci. U.S.A.*, **75,** 4155–4159.
39. Schlegel, W., Cooper, D. M. F., and Rodbell, M. (1980). *Arch. Biochem. Biophys.*, **201,** 678–682.
40. Schlegel, W., Kempner, E. S., and Rodbell, M. (1979). *J. Biol. Chem.*, **254,** 5168–5176.
41. Sternweis, P. C., Northup, J. K., Hanski, E., Schleifer, L. S., Smigel, M. D., and Gilman, A. G. (1981). *Adv. Cyc. Nuc. Res.*, **14,** 23–36.
42. Ohisalo, J. J., and Stouffer, J. E. (1979). *Biochem. J.*, **178,** 249–251.
43. Levine, M. A., Downs, R. W., Jr., Singer, M., Marx, G. D., Aurbach, G. D., and Spiegel, A. M. (1980). *Biochem. Biophys. Res. Comm.*, **94,** 1319–1324.
44. Hanoune, J., Lacombe, M.-L., and Pecker, F. (1975). *J. Biol. Chem.*, **250,** 4509–4513.
45. Kimura, N., and Nagata, N. (1977). *J. Biol. Chem.*, **252,** 3829–3835.
46. Cooper, D. M. F., Londos, C., Gill, D. L., and Rodbell, M. (1982). *J. Neurochem.* (in press).

14 *Synthetic inducers of interferons*

P. F. Torrence

1 MOLECULES THAT STIMULATE INTERFERON PRODUCTION: A SURVEY OF STRUCTURAL DIVERSITY

In addition to viruses and certain intracellular microbes, a wide variety of different substances stimulate interferon production in cell culture or animals. The diversity of structural classes involved, coupled with the fact that different types of interferon may be produced, has thus far precluded any attempt to construct a unifying hypothesis regarding their mode of action. The various classes of molecules that induce interferon are:

   (i) double-stranded RNAs;
  (ii) low molecular weight compounds;
 (iii) polycarboxylates;
  (iv) endotoxins;
   (v) T or B lymphocyte stimulating agents.

In this overview, no attempt will be made to cover all these inducer classes; instead, the focus will be on the two classes which currently show the most promise for drug development because of their potent activity and because they are readily modified in structure. Various reviews[1,2] have dealt, more or less comprehensively, with the other classes of inducers; T or B lymphocyte stimulating agents;[3] endotoxin induction;[4] polycarboxylates;[1] viruses.[5]

2 DOUBLE-STRANDED RNAS

In general, double-stranded RNAs (dsRNAs) are the most potent, least toxic and most widely active of all the different classes of interferon inducers. One obvious possible explanation of their potency and wide spectrum of effectiveness is that dsRNA may be the actual intermediary of interferon induction during virus infection. Thus, for dsRNA viruses like

335

reovirus, the genome RNA would be the inducing molecule; For single-stranded RNA viruses,† like Mengovirus, the replicative intermediate or replicative form would act as the interferon inducer; for DNA viruses, like vaccinia virus or adenovirus, the hypothesis is more difficult to argue: Colby and Duesberg[6] presented evidence that interferon induction by vaccinia virus depended upon the generation of dsRNA intermediates of viral replication. Such dsRNA might arise during symmetrical transcription as has been reported with vaccinia virus, polyoma virus, simian virus 40, herpesvirus and adenovirus.[7] That dsRNA is the 'natural' inducer of interferon in viral infections remains an attractive, but unproven hypothesis.

Isaacs and his co-workers originally proposed that 'foreign' nucleic acids, that is, those heterologous to the cell in question, might be inducers of interferon.[8] This concept had to be abandoned since the experiments that supported such a concept could not be repeated by Isaacs' group or by others. Then, for a time after this, polysaccharides were considered as the actual interferon inducers,[9] and this idea was furthered when the active component of the antiviral substance statolon, a fermentation product of *Penicillium stoloniferum*, was erroneously identified as an anionic polysaccharide.[10] In 1967, polynucleotides became firmly established as potent interferon inducers when workers at Merck and Co. showed that both poly(I)·poly(C) and poly(A)·poly(U) were able to produce high titers of interferon in cell cultures and in animals,[11] moreover, it was also established that the active interferon inducing agent of the mycophage of *P. funiculosum* is dsRNA.[12] Reinvestigation of statolon showed that its active antiviral principle could also be attributed to dsRNA, derived from the mycophage that infected the *P. stoloniferum* cultures.[13]

To induce interferon, however, dsRNA need not be of phage or viral origin. Certain 'killer' strains of *Saccharomyces cereviseae* possess a dsRNA plasmid,[14] and isolated dsRNA from this plasmid is an active interferon inducer in mouse L cells.[15] The source of the interferon inducing dsRNA also may be eukaryotic cells: thus, Stern and Friedman,[16] Kimball and Duesberg[17] and De Maeyer *et al.*[18] isolated dsRNA from normal, apparently uninfected cells and found it induced interferon. In the latter instance, De Maeyer *et al.* demonstrated that the isolated dsRNA could confer protection on the same cells from which it was derived. This finding is contrary to the hypothesis of Isaacs and Lindenmann that interferon production is the cell's response to 'foreign' nucleic acid.

---

† For negative strand viruses, like VSV, an intriguing hypothesis is suggested. (Marcus, P. I., and Sekellick, M. J., *Nature*, **260**, 815–819 (1977).) A defective interfering (DI) particle from vesicular stomatitis virus has been shown to be a highly efficient interfering particle. This DI contains a single molecule of dsRNA. Might induction of interferon by VSV be due to the presence of a very small percentage of such DIs?

The discovery that a synthetic dsRNA, poly(I)·poly(C), could induce interferon[11] was cause for a great deal of excitement in the late 1960s. At that time interferon in sufficient quantities for clinical studies was nearly impossible to come by; thus, the hope developed that poly(I)·poly(C), or a similar molecule, might be used to induce interferon in humans. Studies with poly(I)·poly(C) in rodents and mice have been extremely promising (for review see Reference 19). For instance, poly(I)·poly(C) can protect mice against the following experimental virus infections: herpes simplex virus-, vaccinia virus-, encephalomyocarditis virus- and West Nile virus-induced encephalitis, progressive pneumonia and rabies. Poly(I)·poly(C) can also inhibit the growth of experimental tumours in mice, rats, rabbits and hamsters. The tumours may be virus-induced, chemically-induced, x-ray induced, transplanted or spontaneous. Poly(I)·poly(C) can also prevent viral oncogenesis. Intranasal administration of poly(I)·poly(C) in man resulted in a small reduction in the number and severity of common colds caused by rhinovirus challenge. Poly(I)·poly(C), is, however, a very poor inducer of interferon in man and shows no detectable antiviral or antitumour activities.

Several different approaches have evolved in the attempt to produce a 'superior' interferon inducer to poly(I)·poly(C).

The first approach is based on the observation that human (and primate) serum contains a nuclease activity that quickly hydrolyses and inactivates poly(I)·poly(C).[20-22] The hypothesis has been advanced that a nuclease resistant derivative of poly(I)·poly(C) might be a more effective interferon inducer. Two different substances have been developed for this purpose. The first is a complex of poly(I)·poly(C) with poly-L-lysine in carboxy-methylcellulose (poly ICLC).[23] The second is a chemically modified analogue of poly(I)·poly(C); namely, poly(I) complexed to poly(2-thio-cytidylic acid) [poly(I)·poly(s^{2}C)].[24] This analogue is resistant to degradation by RNase A and human serum. The RNase A resistance is due to the substitution of sulphur for oxygen at the 2 position of the pyrimidine ring. No studies in man with this latter poly(I)·poly(s^{2}C) complex have yet been reported, but extensive studies in primates and a phase I clinical trial in man has been reported with poly(ICLC).[25] Poly(ICLC) has been reported to protect rhesus monkeys against simian hemorrhagic fever,[26] Venezuelan equine encephalomyelitis[27] and rabies.[28] Poly(ICLC) also induces significant levels of interferons in monkeys, chimpanzees and, most significantly, in humans, but not without considerable toxic reactions.

A second approach to develop a superior inducer of interferon is based on the hypothesis that the interferon-inducing capacity of poly(I)·poly(C) does not require the prolonged maintenance of an intact double-helix; moreover, this hypothesis argues that the toxic responses to dsRNA are due to the persistence of the double-helical structure. The suggestion has thus been advanced that polynucleotide structures be designed that have very short lifetimes in human serum, lifetimes long enough to induce interferon, but not long enough to trigger toxic responses.[29,30] To fulfil these requirements, 'mismatched' analogues of poly(I)·poly(C) have been prepared. These analogues have an identical structure to poly(I)·poly(C) except for an occasional interruption of one strand with a base that cannot hydrogen bond to its opposing neighbour in the helix; for instance, the G-I opposition or the I-U opposition leads to 'mismatching'. Evidence has been presented that such 'mismatched' complexes are less toxic than poly(I)·poly(C) but are still able to induce high levels of interferon in experimental animals. No studies in man have been reported. This approach is directly opposite to the aforementioned approach of preparing a long-lived nuclease resistant derivative of poly(I)·poly(C). This diametric opposition of theoretical approaches underlines our ignorance of the interferon induction system, especially in *Homo sapiens*.

The third and last approach is more classical in nature and rephrases a longstanding question: are the receptors for polynucleotide interferon-inducers, on the one hand, and the toxic responses to polynucleotide interferon-inducers, on the other, sufficiently different in their structure-activity requirements to permit a separation of the beneficial and toxic responses by modification of polynucleotide structure? Unfortunately, thus far significant information regarding the polynucleotide structural requirements for interferon induction alone has been obtained. The next several paragraphs summarize current knowledge in this area.

2.1 Polynucleotide interferon-inducers must be double-stranded

Neither single- or triple-stranded polynucleotides are capable of interferon induction. The inactivity of triple-helical nucleic acids can be illustrated rather dramatically: poly(A)·poly(U) is an excellent inducer of interferon in primary rabbit kidney cell culture 'superinduced' with cycloheximide and actinomycin. A typical interferon titer for poly(A)·poly(U) in this system is 3000 units/ml; the addition of poly(U) or poly(I) together with the poly(A)·poly(U) gives a titer of ≤ 10 units/ml. This abolition of the interferon-inducing ability of poly(A)·poly(U) is due to the formation of a triple-helix:

$$\text{poly(A)·poly(U)} + \text{poly(U)} \rightarrow \text{poly(A)·2 poly(U)}^{31}$$

$$\text{poly(A)·poly(U)} + \text{poly(I)} \rightarrow \text{poly(A)·poly(U)·poly(I).}^{32}$$

The report[33,34] that certain preparations of poly(I) could induce titers of interferon *in vitro* and *in vivo* only slightly less than that induced by poly(I)·poly(C) does not constitute an exception to the requirement for double-strandedness. First of all, the polymers in question were prepared in an unusual manner by synthesis on an insolubilized enzyme as opposed to solution synthesis; secondly, reproducibily active preparations could not be prepared; third, these active poly(I) preparations reacted poorly to antibody to ('normal') poly(I); finally, 1 per cent of the total sample of these active poly(I) preparations reacted with antibody specific for dsRNA.[35] The hypothesis was advanced that the interferon-inducing poly(I) molecules contained local regions of double-helical configuration. These results suggested that a continuous high molecular weight base-paired structure is not an absolute requirement for a polynucleotide interferon inducer. Additionally, in regard to these anomalous poly(I) preparations, it is to be noted that poly(I) can assume multistranded structures depending on environmental conditions. Low potassium ion concentrations are sufficient to induce considerable strandedness in poly(I) solutions.[36]

2.2 The melting temperature (T_m) (or temperature at which one-half of the helix base-pairs have been disrupted) of the double-helix plays an important role in determining whether or not a given nucleic acid complex can induce interferon

This is a necessary outcome of the requirement that nucleic acid interferon inducers be double-helical; i.e. they must be intact under assay conditions. Generally, the most active interferon inducers have T_ms of $\approx 60°$ or greater (measured in phosphate buffered saline), but complexes with lower T_ms can be active inducers. For instance, poly(c^7I)·poly(I) has a T_m of $49°$ and is an active inducer *in vitro* and *in vivo*.[37] The hypothesis has been advanced that for some polynucleotide complexes both the T_m for helix to coil and the T_m for rearrangement to different helical forms are important. For instance, poly(A)·poly(U) and its analogues rearrange under certain conditions of temperature and ionic strength to a triply-stranded complex

$$2 \text{ poly(A)·poly(U)} \rightarrow \text{poly(A)·2 poly(U)} + \text{poly(A)}.$$

The potential rearrangement is a factor to be considered for all duplexes which are capable of triple-helix formation.[38] Finally, it is of interest that for a large number of helix-coil displacement reactions, the reaction proceeded in the direction of the helix with the higher T_m and greater interferon-inducing capacity.[39] For instance,

$$\text{poly}(c^7\text{A})\cdot\text{poly(rT)} + \text{poly(A)} \rightarrow \text{poly(A)}\cdot\text{poly(rT)} + \text{poly}(c^7\text{A})$$

T_m	47°	—	72°	—
IFN titer	<10	<10	3000	<10

2.3 Good polynucleotide interferon inducers are resistant to the action of nucleases

The T_m requirement may, to some extent, reflect this requirement; that is, the helix must survive intra- and/or extra-cellular nucleases in order to reach its receptor site. Double-helices with low T_ms 'breathe' extensively; these transient single-stranded regions are open to attack by nucleases with specificity toward single-stranded nucleic acids. This relationship between interferon-inducing ability and nuclease resistance is not a simple one, however. Complexes [like poly(A)·poly(dUz)[40]] with absolute resistance to nucleases, such as RNase A, are inactive as interferon inducers. On the other hand 'mismatched' complexes, with enhanced sensitivity to RNase A, are just as efficient inducers as poly(I)·poly(C).[29,30]

2.4 Interferon induction depends on the molecular weight of the nucleic acid complex and, where applicable, on the molecular weights of the individual homopolymers

The relationships and data obtained with poly(I)·poly(C) have been well-reviewed.[1] For example,[41] different sized poly(I)·poly(C) complexes were prepared from two different batches of poly(I) ($S_{20} = 12.5$ and 2.5) and two different batches of poly(C) ($S_{20} = 13.2$ and 3.0). (The different sizes of poly(I) and poly(C) will be referred to as high (H) and low (L), respectively.) Although the complex poly(I_H)·poly(C_L) induced 300 units IF/ml in interferon-primed mouse L cells, poly(I_L)·poly(C_L) produced <30 units IF/ml. The complex poly(I_H)·poly(C_L) and the complex poly(I_L)·poly(C_H) were more effective inducers than the preparation in which both strands were of low molecular weight, but, giving an interferon titer of 30 units/ml, they were significantly less active than the duplex constituted from the high molecular weight samples of poly(I) and poly(C).

A similar and more extensive study has been carried out with differently sized samples of poly(A) and poly(U).[42] The results of this study are summarized as follows:

(a) Interferon-inducing activity increased with the chain length of the uninterrupted double-helical portion in the polynucleotide complex; the number of such double-helical segments per molecule played a minor role. For instance, $(A)_{1225}·(U)_{1575}$ was more active than $(A)_{1225}·(U)_{350}$ which was more active than $(A)_{1225}·(U)_{100}$.

(b) At a constant concentration of constituting nucleotide residues, interferon-inducing activity increased with the number of double-stranded molecules available to the cell. For instance, $(A)_{90}·(U)_{100}$ was found to be more active than $(A)_{90}·(U)_{1575}$ or $(A)_{1225}·(U)_{100}$.

(c) When complexes have a $(U)_n$ strand that is much longer than the $(A)_n$ strand, inactive triple-helical structures, inactive as interferon inducers, may form. When a complex of one chain of $(A)_{410}$ and one chain of $(U)_{1575}$ was prepared, it was virtually devoid of interferon-inducing ability.

(d) Finally, in contrast to the situation for poly(I)·poly(C), for poly(A)·poly(U) complexes, it appeared that poly(A) and poly(U) strands, regardless of length, were of equal importance for the activity of the complex; i.e. $(A)_{410}\cdot(U)_{1575}$ was of the same activity as $(A)_{1225}\cdot(U)_{350}$ and $(A)_{1225}\cdot(U)_{100}$ induced as much interferon as $(A)_{90}\cdot(U)_{1575}$.

2.5 Interferon induction is critically dependent on the nature of the ribose-phosphate backbone[1,2] (Figures 1 and 2)

The phosphorothioate group can lead to a dramatic increase in antiviral activity and interferon-inducing ability in the case of the alternating copolymers poly(A-U) and poly(I-C). A similar modification of the poly(I) and/or poly(C) strand of poly(I)·poly(C) does not give any increase in activity. The nature of the group at ribose C2′ is of great importance in determining the interferon-inducing ability of a nucleic acid. For instance, the following complexes are completely inactive as interferon inducers: poly(Am)·poly(Um), poly(I)·poly(Cm), poly(Ae)·poly(U), poly(A)·poly(Ue), poly(A)·poly(dUf), *etc.* Poly(A)·poly(dUz), poly(I)·poly(dCz), poly(dAz)·poly(dUz) and poly(dAz)·poly(U) are all inactive as inducers of interferons, but remarkably, poly(dIz)·poly(C) is an active inducer both in tissue culture and in rabbits.[43]

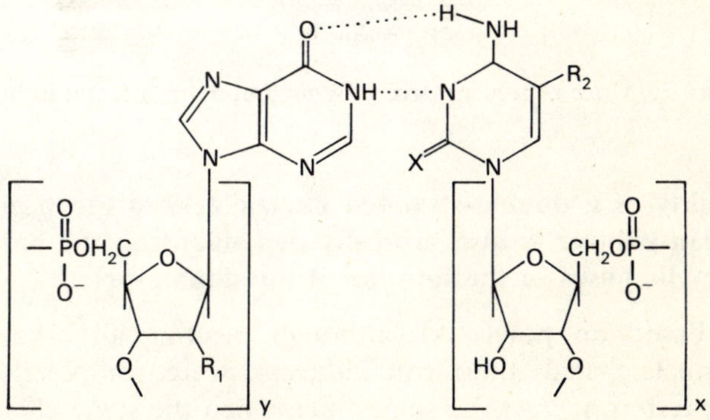

X=O, R_1=OH, R_2=H; Poly(I)·Poly(C)
X=O, R_1=OH, R_2=Br; Poly(I)·Poly(br^5C)
X=O, R_1=N_3, R_2=H; Poly(dIz)·Poly(C)
X=S, R_1=OH, R_2=H; Poly(I)·Poly(s^2C)

Figure 1. Some representative polynucleotide interferon inducers

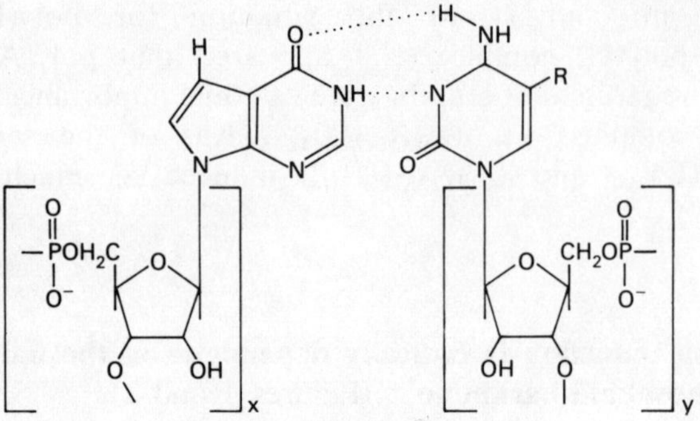

R=H; Poly(c⁷I)·Poly(C)

R=Br; Poly(c⁷I)·Poly(br⁵C)

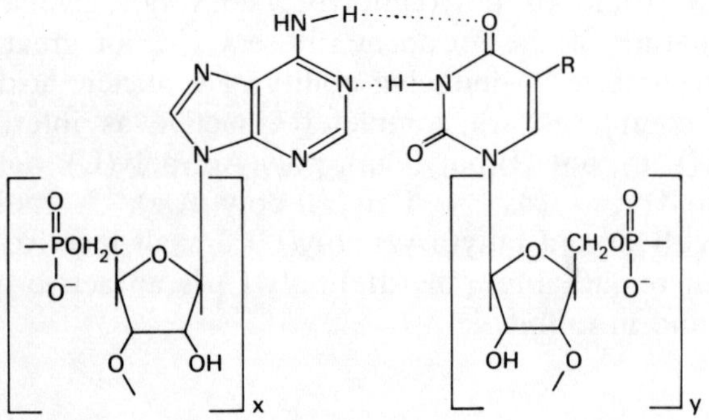

R=H; Poly(A)·Poly(U)

R=CH₃; Poly(A)·Poly(rT)

Figure 2. More representative polynucleotide interferon inducers

2.6 The ability of a double-stranded nucleic acid to function as an interferon-inducer is also critically dependent on the nature of the heterocyclic bases in the interior of the double-helix[1,4]

Duplexes based on poly(c⁷A), although meeting all other established criteria for nucleic acid interferon inducers, were completely inactive as inducers of interferon.[38] At the same time, when the same chemical modification was introduced into the poly(I)·poly(C) molecule, active inducers resulted *in vitro* and in vivo.[37] Other complexes, based on polylaurusin,[44] polyxanthylic acid[45] and poly(3-isoadenylic acid),[46] also failed to induce interferon even though they met all previously established requirements for polynucleotide inducers. Complexes derived from poly(3-deazaadenylic acid) and poly(3-deazainosinic acid) also failed to induce interferon, but, in these instances, thermal stability of the complexes was probably the critical determinant.[47] Similar considerations hold for complexes derived from poly-(2-azaadenylic acid) and poly(2-azainosinic acid).[48]

The fact that quite a variety of structural changes at various locations of the monomeric nucleotide structure can lead to an abolition of a polynucleotide's ability to induce interferon suggested that the overall conformation, i.e. the spatial and steric configuration, of the polynucleotide rather than specific site recognition, plays a determinant role in the interaction of the nucleic acid inducer with the putative interferon inducer receptor.[37,38,44] Studies with antibody to dsRNA showed that the same polynucleotide structural features that substantially alter recognition by antibody also lead to major changes in the interferon-inducing ability of the nucleic acid.[49] For example, alteration of one or both strands of the double-helix by replacement of 2'-hydroxyl to give 2'-deoxy, 2'-O-methyl, 2'-fluoro (in poly(U) or poly(C))-substituted polymers effected a dramatic drop in reactivity with antisera to poly(A)·poly(U). To this behaviour, there was one notable exception: poly(dIz)·poly(C), a good interferon inducer, reacted like poly(I)·poly(C).[43] These and other similarities provided circumstantial evidence to support the hypothesis, previously offered by Colby and Chamberlin,[50] that the putative dsRNA interferon inducer receptor site may be protein in nature.

The role of nucleic acid conformation in interferon induction was further indicated from circular dichroism studies of the non-interferon-inducing duplexes based on poly(7-deazaadenylic acid) and the interferon-inducing duplexes derived from poly(7-deazainosinic acid).[51] Substitution of CH for purine N7 of poly(A) or poly(I) affected the resultant duplex conformation of the two series in different ways. The change poly(A) → poly(c^7A) resulted in an increase of the positive base tilt relative to poly(A)·poly(U) whereas the change poly(I) → poly(c^7I) caused the resultant duplexes to undergo a decrease in the positive base tilt relative to poly(I)·poly(C). Thus, the two analogues, poly(c^7A) and poly(c^7I), underwent conformational changes in different directions compared to their parent polynucleotides. This departure in at least one conformational parameter was accompanied by a radical difference in interferon-inducing ability.

More recently, theoretical calculations on the conformational properties of adenosine, 7-deazaadenosine, inosine, 7-deazainosine and 2'-deoxyadenosine have suggested that the energetic accessibility of certain glycosidic orientations may determine the ability of a nucleic acid to induce interferon.[52] Specifically, the purine glycosidic bond angles in the vicinity of 20°, 80° and 160° may be necessary to confer interferon-inducing activity to polynucleotides. Inosine and 7-deazainosine were determined to possess roughly the same conformational stability in all three regions near 20°, 80° and 160°. Removal of the 2'-OH group destabilizes the high anti-region near 160°. Adenosine and 7-deazaadenosine differ markedly in their stability in the 60°–80° region. This divergence in conformational stabilities at the nucleoside level may be reflected in the divergent interferon-inducing abilities at the polynucleotide level.

The argument has been presented that, as a consequence of the foregoing considerations, the inactivity (or activity) of 2′-modified polynucleotides may have different origins. Consider that the putative interferon-inducer receptor for nucleic acids possesses a cleft or other recognition site that can accommodate the 2′-hydroxyl group, as Colby suggested. Replacement of the hydroxyl group by hydrogen changes the helix conformation from the A family of nucleic acids to the B (DNA) family. Thus the 2′-substituent is no longer in physical proximity to the recognition site. If 2′-hydroxyl is replaced by 2′-O-methyl, the conformation of the resulting duplex remains the same, but now the steric bulk of the methyl group would prevent bonding to the receptor site. This view does not limit active interferon inducers to polynucleotides with 2′-hydroxyl groups alone. It is reasonable to expect that other groups may be found that could mimic the hydroxyl group in binding to the receptor. Thus the interferon-inducing ability of poly(dIz)·poly(C)[43] is consistent with the hypothesis that interferon induction depends on a particular spatial and steric configuration of the double-helix. Incidently, the activity of poly(dIz)·poly(C) does not require that all other 2′-azido substituted polymers be active, since as we have seen, the overall conformation depends on a number of factors including the nature of the heterocyclic base.

The hypothesis that interferon induction is dependent on the conformation of the nucleic acid emphasizes the importance of the information conveyed by the two complementary helical strands acting as a single determinant. The observations[53–56] that interferon of several sources can bind to poly(I)-Sepharose, poly(U)-Sepharose and poly(A)-Sepharose and that poly(U) and poly(I) can displace interferon from columns of blue dextran-Sepharose, has led to the suggestion that direct interaction of interferon with RNA sequences might be important in the induction process. That poly(I) and poly(U) each constitute one chain of the inducers, poly(I)·poly(C) and poly(A)·poly(U), and also bind well to interferon, has been taken to implicate single-stranded structure in the induction process.[56] This hypothesis raises many questions: does poly(I)·poly(C), poly(A)·poly(U) or other established inducers bind interferon? They do not desorb interferon from blue dextran-Sepharose columns. Why does yeast tRNA not induce interferon? It desorbs interferon from blue dextran-Sepharose and, when bound to sepharose, it binds interferon. How does this hypothesis explain induction by natural dsRNAs or alternating ribocopolymers, like poly(A-U), which do not have poly(I) or poly(U) chains? More recently, it has been suggested that the polynucleotide binding site may be related to the polypeptide domain responsible for species specificity.[53]

Tan and Berthold have advanced the notion that inhibition of macromolecular synthesis is associated with ability to induce interferon.[57] They found that cycloheximide, actinomycin D, 2-[4-methyl-2,6-dinitroanilino]-*N*-methyl-propionamide and 1-(β-D-ribofuranosyl)-5,6-dichlorobenzimidazole could induce low levels (usually a few per cent of that induced by poly(I)·poly(C)) of interferon in certain lines of human fibroblasts. Their findings, coupled with established fact that many viruses as well as poly(I)·poly(C) can inhibit cell macromolecule synthesis, suggested that induction of interferon might be effected by reduction in the concentration of an interferon gene repressor. At least four dramatic exceptions to this correlation can be noted, however. Poly(I)·poly(br^{5}C), poly(c^{7}I)·poly(C), poly(c^{7}I)·poly(br^{5}C) and poly(I)·poly(s^{2}C) all are good to excellent interferon inducers in several cell lines as well as animals. None of these four duplexes can inhibit protein synthesis in cell-free extracts of interferon-treated L cells[41] and at least two of them (poly(I)·poly(br^{5}C) and poly(I)·poly(s^{2}C)) do not inhibit protein synthesis in rabbit reticulocyte lysates under conditions where poly(I)·poly(C) is a protein synthesis inhibitor.[58]

Very little is known of the putative nucleic acid interferon inducer receptor site itself. A number of attempts to discern the location (intracellular or in or on the cell membrane) have given equivocal answers since it has not been possible to completely eliminate the possibility of leakage of poly(I)·poly(C) from insoluble supports to where it has been bound.[59] Membrane changes can apparently alter the cell's response to poly(I)·poly(C). Concanavalin A does not change gross binding of poly(I)·poly(C) to 3T3 cells, but it does abolish interferon production in response to poly(I)·poly(C).[60] Neuraminadase or phospholipase C treatment of cells can block the antiviral effect of poly(I)·poly(C) without effecting its binding to the cell.[61] These and similar experiments suggest involvement of the cell membranes at some stage of the induction process, but it is not possible to know which step is affected and if the effect is specific. Moreover, it is quite likely that gross binding of nucleic acid to the membrane is not correlated with interferon induction, and only a small percentage (if any) of the bound molecules are involved in induction.

A few experiments have been directed toward understanding something of the nature of the nucleic acid interferon inducer receptor. Polynucleotides that do not induce interferon can be placed in two classes: (i) those which can inhibit the induction of interferon by poly(I)·poly(C) on other active polymers and (ii) those which have no effect on interferon induction by active nucleic acid inducers. For the first inhibitor class, three separate mechanisms of action have been advanced:[1] (a) inhibition of induction by specific interaction of the inhibiting polynucleotide with the inducing molecule; for instance, poly(U) or poly(I) inhibition of induction by poly(A)·poly(U) through formation of the inactive poly(A)·2 poly(U)[31] or poly(A)·poly(U)·poly(I) triplexes,[32] respectively. (b) Inhibition of induction through the antimetabolic activity of the interfering polynucleotide; for example, poly(c⁷A) can be degraded to the nucleoside antibiotic tubercidin which can inhibit RNA synthesis.[38,62] (c) For this third class of nucleic acid antagonists of interferon induction, the precise mechanism of action has not been unequivocally established. For instance, it has been found that certain base-modified duplexes[38,63] but not single-stranded polymers,[64] could reduce the interferon titer of poly(A)·poly(U) or poly(I)·poly(C). The hypothesis was forwarded that the inactivity of single-stranded and inactive double-helical polymers could have different roots. Single-stranded nucleic acids would fail to bind to the putative receptor whereas inactive duplexes could bind, but failed to trigger an interferon response since they could not attain a required conformation.[38,63] In a similar vein, circular dichroism studies[65] of poly(Im)·poly(C), poly(I)·poly(Cm) and poly(Im)·poly(Cm) showed that these duplexes, like poly(Am)·poly(U),[66] possessed nearly the same conformation as poly(I)·poly(C) itself, and these methylated polymers were capable of inhibiting interferon induction by poly(I)·poly(C). At the same time, deoxypolymers, like poly(dI)·poly(rC) which have a DNA like (B) conformation, could not effect interferon induction by poly(I)·poly(C).

Studies with copolymers of poly(Im, rI)·poly(Cm, rC) showed that the interferon-inducing capability of such duplexes correlated closely with the presence of clusters that contained six or more uninterrupted ribosides.[65] This finding, coupled with earlier data, has produced the hypothesis that the interaction of poly(I)·poly(C) with the cell (as far as interferon induction is concerned) may occur in two stages. The first would be a recognition of a large portion of the dsRNA to permit binding to the cellular receptor. The second stage would involve recognition of a smaller region of this dsRNA corresponding to 6–12 base-pairs, and this step would lead to triggering of the message for interferon production.

3 LOW MOLECULAR WEIGHT (LMW) INDUCERS
(Figures 3 and 4)

The discovery[67] that a small relatively non-toxic molecule (tilorone) could induce interferon was, from a purely practical viewpoint, of enormous importance. Rather than dealing with nucleic acids as drugs, a pharmacological concept without precedence, drug development could concentrate on more classical molecules for which the appropriate concepts and tools of drug development and improvement already existed. Moreover, quite a few of the small molecule inducers, tilorone included, possess oral activity, also a

1. Tilorone and Its Analogs

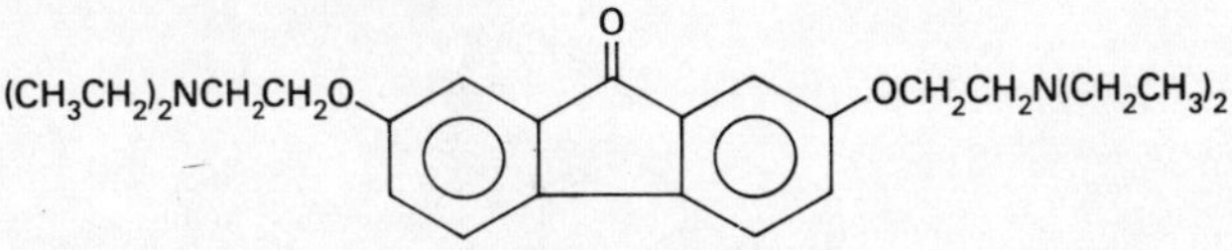

Tilorone (fluorenone ether)

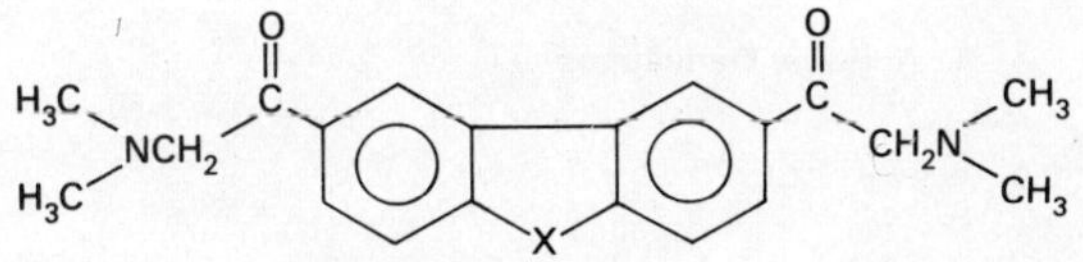

X=O; Dibenzofuran Ketone

X=S; Dibenzothiophene Ketone

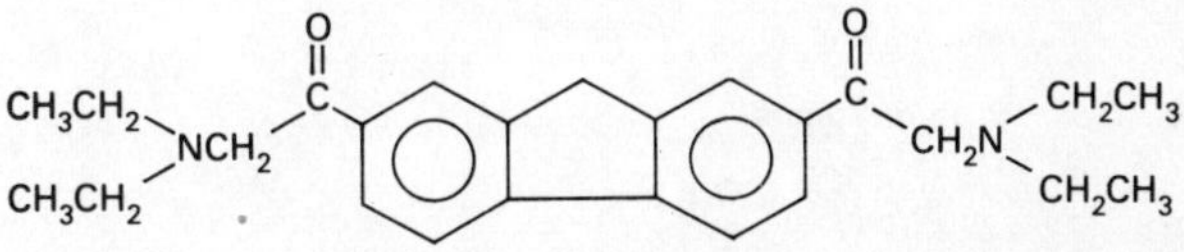

Fluorene Ketone

2. Propane Diamine Derivatives

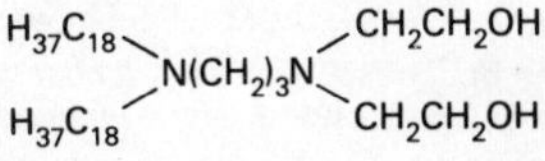

Figure 3. Some representative low-molecular-weitht interferon inducers

boon to drug development. Indeed, tilorone was not the first LMW interferon inducer described. Cycloheximide[68] and kanamycin[69] had earlier been shown to induce interferon; however, the dosages required were quite toxic to the treated animal. For cycloheximide, lethal intraperitoneal doses had to be administered to obtain an interferon response.

Altogether staggering is the difference in structure of the low molecular weight inducers from the dsRNAs. Moreover, the structural differences between the various small molecule classes are nearly as great. No generalizations emerge that relate structure to interferon-inducing ability. Only within narrowly-defined inducer classes is there any semblance of structure-activity relationships. For instance, it has been deduced that for tilorone and its congeners, the molecule should contain a bis-alkylamine moiety with free functional groups on, or adjacent to, a polycyclic aromatic or heteroaromatic nucleus if maximal activity is to be obtained. The implications of this statement and its application to specific analogues has been reviewed in detail elsewhere[70] and will not be expanded on here.

3. Pyrazolo[3,4b]quinolines

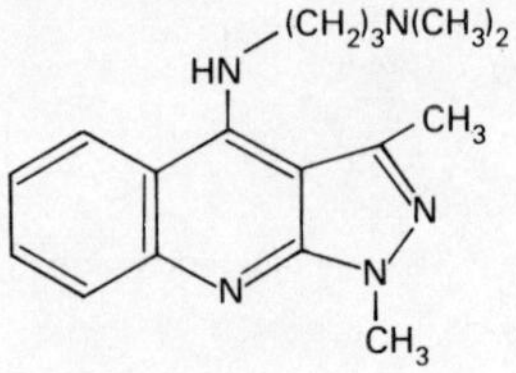

4. Acridine Derivatives

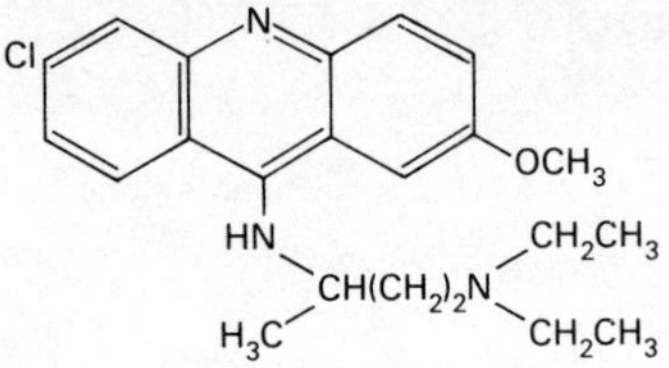

Atabrine

5. Pyrimidines

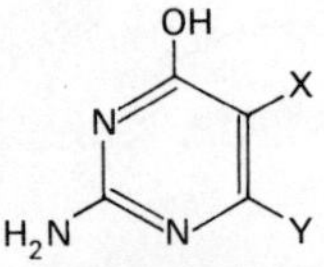

X=Br, Y=CH₃; ABMP, 5-bromo-6-methylisocytosine
X=Br, Y=Ø; ABPP, 5-bromo-6-phenylisocytosine
X=I, Y=Ø; AIPP, 5-iodo-6-phenylisocytosine

6. Mercaptoalkyl Amines

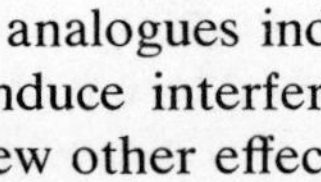

S-(2-aminoethyl)isothiourea

Figure 4. Representative low-molecular-weight interferon inducers (continued)

Tilorone and some of its analogues induce quite high levels of interferon in the mouse, but fail to induce interferon in any other animal, including man. Tilorone has quite a few other effects. It has demonstrated antitumour activity [e.g. against Walker carcinosarcoma 256, Morris hepatoma 7777].[71] In addition, tilorone stimulates both humoural immunity and complement, and has marked anti-inflammatory properties.[72,73]

Two compounds developed by Pfizer laboratories, N,N-dihexadecyl-m-xylylenediamine and N,N'-dioctadecyl-N',N'-bis(hydroxyethyl)propane-diamine, have been shown to induce interferon in nasal washing when administered to man as intranasal sprays.[74-77] A marked reduction in virus-shedding and symptoms from a rhinovirus challenge was obtained when *N,N'*-dioctadecyl-*N',N'*-bis(2-hydroxyethyl)propanediamine was administered as a suitably disperse preparation.[77] Neither of these compounds have been reported to possess activity in man when given by the oral or parenteral routes.

Agents with diverse pharmacological actions have been reported to induce interferon. Several established antimalarial agents, including methylene blue[78] and atabrine (quinacrine)[79] induce interferon, at least in mice. Certain radioprotective substances, such as the mercaptoalkylamine *S*-(2-aminoethyl)isothiourea and the thiophosphate *S*-(2-aminoethyl)thiophosphoric acids, gave low titers of interferon in mice.[80] Apparently, interferon is induced in mice by levamisole[81] which is an anthelmintic drug that also inhibits tumour growth and stimulates humoural and cell-mediated immunological reactions. One report claims interferon induction in mice upon parenteral administration of the biogenic amines serotonin, histamine or mexamine (5-methoxytryptamine).[82]

Several substituted pyrimidine derivatives have been found to induce interferon.[83–86] These include 2-amino-5-bromo-6-methylpyrimidin-4-ol (ABMP), 2-amino-5-bromo-6-phenylpyrimidin-4-ol (ABPP) and 2-amino-5-iodo-6-phenylpyrimidin-4-ol (AIPP). These isocytosines all possess significant antiviral properties in mice and cats[86] and ABPP is active against infectious bovine rhinotracheitis in calves.[87] The antiviral potency and interferon-inducing capacity of these pyrimidines may diverge, however. The iodo analogue (AIPP) was able to protect mice against Semliki Forest virus infection as the bromo analogue, ABPP, but AIPP induced much less interferon ($<$100 units/ml) than ABPP (5000 units/ml).[88] Whether or not these new agents will be active in primates or man remains to be established.

In general, the low-molecular weight inducers of interferon are virtually inactive in fibroblast cell cultures under conditions where viruses or dsRNAs are potent inducers. Several exceptions to this generality have been claimed. Usually near cytotoxic concentrations of inducer were used. Tilorone has been reported to induce low levels of interferon in human and mouse fibroblasts.[89] More extensive reports have documented the ability of these small molecule inducers to produce interferon in cultures of lymphoid origin. Tilorone induced interferon in normal and leukemic human lymphocytes,[90] but, strangely, not in mouse lymphocytes and macrophages of spleen, thymus or peritoneal origins.[91] All three pyrimidine derivatives, ABPP, AIPP and ABMP, induced mouse spleen, thymus or bone marrow cells to produce interferon.[92] The isocytosine, ABPP, induced interferon in human tonsillar tissue co-cultivated with amnion cells but AIPP did not.[86] The pyrazoloquinoline BL-20803 and several analogues have been shown to induce interferon in mouse spleen leukocytes.[93] Tilorone enhanced interferon production by poly(I)·poly(C) and DEAE-dextran in mouse L929 cells, but the mechanism of this enhancement is unknown.[94]

What kind of cells are involved in the interferon response to low molecular weight inducers? Lymphocytes are not likely to be involved in the case of tilorone since tilorone is still active in athymic nude mice and in anti-lymphocyte serum-treated mice.[95,96] For levamisole, on the other hand, T-lymphocytes are believed to be required since no induction could be detected in nude mice.[81] Siminoff has suggested the 'fixed' macrophage of the reticuloendothelial tissues as a target cell of the pyrazolo [3,4-b] quinoline derivatives.[97] In this regard, perhaps the small molecule inducers might share something in common with the dsRNAs and virus interferon inducers. For these latter agents, macrophages may well be the chief target cells both in mouse and man.

Is there any relationship between the mechanism of interferon induction by dsRNAs and the mechanism of interferon induction by small molecules like tilorone? Tilorone, some of its congeners and some other low molecular weight inducers have been shown to interact with nucleic acids. For instance, tilorone can elevate the T_m of calf thymus DNA,[98,99] and shows a specificity for the AT regions of DNA.[98,100–102] This interaction results in inhibition of polymerases that use DNA as a template.[103,104] Fluoranthene analogues of tilorone are good inhibitors of reverse transcriptase.[105] Tilorone and atabrine show anti-plasmid activity.[106,107] The propensity of many low molecular weight inducers to interact with nucleic acids suggests the possibility that this could be related to their mechanism of interferon induction. Could tilorone and related molecules somehow modify the properties of an organism's endogenous nucleic acids to convert them to interferon inducers? Such low molecular weight inducers also could be metabolized to a substance which effects the interaction with the endogenous nucleic acid. Finally, it is possible that the inducing substance is a metabolite that bears little or no resemblance to the drug itself.

Interferons possess a variety of properties in addition to their capacity to induce an antiviral state. Interferons may depress or enhance the immune response, prime cells for interferon production, suppress the growth of uninfected cells, enhance the toxicity of dsRNA, depress or enhance the synthesis of certain enzymes and inhibit tumour growth. These pleiotypic activities of interferon make it plausible to suspect that interferons may also have more than one pathway of induction. In this view, a unifying structure for all interferon inducers would not seem possible.

4 CONCLUSION

From a clinical perspective, interferon inducers are wrought with formidable problems. Their toxicity is certainly a major concern. Reactions such as pyrogenicity, hypotension, leucopenia and myalgia, as recently described for polyICLC, cannot be lightly dismissed. The problem of hyporeactivity (a decreased ability to produce interferon after initial stimulation) is also a possible major limitation. Whether or not this can be overcome remains to be established. A relative lack of activity in humans, compared to other vertebrates, has been a consistent problem with nearly all promising inducers from poly(I)·poly(C) to tilorone. All these considerations notwithstanding, an effective inducer of interferon in humans would be of great interest even in view of the recent bacterial cloning of the interferon gene. For instance, the concurrent administration of exogenous interferon together with a potent inducer could give serum levels of interferon much higher than that attainable with either alone. Such a consideration could be vital in the therapy of either viral disease or malignant growth.

5   ABBREVIATIONS

Standard abbreviations for nucleic acids are used throughout this text. For instance, poly(I)·poly(C) refers to the complex formed when polyinosinic acid (poly(I)) is mixed with an equimolar quantity of polycytidylic acid (poly(C)). Other less familiar abbreviations include poly(s^2C), poly(2-thiocytidylic acid); poly(ICLC), a complex of poly(I)·poly(C) with polylysine and carboxymethylcellulose; poly(br^5C), poly(5-bromocytidylic acid); poly(c^7I), poly(7-deazainosinic acid); poly(c^7A), poly(7-deazaadenylic acid); poly(dIz), poly(2'-azido-2'-deoxyinosinic acid); poly(Am), poly(2'-O-methyladenylic acid); poly(Ue), poly(2'-O-ethyluridylic acid).

REFERENCES

1. Torrence, P. F., and De Clercq, E. (1977). *Pharmacol. Ther.*, **A2,** 1–88.
2. De Clercq, E. (1977). *Texas Rep. Biol. Med.*, **35,** 29–38.
3. Epstein, L. B. (1977). *Texas Rep. Biol. Med.*, **35,** 42–56.
4. Nagano, Y. (1977). *Texas Rep. Biol. Med.*, **35,** 105–110.
5. Jameson, P. (1977). *Texas Rep. Biol. Med.*, **35,** 84–90.
6. Colby, C., and Duesberg, P. H. (1969). *Nature*, **222,** 940–944.
7. Boone, R. F., Parr, R. P., and Moss, B. (1979). *J. Virol.*, **30,** 365–374.
8. Isaacs, A., Cox, R. A., and Rotem, A. (1963). *Lancet*, **2,** 113–116.
9. Isaacs, A. (1965). *Aust. J. Exp. Biol. Sci.*, **43,** 405–412.
10. Kleinschmidt, W. J., and Probst, G. W. (1962). *Antibiot. Chemother.*, **12,** 298–309.
11. Field, A. K., Tytell, A. A., Lampson, G. P., and Hilleman, M. R. (1967). *Proc. Nat. Acad. Sci. U.S.A.*, **58,** 1004–1010.
12. Lampson, G. P., Tytell, A. A., Field, A. K., Nemes, M. M., and Hilleman, M. R. (1967). *Proc. Nat. Acad. Sci. U.S.A.*, **58,** 782–789.
13. Kleinschmidt, W. J., Ellis, L. F., Van Frank, R. M., and Murphy, E. B. (1968). *Nature*, **220,** 167–168.
14. Wickner, R. B. (1976). *Bacteriol. Rev.*, **40,** 757–773.
15. Bevan, E. A., Herring, A. J., and Mitchell, D. J. (1973). *Nature*, **245,** 81–86.
16. Stern, R., and Friedman, R. M. (1970). *Nature*, **226,** 612–616.

17. Kimball, P. C., and Duesberg, P. H. (1971). *J. Virol.*, **7**, 697–706.
18. De Maeyer, E., De Maeyer-Guignard, J., and Montagnier, L. (1971). *Nature New Biol.*, **229**, 109–110.
19. De Clercq, E. (1976). In 'Structure and properties of biopolymers'. *Proceedings of the First Cleveland Symposium on Macromolecules Held at Case Western Reserve University*, Cleveland, Ohio (G. A. Walton, ed.), North Holland Elsevier/Excerpta Medica, Amsterdam, pp. 217–243.
20. Stern, R. (1970). *Biochem. Biophys. Res. Commun.*, **41**, 608–614.
21. Nordlund, J. J., Wolff, S. M., and Levy, H. B. (1970). *Proc. Soc. Exp. Biol. Med.*, **133**, 439–444.
22. De Clercq, E. (1979). *Eur. J. Biochem.*, **93**, 165–172.
23. Levy, H. B. (1977). *Texas Rep. Biol. Med.*, **35**, 91–95.
24. Reuss, K., Scheit, K.-H., and Saiko, O. (1976). *Nucleic Acids Res.*, **3**, 2861–2875.
25. Levine, A. S., Sivulich, M., Wiernik, P. H., and Levy, H. B. (1979). *Cancer Research*, **39**, 1645–1650.
26. Levy, H. B., London, W., Fuccillo, P. A., Baron, S., and Rice, J. (1976). *J. Infect. Dis.*, **133**, A256–A259.
27. Stephen, E. L., Hilmas, D. E., Levy, H. B., and Spertzel, R. O. (1979). *J. Infect. Dis.*, **139**, 267–272.
28. Baer, G. M., Shaddock, J. H., Moore, S. A., Yager, P. A., Baron, S. B., and Levy, H. B. (1977). *J. Infect. Dis.*, **136**, 286–291.
29. Ts'O, P. O. P., Alderfer, J. L., Levy, J., Marshall, L. W., O'Malley, J., Horoszewicz, J. S., and Carter, W. A. (1976). *Molec. Pharmacol.*, **12**, 299–312.
30. Carter, W. A., O'Malley, J., Beeson, M., Cunnington, P., Kelvin, A., Vere-Hodge, A., Alderfer, J. L., and Ts'O, P. O. P. (1976). *Molec. Pharmacol.*, **12**, 440–453.
31. Torrence, P. F., De Clercq, E., and Witkop, B. (1976). *Biochemistry*, **15**, 724–734.
32. De Clercq, E., Torrence, P. F., De Somer, P., and Witkop, B. (1975). *J. Biol. Chem.*, **250**, 2521–2531.
33. Thang, M. N., Bachner, L., De Clercq, E., and Stollar, B. D. (1977). *FEBS Letters*, **76**, 159–165.
34. De Clercq, E., Stollar, B. D., and Thang, M. N. (1978). *J. Gen. Virol.*, **40**, 203–212.
35. Stollar, B. D., De Clercq, E., Drocourt, J.-L., and Thang, M. N. (1978). *Eur. J. Biochem.*, **82**, 339–346.
36. Miles, H. T., and Frazier, J. (1978). *J. Amer. Chem. Soc.*, **100**, 8037–8038.
37. Torrence, P. F., De Clercq, E., Waters, J. A., and Witkop, B. (1974). *Biochemistry*, **13**, 4400–4408.
38. De Clercq, E., Torrence, P. F., and Witkop, B. (1974). *Proc. Nat. Acad. Sci. U.S.A.*, **71**, 182–186.
39. De Clercq, E., Torrence, P. F., and Witkop, B. (1976). *Biochemistry*, **15**, 717–724.
40. Torrence, P. F., Waters, J. A., Buckler, C. E., and Witkop, B. (1973). *Biochem. Biophys. Res. Commun.*, **52**, 890–898.
41. Torrence, P. F., and Friedman, R. M. (1979). *J. Biol. Chem.*, **254**, 1259–1267.
42. Janik, B., De Clercq, E., and Sommer, R. G. (1978). *Biochim. Biophys. Acta*, **517**, 269–273.
43. De Clercq, E., Torrence, P. F., Stollar, B. D., Hobbs, J., Fukui, T., Kakiuchi, N., and Ikehara, M. (1978). *Eur. J. Biochem.*, **88**, 341–349.

44. Torrence, P. F., De Clercq, E., Waters, J. A., and Witkop, B. (1975). *Biochem. Biophys. Res. Commun.*, **62,** 658–664.
45. Torrence, P. F., De Clercq, E., and Witkop, B. (1977). *Biochim. Biophys. Acta,* **475,** 1–6.
46. Uchic, J. T. (1975). *Abstracts of the 170th National Meeting of the American Chemical Society*, abstract MEDI 43.
47. De Clercq, E., Torrence, P. F., Fukui, T., and Ikehara, M. (1976). *Nucleic Acids Res.*, **3,** 1591–1601.
48. De Clercq, E., Huang, G. F., Torrence, P. F., Fukui, T., Kakiuchi, N., and Ikehara, M. (1977). *Nucleic Acids Res.*, **4,** 3643–3653.
49. Johnston, M. I., Stollar, B. D., Torrence, P. F., and Witkop, B. (1975). *Proc. Nat. Acad. Sci. U.S.A.*, **72,** 4564–4568.
50. Colby, C., and Chamberlin, M. J. (1969). *Proc. Nat. Acad. Sci. U.S.A.*, **63,** 160–167.
51. Bobst, A. M., Torrence, P. F., Koruidou, S., and Witkop, B. (1976). *Proc. Nat. Acad. Sci. U.S.A.*, **73,** 3788–3792.
52. Miles, D. L., Miles, D. W., and Eyring, H. (1979). *Proc. Nat. Acad. Sci. U.S.A.*, **76,** 1018–1021.
53. Thang, M. N., Thang, D. C., Chelbi-Alix, M. K., Robert-Galliot, B., Commoy-Chevalier, M. J., and Chany, C. (1979). *Proc. Nat. Acad. Sci. U.S.A.*, **76,** 3717–3721.
54. Wietzerbin, J., Stefanos, S., Lucero, M., and Falcoff, E. (1978). *Biochem. Biophys. Res. Commun.*, **85,** 488–489.
55. De Maeyer-Guignard, J., Tovey, M. G., Greser, I., and De Maeyer, E. (1978). *Nature*, **271,** 622–625.
56. De Maeyer-Guignard, J., Thang, M.-N., and De Maeyer, E. (1977). *Proc. Nat. Acad. Sci. U.S.A.*, **74,** 3787–3790.
57. Tan, Y. H., and Berthold, W. (1977). *J. Gen. Virol.*, **34,** 401–411.
58. Content, J., Lebleu, B., and De Clercq, E. (1978). *Biochemistry*, **17,** 88–94.
59. Pitha, P. M. (1977). *Texas Rep. Biol. Med.*, **35,** 39–41.
60. Harper, H. D., and Pitha, P. M. (1973). *Biochem. Biophys. Res. Commun.*, **53,** 1220–1226.
61. Pitha, P. M., Harper, H. D., and Pitha, J. (1974). *Virology*, **59,** 40–50.
62. De Clercq, E., Billiau, A., Torrence, P. F., Waters, J. A., and Witkop, B. (1975). *Biochem. Pharmacol.*, **24,** 2233–2238.
63. De Clercq, E., Torrence, P. F., Witkop, B., and De Somer, P. (1975). In *Effects of Interferon on Cells, Viruses and the Immune System* (ed. A. Geraldes), Academic Press, New York, pp. 215–236.
64. De Clercq, E., Rottman, F. M., and Shugar, D. (1974). *FEBS Letters*, **42,** 331–334.
65. Greene, J. J., Alderfer, J. L., Tazawa, I., Tazawa, S., Ts'o, P. O. P., O'Malley, J. A., and Carter, W. A. (1978). *Biochemistry*, **17,** 4214–4218.
66. Bobst, A. M., Rottman, F., and Cerutti, P. A. (1969). *J. Mol. Biol.*, **46,** 221–234.
67. Mayer, G. D., and Krueger, R. F. (1970). *Science*, **169,** 1214–1215.
68. Younger, J. S., Stinebring, W. R., and Taube, S. E. (1965). *Virology*, **27,** 541–550.
69. Lukas, B., and Hruskova, J. (1968). *Acta Virol.*, **12,** 263–267.
70. Mayer, G. D. (1980). *Pharmac. Ther.* **8,** 173–192.
71. Adamson, R. H. (1971). *J. Nat. Cancer Inst.*, **46,** 431–434.
72. Hoffman, P. F., and Ritter, H. W. (1971). *J. Ret. Soc.*, **9,** 638–639.

73. Mayer, G. D. and Krueger, R. F. (1980). *Interferon and Interferon Inducers,* volume 17 of *Modern Pharmacology and Toxicology* (ed. D. A. Stringfellow), Marcel Dekker, Inc., New York, pp. 197–221.
74. Hoffman, W. W., Lorst, J. J., Niblack, J. F., and Cronin, T. H. (1973). *Antimicrob. Ag. Chemother.,* **3,** 498–502.
75. Douglas, R. G., Jr., and Betts, R. F. (1974). *Infect. Immun.,* **9,** 506–510.
76. Gatmaitan, B. G., Stanley, E. D., and Jackson, G. G. (1973). *J. Infect. Dis.,* **127,** 401–407.
77. Panusarn, C., Stanley, E. D., Dirda, V., Rubenis, M., and Jackson, G. G. (1974). *New Engl. J. Med.,* **291,** 57–61.
78. Diderich, J., Lodemann, E., and Wacker, A. (1973). *Arch. Ges. Virusforsch.,* **40,** 82–86.
79. Glaz, E. T., Szolgay, E., Stoger, I., and Talas, M. (1973). *Antimicrob. Ag. Chemother.,* **3,** 537–541.
80. Lvovsky, E. A., Levy, H. B., Doherty, D. G., and Baron, S. (1977). *Infect. Immun.,* **15,** 191–196.
81. Matsubara, S., Suzuki, F., and Ishida, N. (1979). *Cell. Immunol.,* **43,** 214–219.
82. Lvovsky, E., Dubuy, H. G., and Levy, H. B. (1977). Abstr. 77th Annual Meeting of the American Society for Microbiology, New Orleans, Louisiana, p. 290.
83. Nichol, F. R., Weed, S. D., and Underwood, G. E. (1976). *Antimicrob. Ag. Chemother.,* **9,** 433–439.
84. Stringfellow, D. A. (1977). *Antimicrob. Ag. Chemother.,* **11,** 984–992.
85. Wierenga, W., Skulnick, H. I., Weed, S. D., and Stringfellow, D. A. (1979). *Abstracts of the 11th International Congress of Chemotherapy and the 19th Interscience Conference on Antimicrobial Agents and Chemotherapy,* Boston, Massachusetts, Abstract No. 262.
86. Stringfellow, D. A., Vanderberg, H. C., and Weed, S. D. (1980). *Abstracts of the 11th International Congress of Chemotherapy and the 19th Interscience Conference on Antimicrobial Agents and Chemotherapy,* Boston, Massachusetts, Abstract No. 264.
87. Hamdy, A. H., and Stringfellow, D. A. (1980). *Abstracts of the 11th International Congress of Chemotherapy and the 19th Interscience Conference on Antimicrobial Agents and Chemotherapy,* Boston, Massachusetts, Abstract No. 263.
88. Weed, S. D., Kramer, G. D., and Stringfellow, D. A. (1980). *Abstracts of the 11th International Congress of Chemotherapy and the 19th Interscience Conference on Antimicrobial Agents and Chemotherapy,* Boston, Massachusetts, Abstract No. 265.
89. Degre, M., and Glaz, E. T. (1977). *Acta Pathol. Microbiol. Scand. [B],* **85,** 189–194.
90. Dennis, A. J., Wilson, H. E., Barker, A. D., and Rheins, M. S. (1972). *Proc. Soc. Exp. Biol. Med.,* **141,** 782–785.
91. De Clercq, E., and Merrigan, T. C. (1971). *J. Infect. Dis.,* **123,** 190–199.
92. Hamilton, R. D., Buthala, D. A., Eidson, E. E., and Tomilo, A. (1980). *Abstracts of the 11th International Congress of Chemotherapy and the 19th Interscience Conference on Antimicrobial Agents and Chemotherapy,* Boston, Massachusetts, Abstract No. 266.
93. Simonoff, P. (1975). *Infect. Immun.,* **12,** 1051–1057.
94. Groelke, J. W., Camyre, K. P., and Mayer, G. D. (1975). *Proc. Soc. Exp. Biol. Med.,* **148,** 1044–1047.

95. Stringfellow, D. A., and Glasgow, L. A. (1972). *Antimicrob. Ag. Chemother.*, **2**, 73–78.
96. Gibson, J. P., Megel, H., Camyre, K. P., and Michael, J. G. (1976). *Proc. Soc. Exp. Biol. Med.*, **151**, 264–266.
97. Siminoff, P. (1976). *J. Infect. Dis.* (*supplement A*), **133**, 37–42.
98. Chandra, P., Zunino, F., Gaur, V. P., Zaccara, A., Woltersdorf, M., Luoni, G., and Gotz, A. (1972). *FEBS Letters*, **28**, 5–9.
99. Chandra, P., Zunino, F., Zaccara, A., Wacker, A., and Gotz, A. (1972). *FEBS Letters*, **23**, 145–148.
100. Chandra, P., and Woltersdorf, M. (1974). *FEBS Letters*, **41**, 169–173.
101. Daune, M., Sturn, J., and Zana, R. (1976). *Stud. Biophys.*, **57**, 139–144.
102. Chandra, P., and Woltersdorf, M. (1976). *Biochem. Pharmacol.*, **25**, 877–880.
103. Chandra, P., Will, G., Gericke, D., and Gotz, A. (1974). *Biochem. Pharmacol.*, **23**, 3259–3265.
104. Chandra, P., and Wright, G. J. (1977). *Topics in Current Chemistry*, **72**, 125–148.
105. Green, M., Rankin, A., Gerard, G. F., Grandgenett, D. P., and Green, M. R. (1975). *J. Nat. Cancer Inst.*, **55**, 433–442.
106. Hahn, F. E., and Ciak, J. (1976). *Antimicrob. Ag. Chemother.*, **9**, 77–80.
107. Hahn, F. E., and Ciak, J. (1977). *Antimicrob. Ag. Chemother.*, **11**, 176–177.

15 The mechanism of action of interferon

J. A. Lewis

Over the last 10 years great strides have been made with regard to our understanding of the ability of interferon to inhibit virus growth. We now know of a variety of mechanisms by which protein synthesis may be inhibited as a result of interferon action but are still ignorant as to which, if any, of these pathways are actually responsible for limiting virus production in the intact, infected cell. This chapter will be concerned with providing an overview of our current understanding of the molecular processes which could account for the anti-viral action of interferon and what little we do know of events occurring in infected cells previously exposed to interferon. Very little is known about the basis for interferon's various other effects (growth regulation, anti-tumour activity, immuno-modulation) and these aspects will undoubtedly become very active fields of research in the near future.

1 ESTABLISHMENT OF THE ANTI-VIRAL STATE

The precise steps involved in establishing an interferon mediated anti-viral state are not yet understood. The fact that interferon exerts a pleiotypic response and appears to require a cell surface receptor for its activity suggests that it probably acts in much the same way as a polypeptide hormone. Two stages can be defined; an initial phase in which extracellular interferon interacts with a plasma membrane receptor and a later phase in which this interaction signals a series of alterations in the intracellular ambient.

A requirement for a cell-surface receptor is indicated by the fairly high degree of species specificity exhibited by most interferons. Somatic cell genetics studies have extended this point by the demonstration that, in mouse–human or hamster–human hybrid cells, the sensitivity to human interferon is correlated with the presence in the hybrid cell lines of human chromosome 21. Furthermore, cell lines trisomic for this chromosome (e.g. fibroblasts from patients with Down's syndrome) were considerably more sensitive to interferon action than haploid or diploid human cell lines.[1] The sensitivity to interferon is probably not determined solely by chromosome 21 since triploid cells (i.e. containing three copies of all chromosomes) are not more sensitive to its action[2] than diploid cells, but this observation probably reflects the existence of regulatory elements on other chromosomes. There is some suggestive evidence that chromosome 21 contains the structural gene for an interferon receptor[2,3] but this point has not been established definitively.

Some direct studies also suggest that interferon must bind to a surface receptor. If interferon is added to cells at 4 °C for a brief period and the cells are then thoroughly rinsed and incubated at 37 °C, an anti-viral state will develop.[4] If the cells are treated with proteases before warming or if anti-interferon sera are added to the cells no anti-viral state develops.[4] Such results would be consistent with the binding of interferon to a specific receptor at 4 °C and a requirement for normal metabolic activity for this interaction to induce an anti-viral state. When low doses of interferon are added to homologous cells no interferon can be recovered from the cells by washing[5] but with increasing amounts of the interferon preparation a point is reached at which some interferon can be recovered.[6,4]

In addition to the specific interferon receptor (protein?) molecule there is a requirement for the presence of certain cell-surface gangliosides and it has been postulated[7] that interferon is first bound to a ganglioside fraction which facilitates interaction with the true receptor. Interestingly, Falcoff[8] has recently presented evidence that gamma interferon does not require gangliosides in order to exert its anti-viral effect and is active in inhibiting the growth of a line of mouse leukemia cells (L1210R) which are resistant to the effects of beta interferon. Thus, although the gangliosides may play quite an important role in the binding of some interferons, they may not be essential for further steps in the induction process. An analogous ganglioside requirement exists in the interaction of cells with cholera, tetanus and botulinum toxins or thyrotropin and human chorionic gonadotrophin hormones and these substances have been shown to prevent the establishment of an anti-viral state by interferon.[9]

There has been some debate as to the need for interferon to penetrate into the cell in order to exert its action. Several studies in which interferon was coupled to large diameter beads and still retained its activity[10,11] can be interpreted in favour of a wholly external function but, owing to the potency of the molecule, the release of a small percentage of the interferon from the beads could account for the results observed. It is extremely difficult to design unambiguous experiments of this type. In addition, even if interferon is active from outside the cell one may ask if, during the normal course of events, it is taken up into the cell as is the case for insulin. Internalization, and possibly degradation, may well be a natural concomitant of receptor binding. Some evidence exists for this view since interferon bound to cells at 4 °C is trypin sensitive whereas interferon can be recovered from the cell sap and membrane fractions of cells incubated at 37 °C in which recoverable interferon is resistant to trypsin.[12,13]

It is generally presumed that interaction of interferon with its receptor results in the production of some kind of signal which will lead to gene derepression. The nature of this signal is unknown. Tovey[14] has reported that cyclic GMP levels rise 2–4 fold within minutes of treating cells with high doses of interferon (several thousand units/ml) and cyclic AMP levels rise several hours later.[15] However, neither cGMP nor cAMP nor agents such as prostaglandins (which modify intracellular concentrations of cyclic nucleotides) cause the establishment of an anti-viral state and it is difficult to understand the specificity of interferon action on this basis. Bourgeade and Chany[16] have reported that drugs which disrupt the cytoskeleton also block development of the anti-viral state and a role for this structure may be inferred.

It has been demonstrated repeatedly that the synthesis of both RNA and protein is required for several hours after treatment with interferon if an anti-viral effect is to be observed.[17,18,19] This obviously suggests that derepression of the cellular genome is required for interferon to exert its action and a requirement for RNA and protein synthesis is frequently used as one of several criteria for defining a substance as interferon. Recently, several groups have succeeded in demonstrating the synthesis of specific proteins in cells treated with interferon. By labelling intact chick embryo fibroblast cells with radioactive amino acids several hours after interferon treatment Ball[20] has observed the synthesis of a new protein of molecular weight 56,000 which is not labelled in control cells; this protein appears to correspond to an enzyme activity which is known to be induced by interferon (see later). Knight and Korant[21] analysed the proteins synthesized by interferon-treated human fibroblast cells by two dimensional gel electrophoresis and identified 4 new proteins in the molecular weight range 44,000 to 68,000. Similar results have been obtained by Gupta et al.[22] who used high resolution polyacrylamide gel electrophoresis to demonstrate that human fibroblasts synthesize at least 5 new proteins (Mr 56,000, 67,000, 80,000, 88,000 and 120,000) after treatment with interferon. De Ley et al. (personal communication, May, 1979) used a similar procedure and identified a protein of Mr 63,000, the synthesis of which was induced by interferon. Farrell et al.[23] have extracted total cellular mRNA from control and interferon-treated cells and translated these fractions in cell-free protein synthesizing systems. After interferon treatment mouse Ehrlich Ascites tumour cells contain at least one new species of mRNA which is translated into a protein of Mr

14,500 whilst HeLa cells contain several interferon induced mRNA species which code for proteins of Mr 15,000 and 50,000–60,000. To date, only the protein demonstrated by Ball[20] can be tentatively correlated with known enzymatic activities but it seems probable that the other species will soon be identified.

Evidence is now accumulating that the induction of these 'anti-viral' proteins, as they have been rather unfortunately termed in the past, leads to the presence in interferon-treated cells of a variety of new enzyme activities (or at least a large stimulation in their level) which are directly responsible for blocking viral replication. The role of these various proteins and enzymes in some of the other interferon related phenomena must now also be studied.

2　WHAT STEP IN VIRUS REPLICATION IS BLOCKED?

This question has been dealt with in many previous reviews[24,25,26] and we can briefly state here that three main events of the replicative cycle may, under appropriate conditions, be the target for interferon activity. The most commonly observed locus is the translation of viral mRNA into virus specific proteins. In a few cases, of which simian virus 40 (SV-40) and vesicular stomatitis virus (VSV) are the most studied examples, transcription of the viral genome may be inhibited. In the case of tumor viruses a late event in the replication cycle has been implicated and this may be due to interference with the release of mature virus from the cell.

Early events such as adorption, penetration and uncoating of viruses do not appear to be the major locus of interferon action since RNA extracted from picornaviruses can no longer cause productive infection of interferon treated cells.[27,28,29,30] Moreover, the infectious cycle of SV-40 can be interrupted by treating infected cells with interferon at a time when these early events have already taken place.[31,32] Under these conditions the transcription of early and late viral RNA species is unaffected by interferon treatment although viral proteins are not synthesized.[32] Similarly, during vaccinia virus infection of cells pre-treated with interferon the synthesis of viral RNA continues unabated (in fact being stimulated as a result of the loss of a control function) even though protein synthesis is inhibited.[33,34,35]

As pointed out elsewhere[36] these types of experiments do not exclude the possibility that some early events may be affected by interferon under some circumstances and there is evidence that viral genome transcription may, in certain cases, be inhibited as a result of interferon treatment. An example of this is provided by SV-40 infection of monkey cells where Oxman and his colleagues[37,38] have shown that accumulation of SV-40 specific RNA is inhibited to an extent comparable with the inhibition of T-antigen forma- tion. Metz *et al.*[38] have ruled out some of the potential causes of such an effect, including reduced uptake of virions into the nucleus of infected cells. Similar findings were reported by Yamamoto *et al.*[39] who showed that interferon pre-treatment inhibits the uptake of viral DNA into nuclei by only 15 per cent whereas T-antigen expression and viral mRNA accumulation are inhibited by 86 per cent and 95 per cent respectively. Interestingly, these investigators found that interferon treatment had relatively little effect (20–40 per cent inhibition) upon viral DNA or mRNA accumulation when infectious SV-40 DNA was used to initiate the replicative cycle at a dose apparently equivalent to 5 pfu/cell. Under these circumstances T-antigen expression was also inhibited only slightly. Although these results have been interpreted to indicate direct inhibition of SV-40 primary transcription, other possibilities have not been eliminated. Studies in which protein synthesis inhibitors were shown not to mimic the effect of interferon on viral RNA accumulation[38] seem to argue against an inhibition of translation exerting some secondary effect upon transcription. The interpretation of such data, however, is not always straightforward as discussed below. Degradation of viral RNA has not been ruled out either, although Metz *et al.*[38] have shown that the reduction in amounts of RNA in cytoplasmic and nuclear fractions was comparable and hence argue that enhanced degradation of untranslated cytoplasmic mRNA is unlikely to explain the decrease in amounts of viral RNA. This interpretation is only valid if nucleolytic activity is restricted to the cytoplasm and there is no evidence to substantiate this condition. On the other hand, Yakobson *et al.*[32] have shown that apparently intact early and late viral mRNA accumulates in the cytoplasm of cells treated with interferon after the onset of SV-40 infection, a finding which argues against the role of ribonucleases in the interferon response. Owing to the current interest in the interferon-specific activation of an endonuclease (see later) degradation of RNA should be further investigated. Some suggestive evidence for a direct effect on transcription comes from a study utilizing nuclei prepared from control and interferon-treated cells which had been infected with SV-40. Metz *et al.*[40] found that synthesis of viral RNA *in vitro* was inhibited as a result of interferon treatment. The results of Yamamoto *et al.*[39] with infectious DNA might suggest that the apparent inhibition of RNA synthesis is actually due to a block in an earlier step such as uncoating which would lower the rate of transcription.

A second case in which interferon is reported to inhibit primary trancription is vesicular stomatitis virus (VSV). Marcus and his colleagues[41] first reported that the rate of primary transcription, as measured in VSV infected cells in the presence of cycloheximide, is inhibited by prior exposure of the cells to interferon. These findings have been extended[42,43,44] although Repik *et al.*[45] using a similar approach with different cell lines concluded that primary transcription of VSV and influenza virus was not affected by interferon. One of the critical problems with this approach is the efficacy of cycloheximide in inhibiting translation and thus preventing secondary transcription. Although the data were not presented the extent of protein synthesis inhibition due to cycloheximide was reported as 95 per cent[41,42] and apparently Sindbis virus, which does not have a virion-associated polymerase, was unable to induce RNA synthesis in cells incubated with similar concentrations of cycloheximide. It is conceivable, however, that the residual levels of protein synthesis of 1–5 per cent may suffice to produce significant levels of newly synthesized polymerase molecules, especially when input multiplicities of 5 to 200 pfu/cell are used (representing possibly 100 to 2000 virus particles). Under these circumstances the additional block of translation by interferon would be expected to lower observed levels of RNA synthesis. To circumvent such arguments Marcus and Sekellick[43] have used a temperature-sensitive mutant of VSV which, at non-permissive temperatures, produces only primary transcripts. When this virus is grown at the non-permissive temperature interferon pre-treatment lowers the rate of RNA synthesis below that observed in the presence of cycloheximide. Weibe and Joklik[46] have also used a temperature-sensitive mutant of reovirus to restrict RNA synthesis to primary transcription and report a 15 to 60 per cent inhibition of RNA synthesis as a result of interferon treatment. Baxt et al.[47] argue that the reduced rate of transcription in interferon treated cells is not due to an effect upon transcription but is related to an inhibition of protein synthesis since, like cycloheximide, interferon suppresses the synthesis of 42S and 20S RNA species in VSV infected cells. These workers show that viral protein synthesis is inhibited despite the presence of normal amounts of viral RNA. Ball and White[48] examined a cell-free system prepared from interferon-treated chick embryo cells which was able to transcribe mRNA from VSV cores and translate the messenger produced. They could find no effect of interferon pre-treatment on the rate of RNA synthesis or on the nature of the mRNA produced even though translation was inhibited by up to 70 per cent.

In most of these experiments it has been supposed that the level of viral RNA present in cells is a reflection of the rate of transcription, but increased degradation of RNA might also account for the results observed. As suggested by Baxt et al.[47] the failure of RNA to associate with ribosomes could lead to a greater susceptibility to ribonucleases. Moreover, it has been demonstrated recently that interferon can lead to an increase in the level of endoribonucleases[49] and the importance of degradation must be reconsidered. In fact very recently, Samuel[50] has reported that the inhibition of reovirus mRNA accumulation in mouse L-cells treated with interferon is accompanied by an enhancement in the rate of viral mRNA turnover as measured by pulse-chase experiments. Similar findings have been reported by Desrosiers and Lengyel.[51] Yet Samuel could find no evidence for enhanced degradation of viral mRNA in SV-40 infected monkey CV-1 cells after interferon treatment nor was accumulation of reovirus mRNA inhibited in these cells. Thus, both virus and host types appear to influence observations on RNA accumulation and degradation.

The debate over the possible effects of interferon on primary transcription will undoubtedly continue for a long time but it appears clear from many studies that the most general site for interferon action is translation[33,52,53,54,46,47,55,35,32] and the mechanism of this effect has been the subject of extensive investigation as described below.

Numerous studies have shown that the synthesis of viral proteins in cells pre-treated with interferon is blocked and the extent of the inhibition is usually correlated with the dose of interferon used. Thus, as seen in Figure 1, synthesis of all the VSV specific proteins is inhibited to the same extent and, at a dose of 1 to 2 interferon units/ml (assayed by 50 per cent inhibition of VSV cytopathic effect), viral protein synthesis is inhibited by approximately 50 per cent. Figure 1 also illustrates the fact that interferon can selectively block viral protein synthesis while permitting cellular protein, in this case actin, to be produced. The rise in cellular protein synthesis with increasing dose of interferon probably reflects a reversal of the host protein synthesis shut-off induced by the virus. It does not appear to be true that all

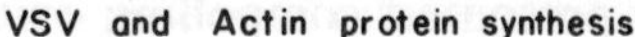

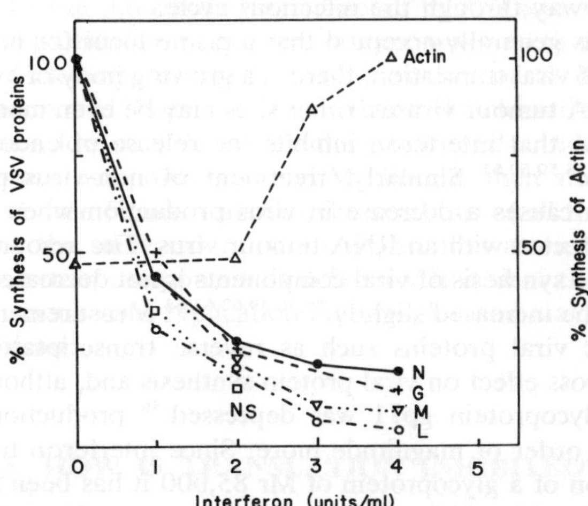

Figure 1. Viral and host protein synthesis in interferon-treated cells. Mouse L-cells were treated with various doses of interferon (specific activity 10^8 reference units/mg protein) for 24 hours and infected with vesicular stomatitis virus (10 pfu/cell). After labeling for 30 min with ^{35}S-methionine at 4 hours post-infection cytoplasmic extracts were prepared and analysed by SDS-PAGE and autoradiography. Synthesis of polypeptides was estimated by densitometric analysis

3.1 Effects on protein synthesis which are independent of dsRNA

Initially some confusion arose as to whether interferon treatment alone induced any alterations in *in vitro* protein synthetic activity. Several groups[70,53,71,72,73,74] found that extracts which had been preincubated to eliminate the translation of endogenous mRNA were defective in their ability to translate various exogenous mRNAs if the cells were pre-treated with interferon. The extent of this impairment appeared to correlate well with the level of anti-viral state induced.[70,71] However, Kerr and his collaborators, preferring to use non-preincubated extracts in order to minimize handling and consequent modification of the cell extracts, were unable to observe any alterations in protein synthetic capacity as a result of interferon treatment alone.[75,76] In fact, even in preincubated extracts Kerr's laboratory did not observe inhibition of translation as reported elsewhere.[76] It is probable that this difference can be ascribed to events occurring during the preparation and handling of cell-free extracts. Thus, Lengyel's group reported[77] that the capacity to charge certain tRNA species, especially those specific for leucine, lysine and serine, was impaired during incubation of extracts of interferon-treated cells. This enhanced impairment was only observed in extracts which were also subjected to Sephadex gel filtration. Falcoff *et al.*[78] reported that the length of time for which extracts were preincubated critically determined whether or not an inhibition of translation was observed. Thus, control and interferon-treated cell extracts preincubated for 60 min had identical capacities to translate Mengo virus RNA but after 90 min of preincubation the interferon-treated cell extract had only 10 per cent of the activity of the control. This effect of preincubation was ascribed to an inactivation of certain tRNA species since the inhibition could be both overcome and prevented by the addition of exogenous tRNA[78] in agreement with previous reports that exogenous tRNA could reverse the translational block.[53,79,80,81,82,83,84]

This effect of tRNA, together with kinetic analyses, had led to the idea that the step inhibited was that of elongation.[85] Experiments in which cell-free extracts were sub-fractionated into cell sap, ribosomal salt wash and ribosomes indicated that the interferon-mediated inhibition of translation was caused by a dominant inhibitor associated rather loosely with ribosomes.[53,70,71,80,73,86] Samuel and Joklik[73] identified a protein of MW 48,000 which was present only in the ribosomal salt-wash fraction of interferon-treated cell extracts. To date the relationship between tRNA inactivation and the ribosome-associated inhibitor has not been clarified nor has the relevance of this type of inhibition observed *in vitro* to the block of viral protein synthesis in intact cells. Apparently cell-free extracts differ from one laboratory to another to such an extent that tRNA inactivation in interferon-treated cells may not be observed (Friedman *et al.*;[76] Lewis, unpublished observations). tRNA extracted from uninfected and virus infected control or interferon-treated cells did not appear to differ in their capacity to be charged with amino acids *in vitrol*[87,88] The role of this particular effect of interferon, therefore, remains to be established.

Although Kerr's group were unable to observe an inhibition of translation in interferon-treated cell extracts, they did obtain a dramatic effect if cells were also infected with vaccinia virus for a short time before preparation of the extracts.[76,89] In this case both non-preincubated and preincubated extracts of interferon-treated, infected cells had impaired ability to translate endogenous and exogenous mRNA. With added encephalomyocarditis virus (EMC) RNA the peptide products were reduced in both size and amount.[76] Curiously, preincubation of the extracts abolished the difference between control and interferon extracts. Inhibitory activity was found to be present in both ribosomal (microsomal) and cell sap fractions but the inhibition could not be overcome by addition of exogenous tRNA.[90] Later studies suggested that the inhibitory activity of sap fractions only reduced the amount of polypeptides formed and not their size. The addition of formylated methionyl tRNA appeared to abolish the inhibitory effect suggesting that a step close to initiation of translation was the target for the interferon mediated inhibition.[89]

In addition to the impairment of tRNA function *in vitro*, interferon-treatment has been shown to cause a reduction in the capacity of cell-free extracts to methylate viral RNA molecules.[77,91] Since methylation of the 5′-termini of mRNA molecules, or 'capping', is essential for the activity of most mRNA species this could lead to the production of inactive mRNA. The impairment is not due to degradation of mRNA and the methylating enzymes of the extract appear to be functional[91] suggesting that the ability of the RNA to interact with the methylating system is affected. Studies in which interferon-treated cells were infected with reovirus[92] or vaccinia virus[93] and the methylation of the residual amount of virus produced was analysed, indicate that no great diminution in 'capping' occurs although the qualitative nature of the cap structures changes slightly. The polyadenylation of the 3′-end of vaccinia mRNA is also reported to be normal in interferon-treated cells.[94]

It seems probable that the inhibition of protein synthesis seen in infected cell extracts is caused by mechanisms distinct from those observed in preincubated extracts of uninfected cells. It is likely that interferon treatment alone induces in the cell a potential to exhibit reduced protein synthetic capacity but that some trigger, e.g. viral infection, is required to unleash this mechanism. Conceivably manipulation of cells, e.g. starvation[90] or of extracts, e.g. pre-incubating them, can also act as a trigger and much interest has focused on the ability of dsRNA to cause such a response *in vitro* since dsRNA may be produced in intact cells which are infected with certain viruses.

3.2 Double-strand RNA-dependent inhibition of translation

Kerr and his co-workers showed that addition of dsRNA to extracts of interferon-treated cells brought about a strong inhibition of protein synthetic activity at concentrations which had little or no effect on control cell extracts[95] as shown in Figure 2. This finding was prompted in part by previous observations that interferon-treated cells are much more sensitive to the cytotoxic effects of dsRNA[96] and partly by the knowledge that dsRNA inhibits translation in a variety of cell-free extracts.[97,98,99] Elucidation of the mechanism by which dsRNA inhibits translation in an extract of interferon-treated cells has led to a series of novel findings which may prove to have a significance far beyond the context of interferon research.[100] A very readable review of dsRNA-dependent effects was published recently.[101]

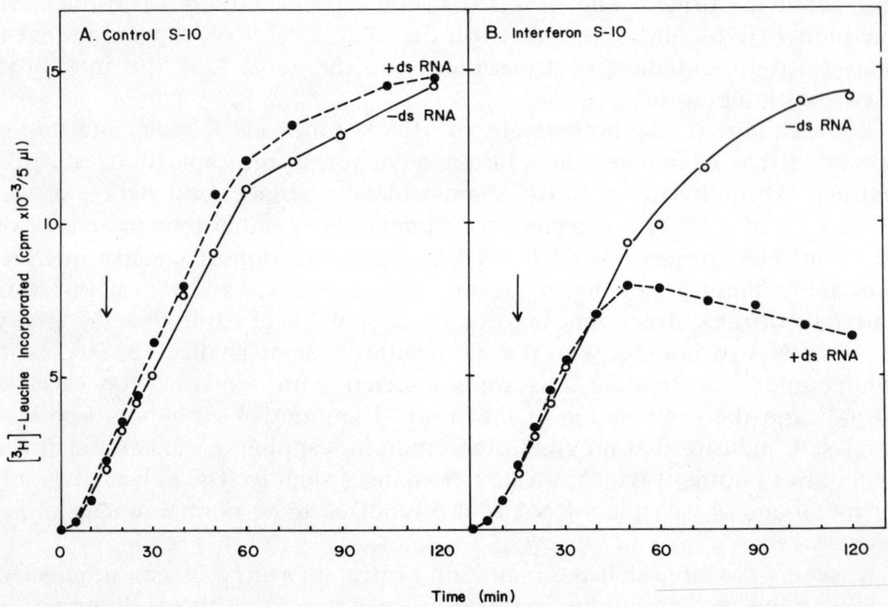

Figure 2. Effect of double-stranded RNA on cell-free protein synthesis. Preincubated cell-free extracts of Ehrlich ascites tumour cells were prepared essentially as described elsewhere (Lewis *et al.*, 1978) after treatment without or with interferon (300 units/ml; 10^8 reference units/mg protein) for 24 hours. Translation of Mengo virus RNA (40 µg/ml) was assayed by measuring incorporation of ^{3}H-leucine into hot trichloroacetic acid insoluble material. Double-stranded RNA (poly I. poly C) was added as shown by the arrows to a final concentration of 100 ng/ml

In part, our understanding of the effects of dsRNA in interferon-treated cell extracts depends on studies of the control of translation in reticulocyte lysates. The group led by Richard Jackson and Tim Hunt found that dsRNA and haemin-deficiency (both inhibitors of protein synthesis in this system) caused a block in the formation of methionyl tRNA·40S ribosomal sub-unit complexes).[102] This obligatory step in the initiation process is mediated by the initiation factor eIf-2[103] and further studies by this[104] and other groups[105,106,107,108] showed that a sub-unit of eIf-2 (MW 38,000) was phosphorylated and it is widely believed that this modification of the native factor in some way incapacitates it, although direct proof is lacking. Several other proteins are phosphorylated too, including, in the case of dsRNA, a protein of MW 68,000 which may be the protein kinase responsible for phosphorylation of eIf-2.[104] Addition of purified eIf-2 to lysates containing dsRNA or deprived of haemin overcomes the inhibition and results in a stimulation of protein synthesis.[109,110]

Following this demonstration of a phosphorylation step in the reticulocyte lysate system, several groups soon showed that addition of dsRNA to extracts of interferon-treated cells also gave rise to an enhanced phosphorylation of proteins of MW 68,000 and 38,000 although no enhancement of phosphorylation could be detected in control cell extracts.[111,112,113,114] The phosphorylation of the protein with MW 68,000 is usually just detectable in extracts of interferon-treated cells but is markedly stimulated by dsRNA (see Figure 3) and the activity resides in both ribosomal salt wash[113] and cell sap fractions.[112] Interestingly, dsRNA mediated stimulation of this peptide also occurs in mouse fibroblasts treated with gamma (immune) interferon prepared by induction of lymphocytes with lectins.[115] Phosphorylation of the protein of MW 38,000 is not always so readily observed in post-mitochondrial supernatants but can be clearly seen in ribosomal salt wash fractions[113] or when exogenous purified eIf-2 is added to extracts.[116,117,118] Various other phosphorylated bands have been observed by polyacrylamide gel electrophoresis, including substances of molecular weight lower than 10,000.[112,119] The protein kinase, which is cyclic AMP independent,[111] will also phosphorylate certain classes of histones.[112,120,121]

By corollary with the reticulocyte lysate system it seems logical to suppose that dsRNA inhibits binding of methionyl·tRNA to native 40S ribosomal sub-units as a result of phosphorylation of eIf-2. Lewis *et al.*[122] have shown directly that the formation of methionyl tRNA-containing pre-initiation complexes is inhibited in interferon-treated cell extracts which were incubated with dsRNA. This inhibition could be overcome by the addition of a crude reticulocyte initiation factor preparation. No inhibition of preinitiation complex formation was observed in the absence of dsRNA, even in extracts which had been pre-incubated to eliminate endogenous translation withconsequent impairment of tRNA function. Ohtsuki *et al.*,[123] however, have

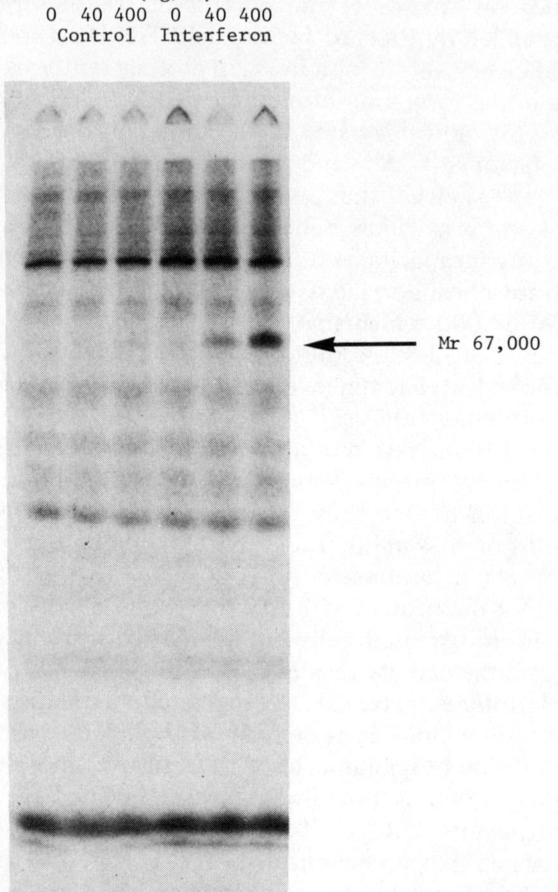

Figure 3. Double-stranded RNA stimulates phosphorylation of a polypeptide (Mr 67,000) in interferon-treated cell extracts. Cell-free extracts of Ehrlich ascites tumour cells (prepared as in legend to Figure 2) were incubated with ^{32}P-ATP (100 μCi/ml; 100 Ci/mol) in the absence or presence of poly I. poly C and analysed by SDS-PAGE and autoradiography

reported that ribosomal salt wash fractions from interferon-treated cells have impaired ability to bind methionyl $tRNA_i^{met}$ in a GTP-dependent reaction. In this report dsRNA was not utilized and it is not clear what relationship this finding bears to the inhibition of methionyl $tRNA_i^{met} \cdot 40S$ sub-unit complex formation reported elsewhere.[122] Cooper and Farrell[124] observed that methionyl tRNA binding to 40S sub-units in reticulocyte lysates could be inhibited by the addition of extracts from interferon-treated cells which had been preincubated with dsRNA. More recently, several groups have reported the partial purification of the protein kinase responsible for eIf-2 phosphorylation[120,125,126,127] and indeed after 1000 fold purification the major protein component is of MW 67,000.[128] Curiously, the dsRNA dependence of this enzyme activity is sometimes lost during purification[126] and this may explain the observations of Ohtsuki et al.[123] of dsRNA-independent inhibition of eIf-2 function since these authors used an ammonium sulphate fractionation step in the preparation of their initiation factors. Using partially purified preparations the protein kinase linked inhibition of translation can be shown to involve two separate steps:[116,120] an initial phase in which the protein kinase is activated by dsRNA and ATP (possibly as a result of autophosphorylation) and a second dsRNA independent step in which eIf-2 acts as a substrate for phosphorylation. Addition of activated, partially purified protein kinase to cell-free systems causes an inhibition of both methionyl tRNA·40S ribosomal sub-unit complex formation and protein synthesis which can be overcome by eIf-2.[116] Revel's group have also detected the presence of a phosphoprotein phosphatase capable of dephosphorylating both the 67,000-Mr protein and eIf-2.[121,127] The role of this enzyme remains to be established but quite clearly a complex system of interdependent enzymes is involved and their inter-relationship in the dynamic state of the intact cell must be considered.

These findings would seem to explain the ATP requirement for the dsRNA inhibitory effect[129] and could, therefore, explain its mechanism of action. However, the effect of dsRNA on extracts of interferon-treated cells has proved to be somewhat more complex. Two further roles of dsRNA have been observed, although they can now be united in a single pathway. Marcus and his co-workers[130] reported that treatment of chick cells with interferon led to the appearance of an alkaline ribonuclease associated with a cell membrane fraction. Soon afterwards Brown et al.[49] observed that reovirus mRNA was degraded much more rapidly in an interferon-treated cell extract than in a control preparation but this enhanced endonuclease activity was only observed when mRNA preparations contaminated with double-stranded reovirus RNA was tested. This dsRNA-dependent endonuclease activity was soon found to be present in many other interferon-treated cell extracts[122,131,132] and is illustrated in Figure 4. Activation of the endonuclease by dsRNA was shown to require ATP.[87,88,133] Although the nuclease showed a slight preference for certain reovirus mRNA species,[133] in general it degrades viral and host mRNA or ribosomal RNA equally well. Samuel, though, has reported a nuclease activity associated with the ribosomal salt-wash fraction of interferon-treated cells and this enzyme does not require dsRNA for activity and is much more effective in degrading reovirus mRNA than poly A-containing cellular RNA. However, Samuel has also

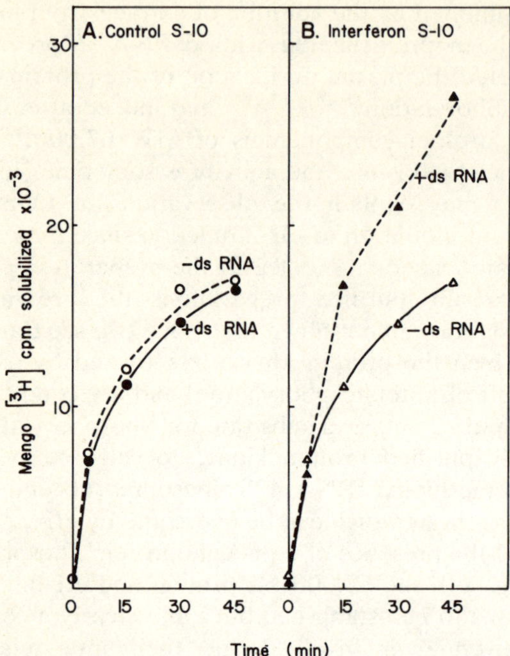

Figure 4. Degradation of Mengo ^{3}H-RNA in extracts of control and interferon-treated cells. Mengo ^{3}H-RNA (40 μg/ml; 90,450 cpm/μg) was incubated with control and interferon-treated cell extracts in the absence or presence of poly I. poly C (100 ng/ml). Liberation of cold trichloroacetic acid soluble radioactivity was assayed

reported that reovirus mRNA will stimulate phosphorylation of eIf-2 and the Mr 67,000 protein by interferon-treated cell extracts[118] and that reovirus mRNA translation by interferon-treated cell extracts is more strongly inhibited than that of cellular mRNA.[73] These observations are contrary to those of other workers and suggest either contamination of the reovirus mRNA by dsRNA or some unusual property of the *in vitro* synthesized viral mRNA used in these studies, such as extensive double-stranded regions.[118]

When cell sap extracts of interferon-treated cells are incubated with dsRNA and ATP a heat stable (90 °C for 5 min) low molecular weight inhibitor of protein synthesis is formed.[112,134] This inhibitor has now been characterized as an oligonucleotide of somewhat unusual structure (pppA2′p5′A2′p5′A) which actually functions as the activator of a latent endonuclease present in cells whether treated with interferon or not. Thus, interferon acts to induce the synthesis of an enzyme responsible for the formation from ATP of pppA2′p5′A2′p5′A or 2′,5′-oligo A and, in fact, the level of 2′,5′-oligo A polymerase in cells is reported to correlate closely with the anti-viral effect of interferon.[135,136] Elucidation of the structure of this oligonucleotide has been carried out principally in Ian Kerr's laboratory and was facilitated by an elegant technique involving retention of the 2',5'-oligo A polymerase to poly I·poly C attached to sepharose.[137] The small columns containing bound enzyme are incubated with ATP and the inhibitor which is formed can be eluted at intervals. The actual structure of the molecule has been determined by paper electrophoresis, DEAE-cellulose and thin-layer chromatography, sensitivity to high pH and sensitivity (or lack of) to various nucleases[134,138,139,140] and by chemical synthesis and NMR analysis.[141] Usually a predominance of tri-nucleotide is formed although small amounts of larger oligomers are also detected;[138] the dimer is apparently ineffective in activating endonuclease activity. Ironically, re-examination of the effect of dsRNA in reticulocyte lysates[142] has shown that here also 2',5'-oligo A is formed, indicating even further enzymological similarity between the reticulocyte and interferon-treated cells. Actually, it now transpires that 2',5'-oligo A synthetase is quite widely distributed in cells which have not been exposed to interferon and its level may vary according to the metabolic state of the cells.[100] In particular, the level of the enzyme appears to increase in both mammary glands and oviducts when the influence of stimulatory hormones is removed. It may be that interferon utilizes pathways commonly used for translational control by other modulators of cellular activity.

Evidence that 2',5'-oligo A mediates the activation of a latent endonuclease has come from studies using purified oligo nucleotide and either crude extracts[143,144,145,146] or partially purified 2',5'-oligo A polymerase.[147,116] When added to reticulocyte lysates 2',5'-oligo A causes degradation of globin mRNA,[143] Mengo and EMC virus RNA, VSV mRNA or L-cell cytoplasmic RNA.[145,146] Similarly, in Hela cell extracts, purified oligonucleotide causes an enhancement in the rate of VSV and cellular mRNA degradation and this effect occurs independently of whether the mRNA is associated with polysomes or not.[144] Moreover, ATP is not required for activation of endonuclease activity by 2',5'-oligo A and extracts from cells which have not been exposed to interferon are just as active in responding to the oligonucleotide as those from interferon-treated cells.[144]

Considerable effort has now been invested in the purification of the various enzymes involved in the oligonucleotide-endonuclease pathway. Three enzyme activities have been characterized and either partially or completely purified. One is the 2',5'-oligo A polymerase which requires dsRNA for activity and utilizes ATP as a substrate.[20,125,128] Lengyel's group have purified this enzyme from interferon-treated mouse cells to apparent homogeneity and it has a molecular weight of 105,000.[128] Ball[20] has obtained a highly purified fraction of polymerase activity from chick fibroblasts which contains one major band of Mr 56,000 which is the only component to be enhanced in amount by interferon treatment. Secondly, there is a latent endonuclease which can be activated by 2′,5′-oligo A in the absence of any other components[147,148] and finally a phosphodiesterase capable of hydrolysing the 2′,5′ linkage of 2′,5′-oligo A.[147,149,150,140] The latent endonuclease and phosphodiesterase appear to be present constitutively[144,150,140] although the level of the latter enzyme increases three-fold after interferon treatment.[136] This system may allow a sensitively balanced response to dsRNA since the 2′,5′-oligo A formed is rapidly degraded by the phosphodiesterase and under these conditions the endonuclease reverts to its inactive form.[150,140,148]

Several recent reports have indicated that purified 2′,5′-oligo A can be introduced into intact cells where it is capable of inhibiting protein synthesis. Williams and Kerr[151] showed that 2′,5′-oligo A inhibited total protein synthesis when introduced into cultured hamster kidney cells by hypertonic treatment. Hovanessian *et al.*[152] used a calcium phosphate-2′,5′-oligo A precipitate to cause penetration of the oligonucleotide into various cell types. These authors observed a transient inhibition of protein synthesis with a secondary effect upon RNA synthesis. In addition, they found evidence for degradation of cellular mRNA and ribosomal RNA and extracts prepared from such cells showed enhanced levels of nuclease activity. Kimchi and Revel (personal communication) and Revel *et al.*[153] found that 2′,5′-oligo A lacking its 5′-triphosphate was taken up by concanavalin A-stimulated lymphocytes and caused an inhibition of DNA synthesis 72 h later, an effect similar to that obtained by treating the cells with interferon.

4 HOW IS THE ANTI-VIRAL EFFECT MEDIATED IN INFECTED, INTERFERON-TREATED CELLS?

Although we can now enumerate several changes in the capacity of cells to translate mRNA *in vitro*, we have little idea as to the events occurring in a virus-infected, interferon-treated cell. An appealing hypothesis which has been suggested before[89,154,101] is that dsRNA formed during a virus infectious cycle may act as an inducer for the protein kinase and endonuclease pathways which have been observed. Yet, even in the case of picornaviruses, there is some dissent as to whether dsRNA actually exists in infected cells even though extracts of such cells can be shown to contain double-stranded forms.[97] Furthermore, it is not obvious why cells infected with DNA viruses should contain dsRNA although vaccinia virus infection is reported to cause an accumulation of such material.[155] It can be argued that the deproteinization steps necessary for analysing RNA species permits annealling of complementary RNA which existed as single-strand forms in the cell. On the other hand, the amounts of dsRNA isolated tend to increase at later stages of infection in a fashion which is not necessarily correlated with the rate of

RNA synthesis, a fact which might suggest that these species are by-products of the replicative process and not artefacts of the extraction procedure. Again, the amount of RNA synthesized in interferon-treated cells infected with picornaviruses is very small and yet viral translation is inhibited even at very early times after infection. Could sufficient dsRNA accumulate under these conditions to trigger the kinase and endonuclease pathways? Finally, how can one explain the apparent specificity of interferon action for viral mRNA translation? The endonuclease does not appear to exhibit any selectivity, being active when tested with viral genome or messenger RNA, cellular mRNA or ribosomal RNA.[131,133] Nor can the protein kinase pathway be expected to permit a discrimination between viral and host mRNA since the step which is inhibited, eIf-2 dependent binding of methionyl tRNA to 40S ribosomal sub-units, precedes binding of mRNA in the currently accepted scheme for protein synthesis.[103] Although Kaempfer et al.[156] report that eIf-2 may interact with mRNA, the recognition is relatively non-specific since both cellular and viral mRNA species bind to the initiation factor. In any case, reduction in the number of methionyl tRNA·40S complexes would be expected to reduce the rate of initiation unless phosphorylated eIf-2 is qualitatively altered in its ability to recognize cellular or viral mRNA species. If competition for pre-initiation complexes by different mRNA species occurs, those with high ribosome binding affinity (i.e. viral messengers) would be favored. The discovery of various factors which appear to modulate the action of eIf-2[157,158] may also require reinterpretation of events which might accompany dsRNA dependent inhibition of initiation. Models for the specificity of interferon action will be further discussed below.

There are essentially three approaches to establish whether any of the mechanisms observed *in vitro* apply to intact infected cells. One is to correlate the *in vitro* activities with the degree of anti-viral state established. In such studies the extent of inhibition of cell free protein synthesis in preincubated extracts produced by various doses of interferon[70,71,76,121] or at different times after interferon treatment[70] followed quite closely the degree to which an anti-viral state was established. On the other hand, no evidence for a deficiency in tRNA activity has been obtained in infected, interferon-treated cells.[87,88] The appearance of protein kinase and 2′,5′-oligo A polymerase activities has also been correlated with anti-viral effect at different interferon doses and at various times after treatment. Whereas Kimchi et al.[136] found that induction of protein kinase and 2′,5′-oligo A polymerase followed similar kinetics when mouse L cells were treated with 200 units/ml of interferon, Baglioni et al.[135] using Hela cells treated with 100 units/ml of interferon find that the protein kinase is not induced until at least 10 hours after treatment, at which time 2′,5′-oligo A polymerase activity is well developed. In their hands establishment of the anti-viral state correlated with the production of 2',5'-oligo A polymerase but not the protein kinase and this would argue against a role for this enzyme in limiting viral replication.

A second approach is to analyse the progress of virus replication in intact, infected cells. Metz *et al.*[54] obtained evidence that the rates of both initiation and elongation of protein synthesis were inhibited in vaccinia virus infected, interferon-treated mouse L-cells. Yau *et al.*[159] have also concluded that a reduced rate of elongation occurs in interferon-treated cells infected with EMC, VSV or murine leukemia virus since inhibitors of elongation such as cycloheximide or anisomycin produce effects which are qualitatively similar to those seen after interferon treatment. According to these authors this effect could account for the specificity of interferon action since the rate limiting step for translation of viral mRNAs may be elongation, whereas that for cellular mRNA translation is initiation; inhibition of elongation would thus have a more severe effect on viral than on host translation and could eliminate the competitive advantages of viral mRNA. Consistent with this model, viral yield was inhibited by approximately 100 fold by cycloheximide treatment at concentrations which inhibited overall protein synthesis by only 50 per cent.

In cells infected with SV-40, interferon treatment initiated after the addition of virus is effective in blocking viral replication.[31] Under these conditions both early and late viral mRNAs are present in the cytoplasm at the same levels as in control infected cells.[32] The size of this RNA and its state of polyadenylation was identical in control and interferon-treated cells, implying that levels of ribonuclease activity were not altered by interferon treatment and infection. Nonetheless, the viral mRNA failed to enter polysomes even though cellular mRNA was associated with ribosomes, and host proteins such as actin and tubulin were synthesized at normal rates. This finding might be explained either by a selective block of viral initiation or a decrease in the ability of viral mRNA to outcompete host mRNA for initiation. In Mengo virus infected, interferon-treated cells the small amounts of viral RNA which could be detected were apparently normal in size as measured by sedimentation on sucrose density gradients and hybridization to a Mengo virus cDNA.[160] It should be remembered, however, that a single nick may labilize RNA species to the activity of cellular nucleases and hence the detection of partially degraded RNA may be technically very difficult.

Recently, Gupta[161] has attempted to detect the phosphorylation of specific proteins (such as the Mr 68,000 and 38,000 protein) in intact cells. By incubating interferon-treated cells with dsRNA in the presence of $[^{32}P]$-PO_4 an enhancement in the labelling of a 65,000 dalton polypeptide was observed. However, infection with either Mengo virus or VSV did not stimulate phosphorylation of this protein. Whether this result is due to technical difficulties or implies that infection does not lead to activation of the protein kinase pathway remains to be established.

A third way of tackling the problem as to which effects of interferon are brought into play during infection is to assay cell-free extracts obtained from interferon-treated, virus infected cells. Extracts of vaccinia-virus infected cells[76] pretreated with interferon were found to be severely limited in their protein synthetic activity and this inhibition was attributed, in part at least, to a block in initiation.[89] Comparison of the profiles of polypeptides synthesized *in vitro* with exogenous viral mRNA indicated that either vaccinia virus[131] or Mengo virus (Lewis *et al.*, unpublished observations) infection of interferon cells produced alterations which were qualitatively similar to those seen when dsRNA was added to cell-free extracts of uninfected but interferon-treated cells. Aujean *et al.*[162] have assayed extracts of Mengo virus infected cells for phosphorylation of protein Mr 68,000 and ribonuclease activity. Both of these parameters were enhanced in extracts of interferon-treated infected cells indicating at least that dsRNA was present under such conditions. Enhanced ribonuclease activity was observed at 15 min of incubation in the absence of ATP which indicated that activation of this pathway occurred in the intact cells. Very recently Williams *et al.*[146] have reported the detection of 2',5'-oligo A in extracts of interferon-treated cells which had been infected for 4 h with EMC virus. The conditions under which these extracts were made essentially preclude the artefactual formation of the oligonucleotide thus making it extremely likely that 2',5'-oligo A was present in the infected cells. It remains to be established whether other viruses are capable of inducing the synthesis of 2',5'-oligo A and thereby activating endonuclease.

How is the specificity of interferon's action determined? Metz[154] has suggested that, since viral mRNAs generally have higher affinity for ribosomes than cellular mRNA species, a possible mechanism would involve an inhibition of the initiation of mRNAs with high ribosome binding affinity. To date there is no known mechanism by which this could occur. Yau *et al.*[159] have modified this proposal slightly and suggested that if the rate of elongation of viral mRNA is limiting rather than the rate of initiation, an inhibition of elongation steps could preferentially block the translation of viral mRNA as opposed to host. There is some evidence[159] that for some viral mRNA species at least, elongation becomes rate limiting rather than initiation which is the usual situation for cellular mRNA. What the block of elongation might be is unclear although, as described above, a reduction in the rate of elongation in interferon-treated, infected cells has been reported.

Evidence is beginning to accumulate for the operation of dsRNA activated pathways in interferon-treated cells infected with picornaviruses (Williams *et al.*;[146] Lewis, unpublished observations). In order to explain the specificity of endonuclease activity, Baglioni *et al.* (1978) have suggestd that endonuclease activation is localized to the picornavirus transcriptional complexes which contain dsRNA since 2′,5′-oligo A polymerase would bind to such complexes. Since 2′,5′-oligo A is rapidly degraded by 2′,5′-phosphodiesterase, activation of endonuclease would occur only in a circumscribed region and degradation of viral RNA being transcribed from the complex could occur with high probability while cellular mRNA at remote sites would be much less likely to be degraded. In support of this model, Nilsen and Baglioni[163] have shown that mRNA attached to dsRNA (e.g. by annealing poly A (+) mRNA to poly U) is preferentially degraded by cell-free extracts from interferon-treated cells.

An alternative model, which need not exclude the operation of the above model, proposes that both the initiation and endonuclease pathways are activated as a result of virus infection.[122] The block of initiation is non-specific and thus stops all protein synthesis. Degradation of RNA occurs through activation of the endonuclease and continues until all traces of the viral genome RNA (and possibly cytoplasmic cellular mRNA) are eliminated. Since activation of these pathways is reversible, once the viral RNA is eliminated the initiation block will cease and the endonuclease will be inactivated. The continuing production of cellular mRNA would permit the return of the cell to its normal state. Some evidence for such a scheme has been reported by Falcoff *et al.*[164] who showed that cellular protein synthesis by Mengo virus infected cells is inhibited for up to 6 h in spite of pretreatment with interferon.

5 CONCLUSION

Our understanding of the mechanisms by which interferon can prevent virus replication has evolved tremendously over the past decade. At present, it remains to be demonstrated which of several possible pathways operate in cells infected with a particular virus. It seems quite likely that different viruses will activate distinct pathways and, indeed, that different steps in the production of mature virus will be affected. Thus, there is evidence that transcription, translation and maturation may be inhibited according to the particular system studied. Although much interest is focused on the dsRNA-mediated activation of RNA degradation and the inhibition of protein synthesis initiation, it remains possible that specific tRNA inactivation or inhibition of mRNA methylation may occur in some systems. The fact that interferon affects such a wide range of enzyme systems makes it unlikely that a universal mechanism will operate for all viruses. Future studies should define how the replication of various classes of virus is blocked and provide an answer to the question of how specificity for viral components is achieved. Finally, it is probable that some explanation of the mechanism by which interferon exerts its other effects (e.g. cell growth inhibition) will shortly be forthcoming.

REFERENCES

1. Tan, Y. H., Schneider, E. L., Tischfield, J., Epstein, C. J., and Ruddle, F. H. (1974). *Science,* **186,** 61–63.
2. Chany, C., Vignal, M., Covillin, P., Van Cong, N., Bove, J., and Boye, A. (1975). *Proc. Nat. Acad. Sci. U.S.A.,* **72,** 3129–3134.
3. Revel, M., Bash, D., and Ruddle, F. H. (1976). *Nature,* **260,** 139.
4. Friedman, R. M. (1967). *Science,* **156,** 1760–1761.
5. Wagner, R. R. (1961). *Virology,* **13,** 23–337.
6. Levine, S. (1966). *Proc. Soc. Exp. Biol. Med.,* **121,** 1041–1045.
7. Besançon, F., and Ankel, H. (1977). In *Texas Reports on Biology and Medicine,* vol. 35. The Interferon System. Ed. S. Baron and F. Dianzani, pp. 282–292.
8. Falcoff, E. (1980). In *Annals of the New York Academy of Sciences— Regulatory Functions of Interferons,* Ed. Vilcek, Merigan and Gresser.
9. Kohn, L. D., Friedman, R. M., Holmes, J. M., and Lee, G. (1976). *Proc. Nat. Acad. Sci. U.S.A.,* **73,** 3695–3699.
10. Ankel, H., Chany, C., Galliot, B., Chevalier, M. J., and Robert, M. (1973). *Proc. Nat. Acad. Sci. U.S.A.,* **70,** 2360–2363.
11. Knight, E. (1974). *Biochem. Biophys. Res. Commun.,* **56,** 860–865.
12. Stewart II, W. E. (1975). In *Effects of Interferon on Cells, Viruses and the Immune System.* Ed. Geraldes, pp. 75–98. Academic Press, New York.
13. Berman, B., and Vilček, J. (1974). *Virology,* **57,** 378–386.
14. Tovey, M. G. (1980). In *Annals of the New York Academy of Sciences— Regulatory Functions of Interferons.* Ed. Vilcek, Merigan and Gresser.
15. Meldolesi, M. F., Friedman, R. M., and Kohn, L. D. (1977). *Biochem. Biophys. Res. Commun.,* **79,** 239–246.
16. Bourgeade, M. F., and Chany, C. (1976). *Proc. Soc. Exp. Biol. Med.,* **153,** 486–488.
17. Lockart, R. Z. (1964). *Biochem. Biophys. Res. Commun.,* **15,** 513–518.
18. Taylor, J. (1964). *Biochem. Biophys. Res. Commun.,* **14,** 447–453.
19. Friedman, R. M., and Sonnabend, J. A. (1965). *J. Immunol.,* **95,** 696–703.
20. Ball, L. A. (1979). *Virology,* **94,** 282–296.
21. Knight, E., and Korant, B. D. (1979). *Proc. Nat. Acad. Sci. U.S.A.,* **76,** 1824–1827.
22. Gupta, S. L., Rubin, B. Y., and Holmes, S. L. (1979). *Proc. Nat. Acad. Sci. U.S.A.,* **76,** 4817–4821.
23. Farrell, P. J., Broeze, R. J., and Lengyel, P. (1979). *Nature,* **279,** 523–525.
24. Sonnabend, J. A., and Friedman, R. M. (1973). In *Interferons and Interferon Inducers,* Ed. Finter, N. B. pp. 201–239, Elsevier, North Holland.
25. Lewis, J. A., Falcoff, E., and Falcoff, R. (1977), *Pathologie–Biologie,* **25,** 9–13.
26. Friedman, R. M. (1977). *Bact. Rev.,* **41,** 543–567.
27. Ho, M. (1961). *Proc. Sco. Exp. Biol. Med.,* **107,** 639–644.
28. Mayer, V., Sokol, F., and Vilcek, J. (1962). *Virology,* **16,** 359–362.
29. Grossberg, S. E., and Holland, J. J. (1962). *J. Immunol.,* **88,** 708–714.
30. Perez-Bercoff, R., Cioe, L., Degener, A. M., Med, P., and Rita, G. (1979). *Virology,* **96,** 307–310.
31. Yakobson, E., Revel, M., and Winocour, E. (1977). *Virology,* **80,** 225–228.
32. Yakobson, E., Prives, C., Hartman, J. R., Winocour, E., and Revel, M. (1977). *Cell,* **12,** 73–81.
33. Joklik, W. K., and Merigan, T. C. (1966). *Proc. Nat. Acad. Sci. U.S.A.,* **56,** 558–565.
34. Metz, D. H., and Esteban, M. (1972). *Nature,* **238,** 385–388.

35. Osterhoff, J., Jager, M., Jungwirth, C., and Bodo, G. (1976). *Eur. J. Biochem.*, **69,** 535–543.
36. Stewart II, W. E. (1979). *The Interferon System*, Springer-Verlag, Wien, New York.
37. Oxman, M. N., and Levin, M. J. (1971). *Proc. Nat. Acad. Sci. U.S.A.*, **55,** 1133–1140.
38. Metz, D. H., Levin, M. J., and Oxman, M. N. (1976). *J. Gen. Virol.*, **32,** 227–240.
39. Yamamoto, K., Yamaguchi, N., and Oda, K. (1975). *Virology*, **68,** 58–70.
40. Metz, D. H., Oxman, M. N., and Levin, M. J. (1977). *Biochem. Biophys. Res. Commun.*, **75,** 172–178.
41. Marcus, P. I., Engelhardt, D. L., Hunt, J. M., and Sekellick, M. J. (1971). *Science*, **174,** 593–598.
42. Manders, E. K., Tilles, J. G., and Huang, A. S. (1972). *Virology*, **49,** 572–581.
43. Marcus, P. I., and Sekellick, M. J. (1978). *J. Gen Virol.*, **38,** 391–408.
44. Thacore, H. R. (1978). *J. Gen. Virol.*, **41,** 421–426.
45. Repik, P., Flamand, A., and Bishop, D. H. L. (1974). *J. Virol.*, **14,** 1169–1178.
46. Weibe, M. E., and Joklik, W. K. (1975). *Virology*, **66,** 229–240.
47. Baxt, B., Sonnabend, J. A., and Bablanian, R. (1977). *J. Gen. Virol.*, **35,** 325–334.
48. Ball, L. A., and White, C. N. (1978). *Virology*, **84,** 496–508.
49. Brown, G. E., Lebleu, B., Kawakita, M., Shalia, S., Sen, G. C., and Lengyel, P. (1976). *Biochem. Biophys. Res. Commun.*, **69,** 114–122.
50. Samuel, C. E. (1980). In *Annals of the New York Academy of Sciences— Regulatory Functions of Interferons*, Ed. Vilcek, Mergian and Gresser.
51. Desrosiers, R. C., and Lengyel, P. (1977). *Fed. Proc.*, **36,** 812.
52. Esteban, M., and Metz, D. H. (1973). *J. Gen. Virol.*, **20,** 111–123.
53. Gupta, S. L., Sopori, M. L., and Lengyel, P. (1973). *Biochem. Biophys. Res. Commun.*, **54,** 777–783.
54. Metz, D. H., Esteban, M., and Danielescu, G. (1975). *J. Gen Virol.*, **27,** 197–213.
55. Jungwirth, C., Horak, I., Bodo, G., Lindner, J., and Schultze, B. (1972). *Virology*, **48,** 59–70.
56. Pitha, P. M., Staal, S. P., Bolognesi, D. P., Denny, T. P., and Rowe, W. P. (1977). *Virology*, **79,** 1–13.
57. Levy, H. B., and Carter, W. A. (1968). *J. Mol. Biol.*, **31,** 561–577.
58. Billiau, A., Edy, V. G., Sobis, H., and Desomer, P. (1974). *Int. Cancer*, **14,** 335–340.
59. Friedman, R. M., and Ramseur, J. M. (1974). *Proc. Nat. Acad. Sci. U.S.A.*, **71,** 3542–3544.
60. O'Shaugnessy, M. V., Easterbrook, K. B., Lee, S. H. S., Katz, L. J., and Rozee, K. R. (1974). *J. Nat. Cancer Inst.*, **53,** 1687–1696.
61. Friedman, R. M., Chang, E. H., Ramseur, J. M., and Myers, M. W. (1975). *J. Virol.*, **16,** 569–574.
62. Chany, C., and Vignal, M. (1970). *J. Gen. Virol.*, **7,** 203–210.
63. Pitha, P., Rowe, W. P., and Oxman, M. N. (1976). *Virology*, **70,** 324–338.
64. Chang, E. H., Myers, M. W., Wong, D. K. Y., and Friedman, R. M. (1977). *Virology*, **77,** 625–636.
65. Chang, E. H., and Friedman, R. M. (1977). *Biochem. Biophys. Res. Commun.*, **77,** 392–398.
66. Chang, E. H., Mims, S. J., Triche, T. J., and Friedman, R. M. (1977). *J. Gen. Virol.*, **34,** 363–367.

67. Chang, E. H., Jay, F. T., and Friedman, R. M. (1978). *Proc. Nat. Acad. Sci. U.S.A.*, **75**, 1859–1863.
68. Wong, P. K. Y., Yuen, P. H., Macleod, R., Chang, E., Myers, M. W., and Friedman, R. M. (1977). *Cell.* **10**, 245–263.
69. Friedman, R. M. (1980). In *Annals of the New York Academy of Sciences—Regulatory Functions of Interferons*, Ed. Vilcek, Merigan and Gresser.
70. Falcoff, E., Falcoff, R., Lebleu, B. and Revel, M. (1972). *Nature New Biology*, **240**, 145–147.
71. Falcoff, E., Falcoff, R., Lebleu, B., and Revel, M. (1973). *J. Virol.*, **12**, 421–430.
72. Gupta, S. L., Graziadei III, W. D., Weideli, H., Sopori, M. L., and Lengyel, P. (1974). *Virology*, **57**, 49–63.
73. Samuel, C. E., and Joklik, W. K. (1974). *Virology*, **58**, 476–491.
74. Hiller, G., Winkler, I., Viehauser, G., Jungwirth, C., Bodo, G., Dube, S., and Ostertag, W. (1976). *Virology*, **69**, 360–363.
75. Kerr, I. M., Sonnabend, J. A., and Martin, E. M. (1970). *J. Virol.*, **5**, 132–144.
76. Friedman, R. M., Metz, D. H., Esteban, R. M., Tovell, D. R., Ball, L. A., and Kerr, I. M. (1972). *J. Virol.*, **10**, 1184–1198.
77. Sen, G. C., Lebleu, B., Brown, G. E., Rebello, M. A., Furuichi, Y., Morgan, M., Shatkin, A. J., and Lengyel, P. (1975). *Biochem. Biophys. Res. Commun.*, **65**, 427–434.
78. Falcoff, R., Falcoff, E., Sanceau, J., and Lewis, J. A. (1978). *Virology*, **86**, 507–515.
79. Gupta, S. L., Sopori, M. L., and Lengyel, P. (1974). *Biochem. Biophys. Res. Commun.*, **57**, 763–770.
80. Content, J., Lebleu, B., Zilberstein, A., Berissi, H., and Revel, M. (1974). *FEBS Letters*, **41**, 125–130.
81. Falcoff, R., Lebleu, B., Sanceau, J., Weissenbach, J., Dirheimer, G., Ebel, J. P., and Falcoff, E. (1976). *Biochem. Biophys. Res. Commun.*, **68**, 1323–1331.
82. Zilberstein, A., Dudock, B., Berissi, H., and Revel, M. (1976). *J. Mol. Biol.*, **108**, 43–54.
83. Weissenbach, J., Dirheimer, G., Falcoff, R., Sanceau, J., and Falcoff, E. (1977). *FEBS Letters*, **82**, 71–76.
84. Mayr, U., Bermayer, H. P., Weidinger, G., Jungwirth, C., Gross, H. J., and Bodo, G. (1977). *Eur. J. Biochem.*, **76**, 541–551.
85. Content, J., Lebleu, B., Nudel, U., Zilberstein, A., Berissi, H., and Revel, M. (1975). *Eur. J. Biochem.*, **54**, 1–10.
86. Samuel, C. E., Farris, D. A., and Eppstein, D. A. (1977). *Virology*, **83**, 56–71.
87. Sen, G. C., Gupta, S. L., Brown, G. E., Lebleu, B., Rebello, M. A., and Lengyel, P. (1976). *J. Virol.*, **17**, 191–203.
88. Sen, G. C., Lebleu, B., Brown, G. E., Kawakita, M., Slattery, E., and Lengyel, P. (1976). *Nature*, **264**, 370–373.
89. Kerr, I. M., Friedman, R. M., Brown, R. E., Ball, L. A., and Brown, J. C. (1974). *J. Virol.*, **13**, 9–21.
90. Kerr, I. M., Friedman, R. M., Esteban, M., Brown, R. E., Ball, L. A., Metz, D. H., Risby, D., Tovell, D. R., and Sonnabend, J. A. (1973). In *Advances in the* (*Habermahland Diefenthal*, eds. Biosciences Vol. 111, pp. 109–123, Pergamon Press, Vieweg.
91. Sen, G. C., Shaila, S., Lebleu, B., Brown, G. E., Desrosiers, R. C., and Lengyel, P. (1977). *J. Virol.*, **21**, 69–83.
92. Desrosiers, R. C., and Lengyel, P. (1979). *Biochem. Biophys. Acta*, **562**, 471–480.

93. Kroath, H., Janda, H. G., Hiller, G., Kuhn, E., Jungwirth, C., Gross, H. J., and Bodo, G. (1979). *Virology,* **92,** 572–577.

94. Klesel, N., Wolf, W., Hiller, G., and Jungwirth, C. (1974). *Virology,* **57,** 570–573.

95. Kerr, I. M., Brown, R. E., and Ball, L. A. (1974). *Nature,* **250,** 57–59.

96. Stewart, W. E., De Clercq, E., Billiau, A., Desmyter, J., and De Somer, P. (1972). *Proc. Nat. Acad. Sci. U.S.A.,* **69,** 1851–1854.

97. Ehrenfeld, E., and Hunt, Y. (1971). *Proc. Nat. Acad. Sci. U.S.A.,* **68,** 1075–1078.

98. Robertson, H. D., and Mathews, M. B. (1973). *Proc. Nat. Acad. Sci. U.S.A.,* **70,** 225–229.

99. Kaempfer, R., and Kaufman, J. (1973). *Proc. Nat. Acad. Sci. U.S.A.,* **70,** 1222–1226.

100. Stark, G. R., Dower, W. J., Schimke, R. T., Brown, R. E., and Kerr, I. M. (1979). *Nature,* **278,** 471–473.

101. Baglioni, C. (1979). *Cell,* **17,** 255–264.

102. Darnbrough, C. H., Legon, S., Hunt, T., and Jackson, R. J. (1973). *J. Mol. Biol.,* **76,** 379–403.

103. Trachsel, H., Erni, B., Schreier, M. H., and Staehelin, T. (1977). *J. Mol. Biol.,* **116,** 755–767.

104. Farrell, P. J., Balkow, K., Hunt, T., Jackson, R. J., and Trachsel, H. (1977). *Cell,* **11,** 187–200.

105. Levin, D. H., Ranu, R. S., Ernst, V., Fifer, M., and London, I. M. (1975). *Proc. Nat. Acad. Sci. U.S.A.,* **72,** 4849–4853.

106. Issinger, O. G., Benne, R., Hershey, J. W. B., and Traut, R. R. (1976). *J. Biol. Chem.,* **251,** 6471–6474.

107. Kramer, G., Cimadevilla, J. M., and Hardesty, B. (1976). *Proc. Nat. Acad. Sci. U.S.A.,* **73,** 3078–3082.

108. Levin, D. H., and London, I. M. (1978). *Proc. Nat. Acad. Sci. U.S.A.,* **75,** 1121–1125.

109. Clemens, M. J., Safer, B., Merrick, W. C., Anderson, W. F., and London, I. M. (1975). *Proc. Nat. Acad. Sci. U.S.A.,* **72,** 1286–1290.

110. Kaempfer, R. (1974). *Biochem. Biophys. Res. Commun.,* **61,** 541–547.

111. Lebleu, B., Sen, G. C., Shaila, S., Cabrer, B., and Lengyel, P. (1976). *Proc. Nat. Acad. Sci. U.S.A.,* **73,** 3107–3111.

112. Roberts, W. K., Hovanessian, A., Brown, R. E., Clemens, M. J. and Kerr, I. M. (1976). *Nature,* **264,** 477–480.

113. Zilberstein, A., Federman, P., Shulman, L., and Revel, M. (1976). *FEBS Letters,* **68,** 119–124.

114. Shalia, S., Lebleu, B., Brown, G. E., Sen, G. C., and Lengyel, P. (1977). *J. Gen. Virol.,* **37,** 535–546.

115. Wietzerbin, J., Stephanos, S., Falcoff, R., and Falcoff, E. (1977). In *Texas Reports on Biology and Medicine,* Vol. 35, *The Interferon System,* Ed. Baron and Dianzani, pp. 205–211.

116. Farrell, P. J., Sen, G. C., Dubois, M. F., Ratner, L., Slattery, E., and Lengyel, P. (1978). *Proc. Nat. Acad. Sci. U.S.A.,* **75,** 5893–5897.

117. Samuel, C. E. (1979). *Virology,* **93,** 281–285.

118. Samuel, C. E. (1979). *Proc. Nat. Acad. Sci. U.S.A.,* **76,** 600–604.

119. Eppstein, D. A., and Samuel, C. E. (1977). *Biochem. Biophys. Res. Commun.,* **79,** 145–153.

120. Sen, G. C., Taira, H., and Lengyel, P. (1978). *J. Biol. Chem.,* **253,** 5915–5921.

121. Revel, M., Gilboa, E., Kimchi, A., Schmidt, A., Schulman, L., Yakobson, E., and Zilberstein, A. (1977). In *Proc. 11th Meeting of FEBS*, Vol. 43 (ed Clark *et al.*), pp. 47–58, Pergamon Press Oxford, New York.
122. Lewis, J. A., Falcoff, E., and Falcoff, R. (1978). *Eur. J. Biochem.*, **86,** 497–509.
123. Ohtsuki, K., Dianzani, F., and Baron, S. (1977). *Nature*, **269,** 536–538.
124. Cooper, J. A., and Farrell, P. J. (1977). *Biochem. Biophys. Res. Commun.*, **77,** 124–131.
125. Zilberstein, A., Kimchi, A., Schmidt, A., and Revel, M. (1978). *Proc. Nat. Acad. Sci. U.S.A.*, **75,** 4734–4738.
126. Hovanessian, A. G., and Kerr, I. M. (1979). *Eur. J. Biochem.*, **93,** 515–526.
127. Kimchi, A., Zilberstein, A., Schmidt, A., Schulman, L., and Revel, M. (1979). *J. Biol. Chem.*, **254,** 9846–9853.
128. Lengyel, P. Pichon, J., Slattery, E., Farrell, P. Broeze, R., Taira, H., Ghosh, N., and Samanta, H. (1980). In *Annals of the New York Academy of Sciences— Regulatory Functions of Interferons*, Ed. Vilcek, Merigan and Gresser.
129. Roberts, W. K., Clemens, M. J., and Kerr, I. M. (1976). *Proc. Nat. Acad. Sci. U.S.A.*, **73,** 3136–3140.
130. Marcus, P. I., Terry, T. M., and Levine, S. (1975) *Proc. Nat. Acad. Sci. U.S.A.*, **72,** 182–186.
131. Kerr, I. M., Brown, R. E., Clemens, M. J., and Gilbert, C. S. (1976). *Eur. J. Biochem.*, **69,** 551–561.
132. Eppstein, D. A., and Samuel, C. E. (1978). *Virology*, **89,** 240–251.
133. Ratner, L., Sen, G. C., Brown, G. E., Lebleu, B., Kawakita, M., Cabrer, B., Slattery, E., and Lengyel, P. (1977). *Eur. J. Biochem.*, **79,** 565–577.
134. Kerr, I. M., Brown, R. E., and Hovanessian, A. (1977). *Nature*, **268,** 540–542.
135. Baglioni, C., Maroney, P. A., and West, D. K. (1979). *Biochemistry*, **18,** 1765–1770.
136. Kimchi, A., Shulman, L, Schmidt, A., Chernajovsky, Y., Fradin, A., and Revel, M. (1979). *Proc. Nat. Acad. Sci. U.S.A.*, **76,** 3208–3212.
137. Hovanessian, A. G., Brown, R. E., and Kerr, I. M. (1977). *Nature*, **268,** 537–540.
138. Kerr, I. M., and Brown, R. E. (1978). *Proc. Nat. Acad. Sci. U.S.A.*, **75,** 256–260.
139. Ball, L. A., and White, C. N. (1978). *Proc. Nat. Acad. Sci. U.S.A.*, **75,** 1167–1171.
140. Minks, M. A., Benvin, S., Moroney, P. A., and Baglioni, C. (1979). *Nucl. Acid Res.*, **6,** 767–780.
141. Martin, E. M., Birdsall, N. J. M., Brown, R. E., and Kerr, I. M. (1979). *Eur. J. Biochem.*, **95,** 295–307.
142. Hovanessian, A. G., and Kerr, I. M. (1978). *Eur. J. Biochem.*, **84,** 14–159.
143. Clemens, M. J., and Williams, B. R. G. (1978). *Cell*, **13,** 565–572.
144. Baglioni, C., Minks, M. A., and Maroney, P. A. (1978). *Nature*, **273,** 684–687.
145. Vaquero, C. M., and Clemens, M. J. (1979). *Eur. J. Biochem.*, **98,** 245–252.
146. Williams, B. R. G., Golgher, R. R., Brown, R. E., Gilbert, C. S., and Kerr, I. M. (1979). *Nature*, **282,** 582–586.
147. Schmidt, A., Zilberstein, A., Shulman, L., Federman, P., Berissi, A., and Revel, M. (1978). *FEBS Letters*, **95,** 257–264.
148. Slattery, E., Ghosh, N., Samanta, H., and Lengyel, P. (1979). *Proc. Nat. Acad. Sci. U.S.A.*, **76,** 4778–4782.
149. Schmidt, A., Chernajovsky, Y., Shulman, L., Federmann, P., Berissi, H., and Revel, M. (1979). *Proc. Nat. Acad. Sci. U.S.A.*, **76,** 4788–4797.

150. Williams, B. R. G., Kerr, I. M., Gilbert, C. S., White, C. N., and Ball, L. A. (1978). *Eur. J. Biochem.*, **92,** 455–462.
151. Williams, B. R. G., and Kerr, I. M. (1978). *Nature*, **276,** 88–90.
152. Hovanessian, A. G., Wood, J., Meur, E., and Montagnier, L. (1979). *Proc. Nat. Acad. Sci. U.S.A.*, **76,** 3261–3265.
153. Revel, M., Shulman, L., Fradin, A. and Kimchi, A. (1980). In *Annals of the New York Academy of Sciences—Regulatory Functions of Interferons*, Ed. Vilcek, Merigen and Gressner.
154. Metz, D. H. (1975). *Cell*, **6,** 429–439.
155. Duesberg, P. H., and Colby, C. (1969). *Proc. Nat. Acad. Sci. U.S.A.*, **64,** 369–403.
156. Kaempfer, R., Hollender, R., Abrams, W. R., and Israeli, R. (1978). *Proc. Nat. Acad. Sci. U.S.A.*, **75,** 209–213.
157. DeHaro, C., Datta, A., and Ochoa, S. (1978). *Proc. Nat. Acad. Sci. U.S.A.*, **75,** 243–247.
158. Dasgupta, A., Das, A., Roy, R., Ralston, R., Majumdar, A., and Gupta, N. K. (1978). *J. Biol. Chem.*, **253,** 6054–6059.
159. Yau, P. M. P., Godefroy-Colburn, T., Birge, C. H., Ramabhadran, T. V., and Thach, R. E. (1978). *J. Virol.*, **27,** 648–658.
160. Jacquet, M., Caput, D., Falcoff, E., Falcoff, R., and Gross, F. (1977). *Biochimie*, **59,** 189–195.
161. Gupta, S. L. (1979). *J. Virol.*, **29,** 301–311.
162. Aujean, O., Sanceau, J., Falcoff, E., and Falcoff, R. (1979). *Virology*, **92,** 583–586.
163. Nilsen, T. W., and Baglioni, C. (1979). *Proc. Nat. Acad. Sci. U.S.A.*, **76,** 2600–2604.
164. Falcoff, R., and Sanceau, J. (1979). *Virology*, **98,** 433–438.

Index